These two volumes are the proceedings of a major International Symposium on General Relativity held at the University of Maryland 27 to 29 May 1993 to celebrate the sixtieth birthdays of Professor Charles Misner and Professor Dieter Brill. Colleagues, friends, collaborators and former students, including many of the leading figures in relativity, have contributed to the volumes, which cover classical general relativity, quantum gravity and quantum cosmology, canonical formulation and the initial value problem, topology and geometry of spacetime and fields, mathematical and physical cosmology, and black hole physics and astrophysics. As invited articles, the papers in these volumes have an aim which goes beyond that of a standard conference proceedings. Not only do the authors discuss the most recent research results in their fields, but many also provide historical perspectives on how their subjects developed and offer individual insights in their search for new directions. The result is a collection of novel and refreshing discussions. Together they provide an authoritative and dynamic overview of the directions of current research in general relativity and gravitational theory, and will be essential reading for researchers and students in these and related fields.

T0192703

# Directions in General Relativity

Proceedings of the 1993 International Symposium, Maryland

VOLUME 2

Professor Dieter Brill

# Directions in General Relativity

Proceedings of the 1993 International Symposium, Maryland

VOLUME 2

### Papers in honor of
# Dieter Brill

B. L. Hu
*University of Maryland*

T. A. Jacobson
*University of Maryland*

**CAMBRIDGE**
UNIVERSITY PRESS

CAMBRIDGE UNIVERSITY PRESS
Cambridge, New York, Melbourne, Madrid, Cape Town, Singapore, São Paulo

Cambridge University Press
The Edinburgh Building, Cambridge CB2 2RU, UK

Published in the United States of America by Cambridge University Press, New York

www.cambridge.org
Information on this title: www.cambridge.org/9780521452670

First published 1993
This digitally printed first paperback version 2005

*A catalogue record for this publication is available from the British Library*

ISBN-13 978-0-521-45267-0 hardback
ISBN-10 0-521-45267-8 hardback

ISBN-13 978-0-521-02140-1 paperback
ISBN-10 0-521-02140-5 paperback

# CONTENTS

# CONTENTS

# CONTENTS

# CONTRIBUTORS

**Arlen Anderson**
Blackett Laboratory, Imperial College, Prince Consort Road, London, SW7 2BZ, UK

**J. David Brown**
Physics Department, North Carolina State University, Raleigh, NC 27695, USA

**Esteban Calzetta**
IAFE, cc 167, suc 28, (1428) Buenos Aires, Argentina

**Riccardo Capovilla**
Centro de Investigacion y de Estudios Avanzados, I.P.N., Apdo. Postal 14-740, Mexico 14, D.F., Mexico

**Pierre Cartier**
Départment de Mathématique et Informatique, Ecole Normale Superieure, 45 rue d'Ulm, F-75005, Paris, France

**Yvonne Choquet-Bruhat**
Mechanique Relativiste, Université Paris 6, 4 Place Jussieu, F-75252 Paris Cedex 05, France

**P. T. Chruściel**
Max-Plank-Institut für Astrophysik, Karl Schwarzschild Str. 1, D8046 Garching bei Munich, Germany

**Jeffrey M. Cohen**
Physics Department, University of Pennsylvania, Philadelphia, PA 19104-6396, USA

**John Dell**
Thomas Jefferson High School for Science and Technology, 6560 Braddock Road, Alexandria, VA 22312, USA

# CONTRIBUTORS

**Stanley Deser**
The Martin Fisher School of Physics, Brandeis University, Waltham, MA 02254-9110, USA

**Cécile DeWitt-Morette**
Center for Relativity and Department of Physics, University of Texas, Austin, TX 78712-1081, USA

**Frank Flaherty**
Department of Mathematics, Oregon State University, Corvalis, OR 97331-4605, USA

**James B. Hartle**
Department of Physics, University of California, Santa Barbara, CA 93106-9530, USA

**Atsushi Higuchi**
Enrico Fermi Institute, University of Chicago, 5640 S. Ellis Avenue, Chicago, IL 60637, USA

**Gary T. Horowitz**
Department of Physics, University of California, Santa Barbara, CA 93106, USA

**Bei-Lok Hu**
Department of Physics, University of Maryland, College Park, MD 20742-4111, USA

**James Isenberg**
Department of Mathematics and Institute of Theoretical Science, University of Oregon, Eugene, Oregon 97403-5203, USA

**Theodore A. Jacobson**
Physics Department, University of Maryland, College Park, MD 20742-4111, USA

**Lee Lindblom**
Department of Physics, Montana State University, Bozeman, MT 59717, USA

**Bahram Mashhoon**
Department of Physics and Astronomy, University of Missouri-Columbia, Columbia, MO 65211, USA

**A. K. M. Masood-ul-Alam**
Department of Physics, Montana State University, Bozeman, MT 59717, USA

**Vincent E. Moncrief**
Department of Physics, Yale University, 217 Prospect Street, New Haven, CT 06511, USA

**Niall Ó Murchadha**
Physics Department, University College, Cork, Ireland

**Errol Mustafa**
Physics Department, University of Pennsylvania, Philadelphia, PA 19104-6396, USA

**Kristin Schleich**
Department of Physics, University of British Columbia, Vancouver, BC, V6T 1Z1, Canada

**Lee Smolin**
Department of Physics, Syracuse University, Syracuse, NY 13244-1130, USA

**Rafael D. Sorkin**
Department of Physics, Syracuse University, Syracuse, NY 13244-1130, USA

**Frank J. Tipler**
Department of Mathematics and Department of Physics, Tulane University, New Orleans, LA 70118, USA

**Robert M. Wald**
Enrico Fermi Institute and Department of Physics, University of Chicago, 5640 S. Ellis Avenue, Chicago, IL 60637, USA

# CONTRIBUTORS

**John A. Wheeler**
Department of Physics, Princeton University Princeton, NJ 08544, USA

**Donald M. Witt**
Department of Physics, University of British Columbia, Vancouver, BC V6T 1Z1, Canada

**James W. York, Jr.**
Physics Department, University of North Carolina, Chapel Hill, NC 27599, USA

**Jong-Hyuk Yoon**
Center of Theoretical Physics and Department of Physics, Seoul National University, Seuol 151-742, Korea

*an International Symposium on*

## DIRECTIONS IN GENERAL RELATIVITY

in celebration of the sixtieth birthdays of
Professors Dieter Brill and Charles Misner

*University of Maryland, College Park, May 27-29, 1993*

| Time | Thursday, May 27 | Friday, May 28 | Saturday, May 29 |
|---|---|---|---|
| 9:00-9:45 | Deser | B. DeWitt | Bennett |
| 9:45-10:30 | Choquet | Horowitz | Sciama |
| 10:30-11:00 | *break* | *break* | *break* |
| 11:00-11:45 | Moncrief | Ashtekar | Matzner |
| 11:45-12:30 | C. DeWitt/ Cartier | Kuchar | Thorne |
| 12:30-2:00 | *lunch* | *lunch* | *lunch* |
| 2:00-2:45 | Wald | Penrose* | Hawking* |
| 2:45-3:30 | York | Sorkin | *Directions in GR* |
| 3:30-4:00 | *break* | *break* | *Disc. leader*: Hartle |
| 4:00-4:45 | *Classical Relativity* | *Quantum Gravity* | Wheeler |
| 4:45-5:30 | *Disc. leader*: Isenberg | *Disc. leader*: Smolin | |

* Unconfirmed as of Mar. 1, 1993.

## Organization

*Advisory Committee* : Deser, C. DeWitt, Choquet, Hartle, Sciama,
Toll, Wheeler
*Scientific Committee*: Hu, Isenberg, Jacobson, Matzner, Moncrief, Wald
*Local Committee*: Hu, Jacobson, Romano, Simon
*Festschrift Editors*: Hu, Jacobson, Ryan, Vishveshwara

## Sponsors

National Science Foundation
University of Maryland

# Papers from Both Volumes
# Classified by Subject

## GEOMETRY, GAUGE THEORIES, AND MATHEMATICAL METHODS

## BLACK HOLES

# RELATIVISTIC ASTROPHYSICS AND COSMOLOGY

# Preface

It is Dieter Brill's gentle insistence on clarity of vision and depth of perception that has so influenced the development of general relativity and the scholarship of his colleagues and students. In both research and teaching, he is always searching for simpler descriptions with a deeper meaning. Ranging from positive energy and the initial value problem to linearization stability, from Mach's principle to topology change, Dieter's unique style has left its mark. The collection of essays here dedicated to Dieter Brill is a fitting tribute and clear testimony to the impact of Dieter's contributions.

This Festschrift is the second volume of the proceedings of an international symposium on *Directions in General Relativity* organized at the University of Maryland, College Park, May 27–29, 1993 in honour of the sixtieth birthdays of Professor Dieter Brill, born on August 9, 1933, and Professor Charles Misner. The first volume is a Festschrift for Professor Misner, whose sixtieth birthday was on June 13, 1992.

Ever since we announced a symposium and Festschrift for these two esteemed scientists in the Fall of 1991, we have been blessed with enthusiastic responses from friends, colleagues and former students of Charlie and Dieter all around the globe. Without their encouragement and participation this celebration could not have been realized. The advisors and organizers of this symposium have shared with us their wisdom and effort. For this volume we would like to thank Professors James Isenberg, Charles Misner, and Herbert Pfister, who joined one of us (TJ) in writing a review of Professor Brill's research. We would also like to offer a special word of appreciation to Professor Stanley Deser, who made valuable suggestions from local problems to global issues.

Our thanks also go to Drs. Joseph Romano and Jonathan Simon for their generous help with the organization of the symposium, to Mrs. Betty Alexander for her patient typing and corrections, and to Mr. Andrew Matacz and Drs. Joseph Romano and Jonathan Simon for their skillful TEXnical assistance in the preparation of manuscripts in both volumes. We also thank Prof. Jordan Goodman for letting us use the VAX facilities and Prof. Charles Misner for helping with some editing and for providing much needed TEXnological assistance.

We gratefully acknowledge support from the National Science Foundation and the University of Maryland.

We wish both Dieter and Charlie many more years of joyous creativity and inspiration.

Bei-Lok Hu
Ted Jacobson

# Dieter R. Brill:
# A Spacetime Perspective

This review of Dieter Brill's publications is intended not only as a tribute but as a useful guide to the many insights, results, ideas, and questions with which Dieter has enriched the field of general relativity. We have divided up Dieter Brill's work into several naturally defined categories, ordered in a quasi-chronological fashion. References [n] are to Brill's list of publications near the end of this volume. Inevitably, the review covers only a part of Brill's work, the part defined primarily by the areas with which the authors of the review are most familiar.

## 1 GEOMETRODYNAMICS—GETTING STARTED

In a 1977 letter to John Wheeler, his thesis supervisor, Brill recalled that after spin 1/2 failed [1] to fit into Wheeler's geometrodynamics program he asked John "for a 'sure-fire' thesis problem, and [John] suggested positivity of mass." Brill's Princeton Ph.D. thesis [A, 2] provided a major advance in Wheeler's "Geometrodynamics" program. By studying possible initial values, Brill showed that there exist solutions of the empty-space Einstein equations that are asymptotically flat and not at all weak. Moreover, in the large class of examples he treated, all were seen to have positive energy. Although described only at a moment of time symmetry, these solutions were interpreted as pulses of incoming gravitational radiation that would proceed to propagate as outgoing radiation.

Brill's examples and results bore on three important questions: (a) the conception of spacetime geometry as a substance or new aether rather than just a background on which real physics was to be displayed, (b) the stability of this aether, and (c) the existence of gravitational radiation in the theory of general relativity.

Question (a) was central to Wheeler's Geometrodynamics and had been previously supported by Wheeler's geons (bundles of electromagnetic radiation bound together by their mutual gravitational attraction) and wormholes (precursors to black holes).

Brill later directed the Princeton senior thesis of Hartle [9] constructing (by important approximations) a spherical geon built from gravitational waves, further reinforcing the idea that interesting structures could be build out of pure geometry. Their approximation technique, which came to be known as "Brill-Hartle averaging", was explored more generally by Isaacson to give a treatment of short wavelength gravitational waves in quite general backgrounds. Another senior thesis (of Graves) [3], which built on the work of Kruskal and Szekeres, determined the causal structure of an electrically charged "wormhole". A summary of the viewpoint being promoted by these researches can be found in [6].

Question (b) was at this time phrased only as the positivity of the (newly formulated) ADM energy in general relativity, but it was soon recognized (by scaling arguments) that negative energy states could not have a lower bound and would therefore allow free unlimited export of positive energy by a region of vacuum. Thus improvements on these positive energy results were pursued for decades with many notable participants.

Question (c) had been provoking much controversy at this time. Bergmann in summarizing the 1957 Chapel Hill conference (GR I) considered "the existence of gravitational waves" to be "the most important nonquantum question" discussed there. Infeld and Plebanski in a 1960 book found in studying equations of motion no suggestion of any radiation except for coordinate waves. The interpretation of exact plane and cylindrical wave solutions was unconvincing due to the non-Minkowskian boundary conditions. Bondi in 1959 expressed doubts that a double star system would radiate gravitationally although expecting nongravitationally driven motions to do so. Into this confusion Brill [2] brought an exact solution (evading doubts that nonlinearity was inadequately treated by approximations) that had a Schwarzschild asymptotic behavior and Euclidean topology, and yet contained no matter or nonmetric field. It was hard to call this anything but a gravitational wave.

## 2 POSITIVE ENERGY

Brill's positive energy results were extended using similar techniques by Arnowitt, Deser, and Misner to a different large class of initial conditions, and they also supplied a physical argument why the negative gravitational binding energies should not drive the total energy of a system negative. (Write the Newtonian energy formula heuristically, in view of relativity, as $E = M_0 - E^2/R$.) These two special cases identified two geometrical problems to which Brill and others gave productive attention in subsequent years: Controlling a manifold's scalar curvature $^{(3)}R$ by conformal transformations, and the existence of maximal ($K = 0$) hypersurfaces in generic solutions of Einstein's equations. These developments were pursued by many people and eventually led to a proof by Schoen and Yau of the positive energy theorem.

Along the way to this Brill and Deser [14, 17] (also with Fadeev [15]) introduced a quite different line of attack on the problem, arguing first that flat space is the unique critical point, and a *local* minimum of the energy functional. This approach motivated many relativists to learn the rigorous tools for treating infinite dimensional manifolds, and recruited many who knew these methods to the positive energy problem, as also to the area of linearization instability initiated by Brill and Deser [21] where such methods were equally important. The previous line of attack was then renewed also with these more rigorous methods, e.g., [24, 25, 26, 27]. A review of these developments written just before the Schoen and Yau proof is found in Brill and Jang [15$^B$]. Deser and Teitelboim had earlier noted that the Hamiltonian of quantum supergravity theory is a formally positive operator, and Grisaru later suggested that taking the $\hbar \to 0$ limit of the (possibly nonexistent) quantum supergravity theory should yield a classical positive energy theorem. Shortly after the Schoen and Yau result, E. Witten found a strikingly different proof while attempting to take such classical limit.

## 3 MAXIMAL AND CONSTANT MEAN CURVATURE HYPERSURFACES

Very soon after it was shown by Y. Choquet-Bruhat that Einstein's field equations constitute a well-posed system for generating solutions from specified initial data, one of the main difficulties with this approach was recognized: in a generic spacetime there is no a priori preferred choice of "time" (i.e., no preferred choice of spacetime foliation by space-like Cauchy surfaces). The choice of time which one makes, either in exploring an existing spacetime or in constructing a new one, is important since the portion of the spacetime that one can explore or build, and the apparent physics that one can examine in that portion, often depends crucially on this choice. While some spacetimes (like the Minkowski and the Friedmann-Robertson-Walker solutions) do have certain families of Cauchy foliations picked out by the spacetime geometry, generic spacetimes do not.

To deal with this problem, various relativists had suggested that maximal (zero mean extrinsic curvature) hypersurfaces in asymptotically flat spacetimes, and constant mean curvature ("CMC") hypersurfaces in spatially-closed spacetimes, might provide a good choice of time in generic solutions. While such foliations were known to simplify significantly the representation of the constraints (Brill was one of the first to recognize this fact, as well as exploit it [A]), and while relativity folklore had it that these foliations led to the cleanest representation of the gravitational physics (even near singularities), little was known about existence and uniqueness of maximal and CMC hypersurfaces until the work of Brill and F. Flaherty during the 1970's. Using variational techniques (including second order and higher variations of the volume form), they were able to show that in a nonflat spatially-closed solution of Einstein's equations (either vacuum or with other fields satisfying the strong energy condition), a maximal hypersurface, if it exists, is unique [24]. Further, they show that in that same spacetime, a foliation by CMC hypersurfaces, if it exists, is also unique [25]. Moreover, it follows from their calculations that the mean curvature behaves monotonically as a function on the leaves of a CMC foliation: it either increases or decreases steadily as one moves from slice to slice into the future or the past [16$^B$].

These results of Brill and Flaherty have lent considerable support to the belief that a CMC foliation provides a preferred, standard choice of time in a given (spatially closed) spacetime, and they have been used as an important ingredient in a number of theorems of mathematical relativity (e.g., the early versions of the positive energy theorem, and the slice theorem for the action of the diffeomorphism group on the space of solutions). However they do not resolve the issue of *existence* of CMC foliations. Brill has done important work on this question, but interestingly his results establish *nonexistence* in certain cases, rather than existence. In an early paper [6$^B$], Brill made the important observation that in a nonflat spacetime which has Cauchy surfaces that are topologically 3-tori and satisfies the Einstein equations (with the strong energy condition), there can be no maximal hypersurface. This result, later extended as a consequence of the work of Gromov and Lawson to a much wider class of topologies, shows that such spacetimes, once expanding, must continue to expand forever; there can be no turnaround. The absence of a maximal hypersurface does not preclude a CMC hypersurface or even a CMC foliation. However in [20$^B$], Brill provided relativists with a rather shocking collection of asymptotically flat solutions containing no maximal or CMC hypersurfaces at all. And soon thereafter Bartnik used Brill's work to show that there are solutions of Einstein's equations with closed Cauchy surfaces which also contain no CMC hypersurfaces. Thus, while a CMC or maximal foliation provides a good choice of time in spacetimes which admit them, we now know that for some solutions neither can be used. There is much research presently being carried out to determine if Brill's and Bartnik's examples are special, or whether perhaps there are large families of solutions with no CMC or maximal slices.

Besides this theoretical work on CMC and maximal foliations, Brill and some coworkers (J. Cavallo and J. Isenberg) made some explicit constructions of constant mean curvature slices in the Schwarzschild [26] and in the Reissner-Nordström spacetimes [18$^B$]. Working from a variational principle, they showed that the spherically symmetric CMC hypersurfaces in these spacetimes correspond to the 2-parameter family of solutions of a particle mechanics-type ordinary differential equation, which is easily studied numerically. As an interesting application of these studies, Brill, Cavallo, and Isenberg showed how to use CMC hypersurfaces in Schwarzschild and Reissner-Nordström spacetimes to construct "lattice" cosmological models (generalizing the lattice models of Lindquist and Wheeler), which match the large scale behavior of open as well as closed Friedmann–Robertson–Walker spacetimes.

## 4 SCALAR CURVATURE AND SOLVING THE CONSTRAINTS

As noted above, one of the virtues of the maximal and constant mean curvature hypersurfaces is that the Einstein constraint equations are considerably simplified when examined on them. Indeed, using a method of field decomposition that was pioneered by A. Lichnerowicz, and developed and extended by Y. Choquet-Bruhat, Brill [A,2], J. York, and others, one can effectively transform the four constraints on a maximal or CMC hypersurface $S$ into a single quasilinear elliptic equation for a positive function $\phi$, which is the conformal factor for the 3-geometry on $S$. This equation involves the conformal geometry $h$ of $S$, and certain parts $p$ of the extrinsic geometry of $S$. The idea is to pick $h$ and $p$ freely on $S$ and then solve the equation for $\phi$.

One cannot pick $h$ and $p$ completely freely, however. For some choices of $h$ and $p$ it turns out that the equation for $\phi$ has no solution. While much was known by the mid 1970's about which $h$ and $p$ work and which do not on *closed* initial data hypersurfaces, the first big breakthrough on this question for *asymptotically flat* initial data hypersurfaces came in 1979 when M. Cantor showed that one can solve the equation if and only if $h$ is conformally related to a metric with vanishing scalar curvature $^{(3)}R$. Thus, one wants to understand which asymptotically flat Riemannian metrics have this property.

Brill, in his thesis [A] and in subsequent work, had studied the question of the sign of $^{(3)}R$ for conformally related metrics. After Cantor's result came out, Brill and Cantor began to look for conditions on $h$ directly that could determine if indeed there is a conformally related metric with vanishing scalar curvature. They found such a condition (involving certain integrals of the scalar curvature of $h$) and reported it in [27]. In addition, they provided in [27] explicit examples of metrics that violate this integral condition. As an interesting mathematical note, they showed (much to the surprise of those more familiar with the scalar curvature of conformal classes of Riemannian metrics on closed manifolds) that if an asymptotically flat Riemannian metric has scalar curvature zero, there are always conformally related asymptotically flat Riemannian metrics with positive scalar curvature, and others with negative scalar curvature.

## 5 LINEARIZATION STABILITY

In the early 1970's, Brill noticed a curious feature of flat $T^3 \times R$ spacetimes: they can not be freely perturbed! More specifically, he found [6$^B$] that some of the solutions of the field equations linearized about such a spacetime are *not* tangent to a curve of exact (inequivalent) solutions. He remarked that, in a certain sense, solutions with this property are "isolated". By contrast, it had been shown earlier by Choquet-Bruhat and Deser that all perturbations of flat $R^4$ *are* tangent to curves of solutions. This work excited much interest, both in the physics community where solution perturbations often play an important role, and in the mathematics community, where it was not clear why solutions with this property should exist. Interest was intensified when Brill's work with Deser [21] showed that a much wider range of solutions of Einstein's equations share this property of having their perturbations restricted.

In the years following the work of Brill and Deser, a large amount of effort was expended—chiefly by J. Marsden, A. Fischer, J. Arms, and V. Moncrief—in attempts to understand this perturbation-restriction phenomenon, which came to be known as "linearization instability". The phenomenon can be characterized in terms of the space of all solutions of the Einstein equations (on a fixed spacetime manifold): A solution $g$ is defined to be linearization stable (and has no perturbation problems) if and only if this space of solutions is a *manifold* around the point $g$. It was shown that a solution of Einstein's equations with closed Cauchy surfaces is linearization stable if and only if the solution has no Killing vector fields; it was also shown (as suggested by Brill and Deser) that asymptotically flat solutions are generally linearization stable, unless one fixes the mass or some other quantity defined "at infinity." Brill provides a very readable account of all these ideas, and some of the machinery behind them, in [17$^B$].

Brill came back to the issue of linearization stability some years later. In work with Vishveshwara [29], he clarified a puzzle involving linearization stability which had been raised in a paper by Geroch and Lindblom. Brill and Vishveshwara showed that the puzzle could be understood in terms of what they called "joint linearization stability," which is essentially the usual linearization stability with additional conditions added to the Einstein equations (as occurs if one works, say, with a minisuperspace of solutions). Then, in work with O. Reula and B. Schmidt [30], Brill examined a localized version of linearization stability: What if one looks at solutions and their perturbations not globally (as with asymptotically flat solutions and solutions with closed Cauchy surfaces) but rather in local (compact, with boundary) spacetime regions? If the boundary conditions are not restricted, then such local solutions turn out to be always linearization stable.

## 6 MACHIAN "FRAME-DRAGGING EFFECTS"

The question whether general relativity realizes Machian ideas has always been a controversial one. This is partly because, beginning with Einstein's 1918 statement of the so called "Mach principle," it has never been entirely clear what the Mach principle is. One well-defined Machian idea is the dragging of inertial frames by accelerated masses. Indications of this sort of effect were first found by Einstein in 1912-1913 in pre-general-relativity gravitation theories, with the help of an ingenious model of an accelerated mass shell. Later Thirring showed in the weak-field approximation that a slowly rotating mass shell drags along the inertial frames within it a little bit. Besides some inconsistencies, mainly with respect to his analysis of the 'centrifugal force', Thirring's results are limited by the weak-field approximation which is never valid in a cosmological context. This is a serious limitation if one wants to address the (less well-defined) original question raised by Mach: are the local inertial frames somehow "determined" by the motions of all of the matter in the universe?

Improving on Thirring's work, Brill and Cohen [12] consider a spherical shell whose mass to radius ratio can reach the collapse limit (as indicated by diverging stresses) and thereby can simulate—in an idealized way—all the matter of the universe. Perturbing the metric of a nonrotating shell, they calculated to first order in the shell's angular velocity and confirmed the Machian expectation that the angular velocity of the inertial frames in the (flat!) interior of the shell approaches the angular velocity of the shell in the collapse limit. That is, the mass shell totally 'shields' the interior from the asymptotic inertial frames. This result has entered standard text books as the single most convincing example for realization of Machian ideas in general relativity. Brill [13] subsequently analyzed the rotating shell model in the Jordan-Brans-Dicke theory of gravity, where he was able to quantify the effect that a change in the cosmic matter density has on the frame-dragging induced by a rotating mass. Recently, it has been shown by Pfister and Braun that, by allowing for a nonspherical mass shell, Brill and Cohen's analysis can be consistently extended to higher orders of angular velocity.

Brill and collaborators later succeeded in transferring the main results of [12] to more realistic models, i.e. an expanding and recollapsing dust cloud at the turning point, an incompressible fluid sphere, and a collapsing dust shell [16,23]. The latter model also illustrated the highly nonlocal character of Machian effects. Even when the angular velocity of the shell is time-dependent, the dragging, as determined by observation at infinity of light beamed from the center of the shell, does not depend on the angular velocity of the shell at times other than the moment the beam crosses the shell.

## 7 QUANTUM GRAVITY AND TOPOLOGY CHANGE

In 1960, Brill and Graves [3] discovered some rather remarkable properties of the "electromagnetic wormhole", i.e., the extended Reissner-Nordström (RN) solution [36] for a spherically symmetric electrovac gravitational field. They determined the global structure of this spacetime by making coordinate transformations to two overlapping coordinate patches, and concluded that due to electromagnetic "cushioning", the wormhole does not pinch off but contracts to a minimum, and then re-expands, with an oscillation period $2\pi M$, where $M$ is the mass of the body. They showed further that the singularity is timelike, and that no timelike geodesic (or charged particle trajectory for a particle with charge/mass ratio less than 1) hits it, although light rays do. The paper includes the phrase:

> "To visualize the manifold it is natural, in general relativity, to consider it
> as a succession of spacelike surfaces."

This is a theme to which Brill would return again and again. We mention this early paper here not only because it marks the beginning of Brill's work on wormholes, but also because, as it turns out, the RN solution has repeatedly figured in many facets of his subsequent research, not least in his investigations of topology change—albeit in a quantum mechanical context.

Evidence of Brill's interest in quantum gravity first appears with the review [18], written with Gowdy and published in 1970. Subsequently Brill published a paper called "Thoughts on Topology Change" [8$^B$] in Wheeler's sixtieth birthday Festschrift (*Magic Without Magic*). This essay is a discussion of the following puzzle: How can quantum gravity, using the Wheeler-DeWitt equation, describe or predict topology changing processes, given the fact that regions of superspace corresponding to topologically distinct 3-manifolds appear to be *disconnected* from each other. Brill suggests an approach in which topology change is associated with the existence of points where the 3-metric is degenerate, and in which singular coordinate transformations are allowed at such points. This produced the concept of a "hybrid manifold", which could have more than one topology. The idea was that, by this device, the manifold structure of superspace may be generalized in such a way that 3-manifolds of different topology would be connected and the theory would have a chance of describing topology change. The paper concludes by saying that it should be possible to work some of this out and its consequences for topology change in minisuperspace calculations. As far as we know, this interesting idea has not been pursued *so far*, but in a recent paper [25$^B$] Brill states his intention to do so using the configurations associated with the wormhole-splitting instanton he discovered in [35].

In the late 1970's Hawking and others developed, by analogy with other quantum field theories, the Euclidean approach to quantum gravity, in which the description of topology changing processes becomes, in the "WKB approximation", a question of the existence and properties of various types of "gravitational instantons"—Riemannian solutions to the Einstein equations. Brill's paper [28=22$^B$], published in Wheeler's seventy-fifth birthday Festschrift, marks the beginning of his interest in the treatment of topology changing processes using the Euclidean approach to quantum gravity. After recalling that the vacuum of 4-dimensional Minkowski space is stable he goes looking for trouble:

> "It may come as a relief to know that at least the vacuum is stable, but in order really to enjoy this fortune one would like to know what fate one has escaped—what might have happened if the vacuum were unstable."

The paper concerns the decay of the Kaluza-Klein (KK) vacuum $M_4 \times S^1$, as given by Witten's "bubble" instanton. This instanton is the five dimensional Euclidean Schwarzschild solution, whose asymptotically flat region contains a flat $R^3 \times S^1$ coinciding with a spatial slice of the KK vacuum, and which contains also a totally geodesic $R^2 \times S^2$ corresponding to the four dimensional space with a bubble.

Brill gives in [28] a particular foliation by 4-spaces that enables one to visualize the change in topology mediated by the bubble instanton. To determine the result of the tunnelling process one uses the totally geodesic $R^2 \times S^2$ as initial data for a Lorentzian solution containing an expanding "bubble". Brill analyzes the behavior of geodesics in this spacetime, showing that "the expanding bubble reflects light like a moving mirror." In [31] with Matlin, he extends the analysis to timelike geodesics, showing that radial trajectories will repeatedly collide with the bubble as it expands, eventually moving with the bubble surface. They point out that this indicates that the back reaction of surrounding matter on an expanding bubble may slow the bubble down. However, since the total energy in KK theory is not bounded below, they remark that it is not clear whether or not the back reaction would eventually stop the bubble's expansion.

The presence of negative energy in KK theories and its connection to topology changing processes became the focus of the next phase of Brill's research. He first developed [28] a generalization of Witten's time symmetric initial data, constructing a class of data with positive and negative energies (although he later showed [32] that all of the negative energy solutions found in [28] possess curvature singularities.) He also discussed [28] the extension of these ideas to higher dimensional KK theories in which the internal space has curvature, and proved that for one such KK vacuum (due to Freund and Rubin), no bubble-type solutions exist.

Brill and Pfister [32] later undertook a systematic examination of spherically symmetric, vacuum initial data in five-dimensional KK theory. They found that there are indeed non-singular solutions with negative energies, and that the energy is un-bounded below in this topological sector, even when the asymptotic compactification radius is held fixed. Brill and Horowitz [34] later showed that the energy is unbounded below even if the minimum bubble area is also held fixed.

In [34] Brill and Horowitz extended the negative energy results of [32] to some 10 dimensional spacetimes that could serve as candidate vacua for superstring theory, in particular, those which are asymptotically $M_4 \times T^6$. First they appropriated the results of [32] by forming orthogonal and tilted products with a flat $T^5$ to obtain 9-dimensional initial data with negative energy, in particular making use of the Eu-clidean 4-dimensional Reissner-Nordström metric. Next they consider the addition of (vector and tensor) gauge fields (present in the low energy string spectrum) that are constant at infinity to determine if the presence of such gauge fields can stabi-lize the vacuum. They find that for each asymptotic "vacuum" configuration, there exist arbitrarily negative energy solutions. The physical implications of the negative energy solutions are not at present clear however, since for small compactification radius (compared with the Planck length), large curvatures occur and the classical analysis is not valid, whereas for large compactification radius, the decay process is presumably exponentially suppressed due to the correspondingly large action of the instanton. In the zero-energy case, they find a generalization of the Witten-bubble instanton using the charged, dilaton black hole solution found by Gibbons and Maeda.

The bubble instantons involve topology change via a decay mode of flat spacetime. Another type of topology change expected in quantum gravity is the fluctuations of wormhole topology envisioned originally by Wheeler. For an approach to this sort of topology change, Brill turned [35] to the Reissner-Nordström wormhole he had discovered with Graves [3] in 1960. Paper [35] begins by noting that in general the Riemannian section of a spacetime containing a (massive) wormhole will not be asymptotically flat, on account of the required periodic identification of the analytic continuation of the Lorentzian time coordinate, this being related to the fact that the classical wormhole is not static but will collapse. An exception is the extremal Reissner-Nordström (RN) solution, whose charge is equal to its mass, for which the wormhole (actually an infinitely long hole) is stable and no periodic identification is required.

Deep in the wormhole the extremal RN solution becomes the Bertotti-Robinson (BR) solution $S^2 \times H^2$, which is the product of a 2-sphere threaded by a uniform magnetic flux, with a 2-dimensional hyperbolic Lorentzian spacetime of constant curvature. By a reinterpretation of the "conformastatic" solutions with different boundary con-ditions on the metric and potential function, Brill obtains a solution describing a universe with $n$ asymptotically BR regions with fluxes $M_i$, and one with flux $\sum M_i$. He interprets the Euclidean version of this solution as an instanton describing quantum fluctuations in which a single wormhole splits into many. The instanton does not describe a tunneling event, since there is no extremal surface in the interior to serve as a "turning point" at which a real classical evolution could be matched on. Rather, all "turning points" are infinitely far away, so it takes infinite Euclidean time for the process to occur. Brill interprets this as analogous to a quantum system with degenerate minima of its potential. In this interpretation, the action of the instanton is related in the WKB approximation to the energy splitting between the ground states and therefore to the frequency with which the system fluctuates between the degenerate minima.

Brill computes the action $S_E$ of his instanton and finds that it is finite and equal to minus one-eighth the change in total area from the "initial" $S^2$ to the "final" asymptotic $S^2$'s. This is equal to minus half the change in the Beckenstein-Hawking entropy of the black holes whose asymptotic interiors are approximated by the BR universes, so $\exp(-2S_E)$ agrees with the probability associated by statistical mechanics with a fluctuation involving such a decrease in entropy from its equilibrium value. Brill's wormhole splitting instanton has generated a lot of interest and stimulated other recent work on the idea of bifurcating quantum black holes in general relativity and in its string-theoretic generalizations.

The visualization of geometrodynamic structures has always been a motivating factor, a powerful tool, and a stylistic feature of Brill's work. When it comes to topology changing processes one sees this characteristic in play ever more so. To visualize the RN wormhole splitting process [35] Brill slices the instanton into a sequence of geometries and fields that are "most likely to be present" during the quantum fluctuations. He uses equipotential surfaces in the Euclidean section to define a natural slicing, in which one can see one wormhole pinch off into two or more, much as loops of equipotential lines for charges on a plane do. In an effort to develop some intuition about the nature of the topology changing Maxwell-Einstein instantons in general, Brill points out in [25$^B$] that the Euclidean Maxwell field (in four dimensions) obeys an "anti-Lenz law", which is crucial to understanding the behavior of the electromagnetic field in these solutions. Armed with this observation, he describes how one can visualize and understand the Euclidean "time development" of some instantons. The paper also contains a new class of instantons constructed by cutting and identifying regions of the Euclidean Bertotti-Robinson solution described above. One of these describes a splitting of one $S^2 \times S^1$ universe into two or, alternatively, the creation of a triplet of $S^2 \times S^1$ universes from "nothing".

In a recent paper with Pirk, "A Pictorial History of Some Gravitational Instantons" [26[B]] (contributed to the Misner Festschrift in Volume I of this proceedings), Brill invokes the aid of computer graphics in his ongoing quest to perceive and understand fluctuations in topology. After warming up by depicting instantons associated with multidimensional tunneling and fluctuation processes in ordinary quantum mechanics, Brill and Pirk construct embedding diagrams, some stereoscopic, of spacetimes, slices, and most probable paths in various topology-changing processes. Included are Vilenkin's tunneling of the universe from nothing into a deSitter space, Witten's bubble nucleation, the Garfinkle-Strominger pair-creation of extremal magnetically charged black holes from an initial Melvin universe, and Brill's extremal Reissner-Nordström wormhole splitting process [35]. These pictures are spectacular (even if you can't manage to see them stereoscopically), and they certainly indicate that Dieter has "seen the light" [B]. Happily for Dieter, there is always more to see!

## 8 PEDAGOGICAL CONTRIBUTIONS

Of course there is much that has been left out of this already long review, but before closing we would particularly like to mention two items of a pedagogical nature. First, over the years, Brill has generously contributed a number of useful review articles on general relativity, both experimental [2[B],9[B]] and theoretical [10,18,3[B]]. And second, he devoted much time and effort to producing with D. Falk and D. Stork a wonderful textbook [B,C] for optics courses for non-science majors, *Seeing the Light: Optics in Nature, Photography, Color, Vision and Holography*. The book promotes genuine scientific literacy through a medium of broad interest and common experience. Although mathematics is used sparingly, the authors do not avoid explanations that can be subtle and involved. The book is full of visual material, including photographs, "flip movies", interactive illusions, color stereographs (anaglyphs), an auto-random dot stereogram, and so on. It seems no accident that the same man who has helped so many physicists see and understand the geometry of spacetime has co-authored this remarkable and popular textbook.

Jim Isenberg
Ted Jacobson
Charles Misner
Herbert Pfister

# Thawing the Frozen Formalism:
# The Difference Between Observables
# and What We Observe

*Arlen Anderson* *

## Abstract

In a parametrized and constrained Hamiltonian system, an observable is an operator which commutes with all (first-class) constraints, including the super-Hamiltonian. The problem of the frozen formalism is to explain how dynamics is possible when all observables are constants of the motion. An explicit model of a measurement-interaction in a parametrized Hamiltonian system is used to elucidate the relationship between three definitions of observables—as something one observes, as self-adjoint operators, and as operators which commute with all of the constraints. There is no inconsistency in the frozen formalism when the measurement process is properly understood. The projection operator description of measurement is criticized as an over-idealization which treats measurement as instantaneous and non-destructive. A more careful description of measurement necessarily involves interactions of non-vanishing duration. This is a first step towards a more even-handed treatment of space and time in quantum mechanics.

There is a special talent in being able to ask simple questions whose answers reach deeply into our understanding of physics. Dieter is one of the people with this talent, and many was the time when I thought the answer to one of his questions was nearly at hand, only to lose it on meeting an unexpected conceptual pitfall. Each time, I had come to realize that if only I could answer the question, there were several interlocking issues I would understand more clearly. In this essay, I will address such a question posed to me by others sharing Dieter's talent:

What is the difference between an observable and what we observe?

---

*Blackett Laboratory, Imperial College, Prince Consort Rd., London, SW7 2BZ, England. I would like to thank A. Albrecht and C.J. Isham for discussions improving the presentation of this work.

This question arises in the context of parametrized Hamiltonian systems, of which canonical quantum gravity is perhaps the most famous example. It is posed to resolve the following paradox: For constrained Hamiltonian systems, an observable is defined as an operator which commutes (weakly) with all of the (first-class) constraints. In the parametrized canonical formalism, the super-Hamiltonian $\mathcal{H}$ describing the evolution of states is itself a constraint. Thus, all observables must commute with the super-Hamiltonian, and so they are all constants of motion. Where then are the dynamics that we see, if not in the observables? This is the problem of the *frozen formalism*[1, 2, 3].

In the context of quantum gravity, the problem of the frozen formalism is closely linked with the problem of interpreting the wavefunction of the universe and the problem of time. Two proposed solutions to the problem of time—Rovelli's evolving constants of the motion[2] and the conditional probability interpretation of Page and Wootters[4]—intimately involve observables which commute with the super-Hamiltonian, and each claims to recover dynamics. These proposals have been strongly criticized by Kuchar[3], who notes that there is a problem with the frozen formalism even for the parametrized Newtonian particle.

In this essay, I shall not address the problem of time but will focus on the simpler case of the Newtonian particle. My intention is to reconcile the different conceptions, mathematical and physical, that we have of observables. This will involve a recitation of measurement theory to establish the connection between the physical and the mathematical. Essential features of both the Rovelli and the Page-Wootters approaches will appear in my discussion as aspects of a careful understanding of observables and how we use them.

There is a general consensus that to discuss the wavefunction of the universe one must adopt a post-Everett interpretation of quantum theory in which the observer is treated as part of the full quantum system. I shall take this position for parametrized Hamiltonian systems as well. Insistence that the measurement process must be explicitly modelled will lead to a sharp criticism of the conventional description of measurement in terms of projection operators. A simple model measurement, related to one originally discussed by von Neumann[5], will clarify the role of observables in the description of measurements. No incompatibility between dynamics and observables which are constants of the motion will be found. With further work, I believe that my discussion can be extended to answer some of the criticisms of Kuchar of the Rovelli and the Page-Wootters proposals on the problem of time.

Before beginning the analysis of observables, return to the formulation of the problem of the frozen formalism. To be assured the problem doesn't lie in the assumptions, consider each of the hypotheses leading up to it. The stated definition of an observable is a sensible one as the following argument shows. In a constrained Hamiltonian system, the set of (first-class) constraints $\{C_i\}$ $(i = 1, \ldots, N)$ define a subspace in the full Hilbert space of an unconstrained system. A state $\Psi$ in this subspace satisfies the constraint equations $C_i \Psi = 0$. When $\Psi$ is acted upon by the observable $A$, one requires that the result $A\Psi$ remain in the constrained subspace. The condition for this is $[A, C_i] = f_i(C_1, \ldots, C_N)$ because then

$$C_i A \Psi = -[A, C_i]\Psi = -f_i(C_1, \ldots, C_N)\Psi = 0.$$

If one were to weaken the definition of an observable by not requiring that it commute with the super-Hamiltonian, as is sometimes done[3], then one must deal with the difficult problem of operators whose action takes one out of one's Hilbert space. This is not an adequate strategy for dealing with the problem of the frozen formalism; it trades one problem for a harder one.

If the difficulty is not in this definition of an observable, perhaps it lies in the fact the super-Hamiltonian is a constraint. Constraints are often a consequence of a symmetry underlying the theory. In the ADM canonical quantization of gravity, it is well-known that invariance of the theory under space-time diffeomorphisms makes the super-Hamiltonian a constraint. In the parametrized canonical formulation of quantum mechanics, reparametrization invariance of the theory makes the super-Hamiltonian a constraint. In both cases, the symmetry making the super-Hamiltonian a constraint is a physically motivated symmetry which is not to be given up lightly.

The problem of the frozen formalism is thus a real one, at least in so far as it reflects a weakness in our understanding. It does not however prevent one from just using the familiar machinery of quantum mechanics. For this reason, it is most often consigned to the limbo of "peculiarities of the quantum formalism," and is either dismissed as a problem in semantics or simply not addressed.

There is without doubt a semantic component to the problem. In common usage, the word "observable" has the connotation "something which can be observed." In ordinary quantum mechanics, it is defined as a self-adjoint operator with complete spectrum. In parametrized and constrained quantum mechanics, it is defined as an operator, not necessarily self-adjoint, which commutes with all of the constraints. The task is to distinguish these meanings. In so doing, we shall find that the problem of the frozen formalism is more subtle than confusing one word with three meanings. It will hinge on how we describe physical measurements in the mathematical formalism of quantum mechanics. I will give an explication of the problem by way of a few examples. These will show that there is no problem with working in the frozen formalism: there are both constant observables and dynamics within the wavefunction of the universe.

The essential property of an observable in both its mathematical definitions is that it has an associated (complete) collection of eigenstates with corresponding eigenvalues. The significance of this is that states can be characterized by the eigenvalues of a collection of commuting observables. The eigenvalues are the quantum numbers of the state. In the parametrized formalism, these eigenvalues characterize the state throughout its entire evolution. This is why they are constants of the motion. If an operator does not commute with the super-Hamiltonian constraint, its eigenstates are not in the constrained Hilbert space and are then of no use for representing states in the constrained Hilbert space.

Because the eigenstates of observables are assumed to be complete, one may represent states as superpositions of eigenstates. The coefficients in the superposition will be constant. It is not necessary to know the observables of which the full state is the eigenstate, though they can be constructed if it is desired.

To firmly establish this perspective on observables, consider the parametrized free particle with the super-Hamiltonian

$$\mathcal{H} = p_0 + p_1^2. \tag{1}$$

Physical states $|\Psi\rangle$ are those which satisfy the super-Hamiltonian constraint

$$\mathcal{H}\Psi = 0. \tag{2}$$

The operator $p_1$ commutes with the super-Hamiltonian and is an observable. Its associated eigenstates may be labelled by the eigenvalue $k$, where

$$p_1|k\rangle_1 = k|k\rangle_1,$$

and, in the coordinate representation (assuming the canonical commutation relations $[q_0, p_0] = i$, $[q_1, p_1] = i$), they are

$$\langle q_1, q_0|k\rangle_1 = \frac{1}{(2\pi)^{1/2}} e^{ikq_1 - ik^2 q_0}. \tag{3}$$

The operator $q_1$ does not commute with the super-Hamiltonian and is not an observable. In particular the state $q_1|k\rangle_1$ does not satisfy the super-Hamiltonian constraint.

An operator closely related to $q_1$ which is an observable is

$$q_{1t} = e^{-ip_1^2(q_0 - t)} q_1 e^{ip_1^2(q_0 - t)} = q_1 - 2p_1(q_0 - t). \tag{4}$$

This is the observable which is equal to $q_1$ at time $q_0 = t$. It is one of Rovelli's "evolving constants of the motion"[2]. Its eigenstates are characterized by

$$q_{1t}|x\rangle_1 = x|x\rangle_1.$$

In the coordinate representation, this is

$$\langle q_1, q_0|x\rangle_1 = (4\pi i(q_0 - t))^{-1/2} e^{i(q_1 - x)^2/4(q_0 - t)}. \tag{5}$$

This may be recognized as the Green's function for the free particle, which reduces to $\delta(q_1 - x)$ as $q_0 \to t$. (The states are normalized using the usual inner product with respect to $q_1$, but this won't be discussed here.)

A Gaussian superposition of momentum eigenstates can be formed by

$$|g; \overline{k}, a\rangle_1 = (\pi a/2)^{-1/4} \int dk e^{-(k-\overline{k})^2/a} |k\rangle_1. \tag{6}$$

This has the coordinate representation

$$\langle q_1, q_0 | g; \overline{k}, a\rangle_1 = (2\pi a)^{-1/4}(iq_0 + 1/a)^{-1/2} \exp(-\overline{k}^2/a) \exp\left(\frac{i(q_1 - 2i\overline{k}/a)^2}{4(q_0 - i/a)}\right).$$

An observable of which this state is an eigenstate is found to be

$$G = q_1 - 2p_1(q_0 - i/a), \tag{7}$$

and the state has eigenvalue $2i\overline{k}/a$. Note that $G$ is not self-adjoint in the usual inner product and its eigenvalue is not real. One expects that this means that it is not physically observable, but to confirm this requires a discussion of measurement.

Measurement theory in the foundation of quantum mechanics has been discussed exhaustively over the past sixty years. To put the use of observables as self-adjoint operators in context, it is necessary to reiterate the litany. I want to emphasize the central role of projection operators in the conventional approach. In contrast, I want to draw attention to an argument from a new perspective compelling the use of a post-Everett description of measurement in which both system and observing apparatus appear explicitly.

In ordinary quantum mechanics, observables as self-adjoint operators play a central role, again through their eigenstates. The conventional description of measurement is the following: If one intends to measure a particular observable, one decomposes the state of the system into a superposition of eigenstates of that observable. The eigenvalues of these eigenstates of the observable are interpreted as the possible outcomes of the measurement. Since the observables are self-adjoint, the eigenvalues, and hence the outcomes of measurement, are necessarily real. The probabilities for each of the outcomes are given by the square-modulus of the coefficients in the superposition. When the measurement is complete, the state of the system is in an eigenstate of the observable.

This procedure is so ingrained in our understanding of quantum mechanics that one easily forgets that it is a theoretical construct and not the measurement process itself. The procedure is primarily based on two *assumptions*[6]: 1) measurement of a state gives a particular result with certainty if and only if the system is in an eigenstate of the observable being measured, and the result is the eigenvalue of that eigenstate; 2) from "physical continuity," after a measurement is made, if that measurement is immediately repeated, the same outcome must be obtained with certainty, and, hence, by 1), the measurement must put the system into an eigenstate of the observable. These two assumptions characterize measurements, distinguishing them from other interactions, and are thus the fundamental tie between the physically observed and the mathematically observable, between measurement outcomes and eigenstates of operators. Few would doubt the validity of the assumptions. I do not claim that the procedure does not work, but rather that it works too well.

Let us call this description of measurement "the projection procedure," as one projects the initial state onto the eigenstates of the observable being measured. This projection procedure neatly summarizes the results of measurement, but does so at the cost of neglecting a description of the process by which the measurement is made. It is as if an external agent is able to effect a measurement on the system without need of introducing any apparatus: suddenly, the measurement is done. The description is wholly isolated. Only the system is present, and the measurement has direct access to its state. Unfortunately, we do not share this luxury of direct access to states. By necessity, we must always employ intermediaries to investigate the state of a system.

A question that we are accustomed to ask in quantum mechanics is

"What is the probability density that the momentum of particle-1 in state $|\Psi\rangle_1$ is $k$?"

Suppose the state $|\Psi\rangle_1$ is the gaussian superposition of momentum eigenstates (6) in the example above. The question inquires directly about the state of particle-1, and, in the projection procedure, the question is meaningful and has the familiar answer $(\pi a/2)^{-1/2} e^{-2(k-\bar{k})^2/a}$. This is not however an entirely sensible question in the context of a system described by a super-Hamiltonian constraint. To verify the answer, we must conduct an experiment. The state solving the super-Hamiltonian constraint is the wavefunction of the universe and contains, along with everything else, all measurements and their outcomes. In fact, no measurements were ever made. The question has no truth value because its answer can be neither confirmed nor denied.

To address the question, additional subsystems must be introduced which interact with particle-1 to produce the measurement. For the purposes of theory, these additional subsystems may be hypothetical, as we need not do every experiment we contemplate, but we must augment the hypothesized super-Hamiltonian as if the experiment were to be performed. In the event that it is, we can then expect to confirm or deny our theoretical result. This treatment of the super-Hamiltonian carries an important resonance with Bohr's insistence that reality is determined by the full experimental arrangement[7]: the choice of experiments determines the super-Hamiltonian; the super-Hamiltonian (plus initial conditions) determines the wavefunction of the universe and hence reality.

An essential consequence of this is that, to understand the measurement process properly, one must model the interaction. It is not enough to add apparatus subsystems to the super-Hamiltonian if one continues to treat measurement as a black box which spontaneously changes the combined system and apparatus state from an uncorrelated to a correlated superposition. This is essentially still the projection procedure, albeit without the final selection of a particular term from the correlated superposition.

Before investigating such a model explicitly, consider the characteristics it must possess. Our goal is to understand the relation between observables as self-adjoint operators and physical measurements. As the correspondence between them is made through the assumptions underlying the projection procedure, we desire a model which is as close to the projection procedure as possible while being more specific about the details of the interaction. In particular, we require that a measurement of a chosen observable return a result which distinguishes between different eigenstates of the observable and that it have the property that if the measurement is immediately repeated, the same result will be found with certainty. This type of model was discussed by von Neumann[5] and plays an important role in the Everett interpretation[8]. I will discuss it again to emphasize certain features.

If one has an isolated state being observed without apparatus, as in the projection procedure, the only quantity which distinguishes between eigenstates of an observable are their eigenvalues. This is why a measurement in the projection procedure must return the eigenvalue of the eigenstate. In a more general setting, in which the state of one subsystem interacts with another to perform a measurement, the result need only be a (non-degenerate) correlation of the states of the observing subsystem with the eigenstates of the observable in the observed subsystem. This correlation allows one to infer the state of one system from the state of the other. Since the eigenstates of the observed subsystem are characterized by their eigenvalues, one may say that the measurement has returned the eigenvalue, in the sense that the eigenvalue can be inferred from knowledge of the state of the observing subsystem. This is however an abstraction: the eigenvalue is not an extant physical quantity. The physical result of a measurement is the correlation of the states of subsystems.

The second criterion—that if the measurement is immediately repeated, the same result is obtained with certainty—is a requirement that the measurement be non-destructive[8]. That is, if the observed subsystem is in an eigenstate of the observable, this eigenstate must be preserved after the interaction, so that it may be measured again and found to give the same result. This rules out, as measurements, interactions which correlate the state of the observing subsystem with the state of the observed system before the interaction but leave it disturbed after the interaction. As one might expect, this restricts the interaction terms that may be classified as measurements in the projection procedure sense. This is significant because it reveals that the projection procedure is an idealization of the process of measurement. There are interactions which are considered measurements in experimental practice that are not measurements in this sense.

A further idealization of the process of measurement in the projection procedure is that it is instantaneous. This feature is not retained in the model system: necessarily all measurements implemented by interaction require finite duration. The implications of this regarding observables will be discussed below. I remark here that this is a profound departure from the projection procedure in both its Copenhagen and Everett incarnations. It has been lamented[9, 5, 10] that one of the most serious failings of the quantum mechanical formalism, especially from the perspective of relativity, is the fact that measurements take place at a precise instant of time. This is where this begins to change. Measurements as projections, and as results computed from expectation values, take place at a precise instant of time. Measurements as interactions require duration.

In the post-Everett view, where the outcome of a measurement is a correlation between subsystems, the second criterion is a question of conditional probability. One confirms that it is satisfied by using the Page-Wootters interpretation[4]. One requires two observing subsystems. Sequentially, each interacts with the observed subsystem establishing correlation with the observed subsystem. The question is then posed: given the result of the first of the measurement, is the probability certain that the result of the second is the same? The answer is yes, by construction. When the first observing subsystem interacts with the observed subsystem, it establishes a correlation which distinguishes the different eigenstates of the observed subsystem. In the manner in which one handles conditional probabilities, one discards all the states except for the one whose correlation reflects the given result of the first measurement. The second observing subsystem then interacts only with an eigenstate of the observable, not with a superposition, and establishes a correlation which is the same as that of the first subsystem. The only thing that could go wrong would be if the observed subsystem is not still in an eigenstate of the observable, but the measurement-interaction is chosen so that this cannot happen.

Consider a model of a measurement of the momentum $p_1$ of particle-1 in the example above. We introduce a second free particle, particle-2, which interacts with particle-1 through the measurement-interaction (cf. [5])

$$\mathcal{H}_I = a(q_0)p_1 q_2. \tag{8}$$

Since the interaction couples to the observable, it will preserve the eigenstates of the observable through the measurement-interaction. Here, $a(q_0)$ is a smooth function which vanishes outside the interval $0 < q_0 < T$ and for which $\int_0^T a(q_0')dq_0' = 1$. It can be viewed as a phenomenological summary of a more detailed process by which particle-1 and particle-2 are brought together to interact. The full super-Hamiltonian is then

$$\mathcal{H} = p_0 + p_1^2 + p_2^2 + a(q_0)p_1 q_2. \tag{9}$$

This problem can be exactly solved, using for example canonical transformations [11] (cf. also [12]). Define

$$A(q_0) = \begin{cases} 0 & q_0 < 0 \\ \int_0^{q_0} a(q_0')dq_0' & 0 \le q_0 \le T. \\ 1 & q_0 > T \end{cases} \tag{10}$$

The super-Hamiltonian $\mathcal{H}$ with the interaction term is related to the super-Hamiltonian $\mathcal{H}_0 = p_0 + p_1^2 + p_2^2$ without interaction term by a time-dependent canonical transformation $C_{q_0}$,

$$\mathcal{H} = C_{q_0} \mathcal{H}_0 C_{q_0}^{-1},$$

where

$$C_{q_0} = e^{-ip_1^2 \int_{-\infty}^{q_0} A^2(q_0')dq_0'} e^{-iA(q_0)p_1 q_2} e^{2ip_1 p_2 \int_{-\infty}^{q_0} A(q_0')dq_0'}. \tag{11}$$

The solutions $|\Psi\rangle$ of $\mathcal{H}$ are given in terms of those $|\Psi_0\rangle$ of $\mathcal{H}_0$ by

$$|\Psi\rangle = C_{q_0}|\Psi_0\rangle.$$

Assume that particle-1 is initially in an eigenstate of momentum $p_1$, $|k\rangle_1$, and that particle-2 is in an eigenstate of momentum $p_2$, $|k_2\rangle_2$, so that

$$|\Psi_0\rangle = |k\rangle_1 |k_2\rangle_2.$$

The coordinate representation of the solution $|\Psi\rangle$ is

$$\begin{aligned} \langle q_1, q_2, q_0 | \Psi \rangle &= \langle q_1, q_2, q_0 | C_{q_0} | k \rangle_1 | k_2 \rangle_2 \\ &= \frac{1}{2\pi} \exp(ikq_1 + i(k_2 - A(q_0)k)q_2 - i(k^2 + k_2^2)q_0 \\ &\quad + i2kk_2 \int_{-\infty}^{q_0} A(q_0')dq_0' - ik^2 \int_{-\infty}^{q_0} A^2(q_0')dq_0'). \end{aligned} \tag{12}$$

The state evolves smoothly from $|k\rangle_1 |k_2\rangle_2$ before $q_0 = 0$ to $e^{i\phi(k,k_2)}|k\rangle_1 |k_2 - k\rangle_2$ after $q_0 = T$. A phase $\phi(k, k_2)$ arises in the evolution and, explicitly,

$$\phi(k, k_2) = i2kk_2(c_1 - T) - ik^2(c_2 - T),$$

where

$$c_1 = \int_0^T A(q_0')dq_0'$$

and

$$c_2 = \int_0^T A^2(q_0')dq_0'.$$

The state of particle-2 is correlated with that of particle-1 after the evolution, and the eigenstate of particle-1 has not been disturbed. A measurement has been performed.

If particle-1 were initially in the Gaussian superposition of momentum-eigenstates (6), the measurement would have produced the smooth transition to a superposition of correlated states

$$\left((\pi a/2)^{-1/4} \int dk e^{-(k-\bar{k})^2/a} |k\rangle_1\right) |k_2\rangle_2 \quad \longrightarrow \tag{13}$$

$$(\pi a/2)^{-1/4} \int dk e^{-(k-\bar{k})^2/a} e^{i\phi(k,k_2)} |k\rangle_1 |k_2 - k\rangle_2.$$

Suppose one introduces a second observer, particle-3, in the same initial state $|k_2\rangle_3$ as particle-2, and couples it to particle-1 for an interval after $q_0 = T$ through a term analogous to (8). Given that the result of the first measurement is $k'$, i.e. the correlation $|k'\rangle_1 |k_2 - k'\rangle_2$, the result of the second measurement will be the correlated state $|k'\rangle_1 |k_2 - k'\rangle_2 |k_2 - k'\rangle_3$. The same measurement result is obtained, as required.

Let us now consider the role of observables. In the absence of the interaction term (8), both $p_1$ and $p_2$ commute with the super-Hamiltonian $\mathcal{H}_0$ and are observables. In the presence of the interaction, $p_2$ is no longer an observable. This is consistent with the fact that the initial state of particle-2, which is characterized by its eigenvalue with respect to $p_2$, changes during the interaction. Even though $p_2$ is not an observable, a modification gives an observable

$$\tilde{p}_2 = C_{q_0} p_2 C_{q_0}^{-1} = p_2 + A(q_0)p_1. \tag{14}$$

The full quantum wavefunction (12) over the whole history of the universe has the eigenvalue $k_2$ for $\tilde{p}_2$ and the eigenvalue $k$ for the observable $p_1$. These eigenvalues label the state, and they are constants throughout the evolution of the state. Nevertheless a measurement has been made. There is no loss of dynamics because one has chosen to work in the frozen formalism.

A closer examination of the relation between observables and dynamics will be illuminating. Note that $\tilde{p}_2$ agrees with $p_2$ when $q_0 < 0$. For this restricted portion of the universe, $p_2$ is an observable in the sense that it commutes with the super-Hamiltonian, and it can be used to label states in this region. This suggests that it is useful to distinguish between a restricted observable which commutes with the super-Hamiltonian in some region and a global observable which commutes with the super-Hamiltonian everywhere.

As participants in the universe, we do not of course know the full super-Hamiltonian which describes it. There will be measurements made in the future which we cannot anticipate now. Since we only discover the details of the super-Hamiltonian of the universe as we go along, we cannot know the global observables which commute with the super-Hamiltonian of our universe. When we say that the states of subsystems we observe are in eigenstates of some observables, they are in eigenstates of restricted observables. For some period of time, those observables commute with the super-Hamiltonian of the universe, and their eigenstates are unchanging with respect to eigenstates of other observables that also commute with the super-Hamiltonian.

To elaborate on this further, consider the observable $p_1$ which is being measured. In the example here, it is both an observable in the sense that it commutes with the super-Hamiltonian and in the sense that a correlation with its eigenstates is established during the measurement-interaction with particle-2. I want to emphasize that it is not necessary that $p_1$ commute with the super-Hamiltonian for all $q_0$, so long as it does so in the neighborhood of the period of measurement.

Suppose one considers the measurement of $p_1$ when the state of particle-1 at $q_0 = 0$ is the gaussian superposition (6). One could add a $q_1$-dependent term to the super-Hamiltonian which evolves some initial state of particle-1 into the gaussian superposition and turns off before $q_0 = 0$, when the measurement begins. Or, one could add such a term some time after $q_0 = T$ when the measurement is complete, and the final state of particle-1 in each correlated state of the superposition would evolve away from a momentum eigenstate. In each case, the momentum of particle-1 in the gaussian superposition state at $q_0 = 0$ would still be measured, but $p_1$ would only be a restricted observable. It would not commute with the super-Hamiltonian if there were $q_1$-dependent terms present. Not being a global observable means that the eigenvalue of $p_1$ could not be used as a quantum number for the wavefunction of the universe, but this is not a serious loss. If one's primary concern is with predictions of the outcomes of measurement, restricted observables are more relevant than global ones.

The nature of observables can be still more closely investigated. At each instant $q_0 = t$, the state of particle-2 is instantaneously an eigenstate of the self-adjoint operator $p_2$ with eigenvalue $k_2 - A(t)k$. In the ordinary quantum mechanical sense, $p_2$ is an observable. One can compute expectation values of it at any time $q_0$, and one thinks of these as predictions of the outcomes of possible measurements. Now, $p_2$ is not a global observable, and it doesn't commute with $\mathcal{H}$ at $q_0 = t$ when $0 < t < T$, so it isn't always a restricted observable. Nevertheless, just as $q_1$ at time $q_0 = t$, in the first example, was made into a global observable above by evolving it with the Hamiltonian, $p_2$ can be made a global observable by applying the canonical transformation $C_{q_0} C_t^{-1}$. The observable is

$$p_{2t} = C_{q_0} C_t^{-1} p_2 C_t C_{q_0}^{-1} = p_2 + A(q_0)p_1 - A(t)p_1. \qquad (15)$$

This gives a family of observables $p_{2t}$ which reduce to the operator $p_2$ at time $q_0 = t$ of which particle-2 is instantaneously an eigenstate. As the state of the system evolves through the measurement, the eigenstate of particle-2 changes at each instant as the observable of which it is the eigenstate changes. In ordinary quantum mechanics, when one speaks of the self-adjoint operator $p_2$ as an observable, one is referring to $p_{2t}$.

Incidentally, this answers Kuchar's criticism that the Page-Wootters conditional probability interpretation does not give the correct answer for propagators[3]. The observables for the position at two distinct instants of time are different, as given by (4).

If, at time $q_0 = T$, one wants to predict the probability of finding the particle at some location at a later instant $q_0 = T'$, one must compute the conditional probability that the particle is in an eigenstate of $q_{1T'}$. If one uses $q_{1T}$ as the position observable for all time, the particle will not appear to move, as Kuchar rightly argues.

It is generally true that an operator at an instant of time can be promoted into a global observable, and hence one has a family of observables parametrized by the time. These are Rovelli's evolving constants of the motion[2]. As these observables change, the eigenstates associated with them change as well. This change embodies the evolution of states.

One may ask whether these observables are all physically measurable. That is, can one introduce an observing subsystem that will correlate with the momentum of particle-2 at time $q_0 = t$ for $0 < t < T$? My answer is no. While one may formally calculate expectation values for the momentum $p_{2t}$ at these times, these calculations do not refer to the results of any physical experiment that can be done, in the projection procedure sense. There are two related difficulties. First, all physical measurements require finite duration in order to establish correlations between the observing and the observed subsystems. This is itself a subject requiring further elaboration, but for the moment suffice it to say that, since the eigenvalue of the operator $p_2$ is changing, an attempted measurement can at best measure an averaged value and not the specific momentum at time $q_0 = t$. Moreover, one expects that no coupling exists which will leave the changing value of $p_2$ undisturbed, so that the measurement of $p_1$ is unaffected. Secondly, because $p_2$ is dynamically changing, it is impossible to arrange that a second measurement will find the same result with certainty. One can couple to the observable which corresponds to the instantaneous momentum eigenstate of particle-2 at time $q_0 = t$, but as it will obtain an average result over a different interval than the first measurement, the results will in general be different. This would not then be a measurement in the projection procedure sense.

Thus, only restricted observables can be physically measured in the projection procedure sense. One is led to the conclusion that the assumptions about the nature of measurement that lie at the foundation of the projection procedure are too idealized. By postulating instantaneous non-disruptive measurement, they both exclude physically relevant measurement-interactions and allow computations for the outcomes of experiments that cannot be realized. It is evident that further work on measurement theory outside the projection procedure framework is necessary.

To close one final loose-end, consider whether the non-self-adjoint observable $G$ (7) can be physically observed. Mathematically, the answer would seem to be yes: one uses a coupling analogous to (8) with $p_1$ replaced by $G$. This would establish a correlation between the state of particle-2 and the $G$-eigenstate of particle-1. There is however a difficulty. Since $G$ is a complex operator, it is not evident that there exists a physical device which can realize the proposed coupling. This serves to emphasize a very important point. In the laboratory, we are restricted to a handful of possible interactions. One must bear in mind that these are the building blocks from which we must ultimately build our super-Hamiltonian.

The following picture of dynamics in the frozen formalism can be assembled from the foregoing discussion. The full quantum state representing the "wavefunction of the universe" is fixed once the initial conditions and the super-Hamiltonian are given. This includes all measurements that will be made during the course of the universe. Dynamical evolution is a process that takes place in the form of changes in the decomposition of the full state into subsystem eigenstates. The wavefunction of the universe need not be expressed as a product state of eigenstates of its global observables. It may of course be represented as a superposition of such eigenstates. More generally it may be represented in terms of eigenstates of operators which are observables only in restricted regions of the universe, or in terms of eigenstates of families of global observables parametrized by the time. When the wavefunction of the universe is expressed in such a fashion, one finds that as the collection of observables used to decompose the state change, the superposition of eigenstates change. This is what gives us the impression of dynamical evolution: it is the changing collection of correlations amongst the eigenstates of restricted observables that constitutes what we observe.

The self-adjoint operators that we speak of in ordinary quantum mechanics as observables are members of families of global observables parametrized by the time. Because any measurement made through interactions requires finite duration to establish correlations between the observing and the observed subsystems, only restricted observables which commute with the super-Hamiltonian through the period of measurement are physically measurable, in the projection procedure sense. In particular, this means that one can compute expectation values for many self-adjoint operators which do not refer to the outcomes of physically realizable experiments. If one is interested in physics, care must be taken with the use of expectation values. More importantly, one must appreciate that the projection procedure, which so strongly colors our perception of quantum mechanics, overly idealizes measurement as instantaneous and non-destructive. Recognizing that a proper description of measurements within the quantum formalism requires interactions of finite duration is a first step towards resolving the long-standing conflict over the role of time in quantum mechanics and relativity.

# References

[1] A selection of papers raising the issue of the frozen formalism are, Misner C.W., *Feynman Quantization of General Relativity. Rev. Mod. Phys.* **29** (1957), p. 497; Bergmann P.G. and Komar A.B., *Status Report on the Quantization of the Gravitational Field.* in *Recent Developments in General Relativity*, Pergamon Press, New York, 1962, p. 31; Komar A., *Hamilton-Jacobi Quantization of Gen-*

*eral Relativity. Phys. Rev.* **153** *(1967), p. 1385;* DeWitt B.S., *Quantum Theory of Gravity. I. Phys. Rev.* **160** *(1967), p. 1113;* Brill D. R. and Gowdy R.H., *Quantization of General Relativity. Rep. Prog. Phys.* **33** *(1970), p. 413.*

[2] Rovelli C., *Is There Incompatibility Between the Ways Time is Treated in General Relativity and in Standard Quantum Mechanics.* in Conceptual Problems of Quantum Gravity, eds. A. Ashtekar and J. Stachel, Birkhauser, Boston, 1991, p. 126; *Quantum Mechanics Without Time: A Model. Phys. Rev.* **D42** *(1990),* p. 2638; *Time in Quantum Gravity: An Hypothesis. Phys. Rev.* **D43** *(1991),* p. 442.

[3] Kuchar K., *Time and Interpretations of Quantum Gravity.* in Proceedings of the 4th Canadian Conference on General Relativity and Astrophysics, World Scientific, Singapore, 1992.

[4] Page D. and Wootters W.K., *Evolution Without Evolution: Dynamics Described by Stationary Observables. Phys. Rev.* **D27** *(1983),* p. 2885; Page D.N., *Interpreting the Density Matrix of the Universe.* in Conceptual Problems of Quantum Gravity, eds. A. Ashtekar and J. Stachel, Birkhauser, Boston, 1991, p. 116; Hotke-Page, M.D., C.A. and Page D.N., *Clock Time and Entropy.*, to appear in Physical Origins of Time Asymmetry, eds. J.J. Halliwell, J. Perez-Mercader, and W.H. Zurek, Cambridge Univ. Press, Cambridge (1993).

[5] von Neumann J., *Mathematical Foundations of Quantum Mechanics, transl. by R.T. Beyer.* Princeton Univ. Press, Princeton, N.J., 1955.

[6] Dirac P.A.M., *The Principles of Quantum Mechanics, 4th ed.* Oxford Univ. Press, Oxford, 1958, pp. 35-36.

[7] Bohr N., *Can Quantum-Mechanical Description of Physical Reality be Considered Complete? Phys. Rev.* **48** *(1935),* p. 696.

[8] DeWitt B.S. and Graham N., eds., *The Many-Worlds Interpretation of Quantum Mechanics.* Princeton Univ. Press, Princeton, N.J., 1973.

[9] Schrodinger E., *Die gegenwärtige Situation in der Quantenmechanik. Die Naturwissenschaften* **23** *(1935),* p. 807 as quoted by K. Kuchar in *Canonical Methods of Quantization. Quantum Gravity II: An Oxford Symposium,* eds. C.J. Isham, R. Penrose and D.W. Sciama, Clarendon, Oxford, 1981.

[10] Kuchar K., *Canonical Methods of Quantization. Quantum Gravity II: An Oxford Symposium,* eds. C.J. Isham, R. Penrose and D.W. Sciama, Clarendon, Oxford, 1981.

[11] Anderson A., *Canonical Transformations in Quantum Mechanics. McGill Univ. preprint 92-29, hepth/9205080.*

[12] Kuchar K., *Gravitation, geometry, and nonrelativistic quantum theory.* Phys. Rev. **D22** *(1980)*, p. 1285.

# Jacobi's Action and the Density of States

## J. DAVID BROWN
North Carolina State University, Raleigh

## JAMES W. YORK, JR.
University of North Carolina, Chapel Hill

The authors have introduced recently a "microcanonical functional integral" which yields directly the density of states as a function of energy. The phase of the functional integral is Jacobi's action, the extrema of which are classical solutions at a given energy. This approach is general but is especially well suited to gravitating systems because for them the total energy can be fixed simply as a boundary condition on the gravitational field. In this paper, however, we ignore gravity and illustrate the use of Jacobi's action by computing the density of states for a nonrelativistic harmonic oscillator.

## 1 DEDICATION
We dedicate this paper to Dieter Brill in honor of his sixtieth birthday. His continued fruitful research in physics and his personal kindness make him a model colleague. JWY would especially like to thank him for countless instructive discussions and for his friendship over the past twenty–five years.

## 2 INTRODUCTION
Jacobi's form of the action principle involves variations at fixed energy, rather than the variations at fixed time used in Hamilton's principle. The fixed time interval in Hamilton's action becomes fixed inverse temperature in the "periodic imaginary time" formulation, thus transforming Hamilton's action into the appropriate (imaginary) phase for a periodic path in computing the canonical partition function from a Feynman functional integral (Feynman and Hibbs 1965). In contrast, fixed total energy is suitable for the microcanonical ensemble and, correspondingly, Jacobi's action is the phase in an expression for the density of states as a real–time "microcanonical functional integral" (MCFI) (Brown and York 1993b).

We wish to characterize briefly the canonical and microcanonical pictures. (We shall speak only of energy and (inverse) temperature here, ignoring the other possible conjugate pairs of variables in order to simplify the discussion.) In the canonical picture, with a fixed temperature shared by all constituents of a system, there are no constraints on the energy. This feature simplifies combinatorial (counting) problems for canonical systems and leads to the factorization of the partition function for weakly coupled constituents. For gravitating systems in equilibrium, the temperature is not spatially uniform because of gravitational red and blue–shift effects. In such cases the relevant temperature is that determined at the boundary of the system (York 1986). It can therefore be specified as a boundary condition on the metric (York 1986, Whiting and York 1988, Braden et al. 1990) and used in conjunction with Hamilton's principle, which is the form of the action for gravity in which the metric is fixed on the boundary (Brown and York 1992, 1993a). (The metric determines the lapse of proper time along the boundary.) On the other hand, equilibrium in the canonical picture is not always stable when gravity is present, as is well known. For some pertinent examples, see York (1986), Whiting and York (1988), and Braden et al. (1990).

With its constraint on the energy, the microcanonical picture leads to more robust stability properties. However, the energy constraint can complicate calculations of relevant statistical properties because the constituents of the system share from a common fixed pool of energy. For field theories, with a continuous infinity of degrees of freedom, the energy constraint restricts the entire phase space of the system *unless gravity is taken into account*. For gravitating systems, as a consequence of the equivalence principle, the total energy including that of matter fields is an integral of certain derivatives of the metric over a two–surface bounding the system. Therefore, if we specify as a boundary condition the energy per unit two–surface area, we have constrained the total energy simply by a boundary condition (Brown and York 1993a, 1993b). Thus, the canonical and microcanonical cases differ only in which of the conjugate variables (Brown et al. 1990), inverse temperature or energy, is specified on the boundary. The corresponding functional integrals, for partition function or density of states, differ in which action gives the correct phase, Hamilton's or Jacobi's.

We have recently applied this reasoning to the case of a stationary black hole (Brown et al. 1991a, 1991b, Brown and York 1993b). The MCFI, in a steepest descents approximation, shows that the logarithm of the density of states is one–quarter of the area of the event horizon (that is, the Bekenstein–Hawking entropy) (Brown and York 1993b). In the present paper we shall disregard gravity and obtain the density of states for a nonrelativistic harmonic oscillator. This is a relatively simple situation in which to recall the properties of Jacobi's action and to see the MCFI at work.

## 3 JACOBI'S ACTION
Consider, for simplicity, a particle of mass $m$ with a one–dimensional configuration space. The Lagrangian form of Jacobi's action is (Lanczos 1970, Brown and York

1989)

$$S_E[x] = \int dx \sqrt{2m[E - V(x)]} \,, \tag{1}$$

where $V(x)$ is the potential energy and the energy $E$ is a fixed constant. $S_E[x]$ is extremized by varying the path freely except that the end points are fixed. Now introduce a parameter $\sigma$ increasing monotonically from $\sigma'$ at one end of the path to $\sigma''$ at the other. Denoting $dx/d\sigma$ by $\dot{x}$, we can write the action as

$$S_E[x] = \int_{\sigma'}^{\sigma''} d\sigma\, \dot{x}\sqrt{2m[E - V(x)]} \,, \tag{2}$$

where $x' = x(\sigma')$ and $x'' = x(\sigma'')$ are fixed. Jacobi's action is invariant under changes $\delta x$ induced by changes of parameterization that preserve the end–point values of $\sigma$.

For constructing the MCFI, we employ the canonical form of Jacobi's action. Because of the reparameterization invariance of $S_E[x]$, the corresponding canonical Hamiltonian $\dot{x}(\partial L/\partial \dot{x}) - L$ vanishes identically. Furthermore, the canonical momentum

$$p = \frac{\partial L}{\partial \dot{x}} = [2m(E - V)]^{1/2} \tag{3}$$

is independent of $\dot{x}$ in one dimension and, in general, does not allow one to solve for all the $\dot{x}$'s as functions of the $p$'s. Indeed, from (3) we obtain the "Hamiltonian constraint"

$$\mathcal{H}(x,p) \equiv \frac{p^2}{2m} + V(x) - E \approx 0 \,. \tag{4}$$

Because the canonical Hamiltonian is zero, there are no secondary constraints and $\mathcal{H}$ is then trivially first class. Jacobi's action in canonical form is thus

$$S_E[x,p,N] = \int_{\sigma'}^{\sigma''} d\sigma[p\dot{x} - N\mathcal{H}(x,p)] \,, \tag{5}$$

where $N$ is a Lagrange multiplier. The equations of motion following from variation of (5) are

$$\dot{x} = N[x,\mathcal{H}] = \frac{Np}{m} \tag{6}$$

$$\dot{p} = N[p,\mathcal{H}] = -N\frac{\partial V}{\partial x} \tag{7}$$

$$\mathcal{H} = \frac{p^2}{2m} + V - E = 0 \,. \tag{8}$$

Combining (6) and (8) determines the multiplier as

$$N = \dot{x}[2(E - V)/m]^{-1/2} \,. \tag{9}$$

The interpretation of (9) is that

$$dt = N \, d\sigma \tag{10}$$

is the lapse of physical time, in accordance with the definition of energy.

The canonical statement of reparameterization invariance is that the action (5) is invariant under the gauge transformation given by

$$\delta x = \epsilon[x, \mathcal{H}] \,, \tag{11}$$
$$\delta p = \epsilon[p, \mathcal{H}] \,, \tag{12}$$
$$\delta N = \dot{\epsilon} \,, \tag{13}$$

where $\epsilon(\sigma') = \epsilon(\sigma'') = 0$. With the choice

$$\epsilon(\sigma) = \left(\frac{\sigma - \sigma'}{\sigma'' - \sigma'}\right) T - \int_{\sigma'}^{\sigma} d\alpha \, N(\alpha) \,, \tag{14}$$

where $T$ is the total time

$$T = \int_{\sigma'}^{\sigma''} d\sigma \, N(\sigma) \,, \tag{15}$$

the lapse function is transformed to a constant, namely, $N = T/(\sigma'' - \sigma')$. This shows that every history is gauge related to a history with a constant lapse, and the time $T$ is the gauge invariant part of the lapse function. If the histories under consideration are restricted to those with constant lapse, the gauge freedom of Jacobi's action is removed and (5) becomes

$$S_E[x, p; T] = \int_{\sigma'}^{\sigma''} d\sigma[p\dot{x} - T\mathcal{H}/(\sigma'' - \sigma')] \,. \tag{16}$$

This form of Jacobi's action is a functional of $x(\sigma)$ and $p(\sigma)$ and an ordinary function of the time interval $T$. The classical equations of motion for (16), that is, the conditions for the extrema of (16), are given by (6) and (7) with $N = T/(\sigma'' - \sigma')$ along with

$$0 = \frac{\partial S_E}{\partial T} = -\frac{1}{(\sigma'' - \sigma')} \int_{\sigma'}^{\sigma''} d\sigma \, \mathcal{H} \,. \tag{17}$$

Since (6) and (7) imply that $\mathcal{H}$ is constant, equations (6), (7), and (17) together imply $\mathcal{H} = 0$. It follows that the form (16) for Jacobi's action is classically equivalent to (5), but has no gauge freedom.

## 4 FUNCTIONAL INTEGRAL FOR JACOBI'S ACTION

The functional integral associated with Jacobi's action can be constructed by integrating over all histories $x(\sigma)$, $p(\sigma)$, $T$, with fixed endpoints $x(\sigma') = x'$ and $x(\sigma'') = x''$, where the phase for each history is given by the action (16). Thus, the functional integral is

$$Z_E(x'', x') = \frac{1}{2\pi\hbar} \int dT \int_{x(\sigma')=x'}^{x(\sigma'')=x''} \mathcal{D}x\mathcal{D}p \exp\left\{\frac{i}{\hbar} \int_{\sigma'}^{\sigma''} d\sigma[p\dot{x} - T\mathcal{H}/(\sigma'' - \sigma')]\right\}, \quad (18)$$

where $\mathcal{D}x\mathcal{D}p$ is (formally) the product over $\sigma$ of the Liouville phase space measure $dx(\sigma)dp(\sigma)/(2\pi\hbar)$. The integration measure in (18) can be justified by appealing to a BRST analysis based on the canonical action (5), as is done in the Appendix of Brown and York (1993b).

The functional integral over $x(\sigma)$ and $p(\sigma)$ in (18) has the familiar form of the path integral associated with Hamilton's action, where $\sigma$ plays the role of time and the Hamiltonian is $T\mathcal{H}/(\sigma'' - \sigma')$. This path integral can be written as the matrix elements of the evolution operator $\exp(-iT\hat{\mathcal{H}}/\hbar)$, so the path integral for Jacobi's action becomes

$$Z_E(x'', x') = \frac{1}{2\pi\hbar} \int dT < x''|e^{-iT\hat{\mathcal{H}}/\hbar}|x' > . \quad (19)$$

Hence, taking the integration of $T$ over all real values, we have

$$Z_E(x'', x') = < x''|\delta(\hat{\mathcal{H}})|x' > . \quad (20)$$

Note that $Z_E(x'', x')$ satisfies the time independent Schrödinger equation, namely $\hat{\mathcal{H}}Z_E(x'', x') = 0$ (where $\hat{\mathcal{H}}$ acts on the argument $x''$), since formally $\hat{\mathcal{H}}\delta(\hat{\mathcal{H}}) = 0$.

From (20) it follows that the trace of $Z_E(x'', x')$ yields the density of states

$$\nu(E) = \int dx\, Z_E(x, x) = \text{Tr}\delta(\hat{\mathcal{H}}) = \text{Tr}\delta(E - \hat{H}) , \quad (21)$$

where $\hat{H}$ is the usual Hamiltonian operator. By combining this result with (18) we find that $\nu(E)$ can be written directly as a functional integral, the MCFI:

$$\nu(E) = \frac{1}{2\pi\hbar} \int dT \int \mathcal{D}x\mathcal{D}p \exp\left\{\frac{i}{\hbar} \int_{-\pi}^{\pi} d\sigma[p\dot{x} - T\mathcal{H}/2\pi]\right\} . \quad (22)$$

For later convenience, the endpoint parameter values have been chosen to be $\sigma' = -\pi$ and $\sigma'' = \pi$. The derivation of the path integral (22) for the density of states shows that the integration can be described as a sum over all phase space curves that begin and end at some "base point" $x(\pi) = x(-\pi) = x$, plus an integral over the base point $x$. Then roughly speaking, the density of states is given by a sum over all periodic histories. However, to be precise, it should be recognized that the sum in (22) counts each closed phase space curve a continuous infinity of times because any point on the curve can serve as the base point $x$. Also observe that the integration in (22) is over all real values of the time interval $T$, rather than just positive values. This implies that the functional integral for $\nu(E)$ consists of a sum over *pairs* of histories with members contributing equal and opposite phases (Brown and York 1993b). As a consequence, the density of states so constructed is real.

## 5 DENSITY OF STATES FOR THE HARMONIC OSCILLATOR

We now turn to the evaluation of the MCFI (22) for the density of states of a simple harmonic oscillator with angular frequency $\omega$ and Hamiltonian constraint

$$\mathcal{H} = \frac{p^2}{2m} + \frac{m\omega^2 x^2}{2} - E . \tag{23}$$

The periodic nature of the histories suggests the use of Fourier series techniques (Feynman and Hibbs 1965) for this calculation. Accordingly, write the phase space coordinates as

$$x(\sigma) = a_0 + \sum_{k=1}^{\infty}(a_k \cos k\sigma + b_k \sin k\sigma) , \tag{24}$$

$$p(\sigma) = c_0 + \sum_{k=1}^{\infty}(c_k \cos k\sigma + d_k \sin k\sigma) . \tag{25}$$

The functional integral over $x(\sigma)$ and $p(\sigma)$ is replaced by a multiple integral over the coefficients in the Fourier series (24) and (25) with measure

$$\mathcal{D}x\mathcal{D}p = J \, da_0 \, dc_0 \prod_{k=1}^{\infty}(da_k \, db_k \, dc_k \, dd_k) . \tag{26}$$

Here, $J$ is (formally) the Jacobian of the transformation from $x(\sigma)$, $p(\sigma)$ to $a_0$, $c_0$, $a_k$, $b_k$, $c_k$, $d_k$. The form (24), (25) of this transformation shows (again, formally) that $J$ should be a *real constant*, and should be independent of $T$, $m$, $\omega$, and $E$. ($J$ should depend on $\hbar$, since $\hbar$ appears in the definition of $\mathcal{D}x\mathcal{D}p$.) One of the goals of the present calculation is to determine the real constant $J$ that characterizes the change of integration variables specified by (24) and (25). Note that by integrating freely over all Fourier coefficients $a_0$, $c_0$, $a_k$, $b_k$, $c_k$, $d_k$, we have each closed phase space curve correctly counted a continuous infinity of times. This is because the values of the Fourier coefficients depend on the choice of base point that is assigned the parameter value $\sigma = \pi$ (identified with $\sigma = -\pi$) on a given closed phase space curve.

With the change of variables (24), (25), the density of states (22) for the harmonic oscillator becomes

$$\nu(E) = \frac{J}{2\pi\hbar} \int dT \int da_0 \, dc_0 \prod_{k=1}^{\infty}(da_k \, db_k \, dc_k \, dd_k) \exp\{iS_E/\hbar\} , \tag{27}$$

where the phase is obtained by substituting the Fourier series for $x(\sigma)$ and $p(\sigma)$ into the action (16):

$$S_E = ET - \frac{m\omega^2 T}{2} a_0^2 - \frac{T}{2m} c_0^2$$
$$- \frac{1}{2} \sum_{k=1}^{\infty} \left\{ 2\pi k(a_k d_k - b_k c_k) + \frac{T}{2m}(c_k^2 + d_k^2) + \frac{m\omega^2 T}{2}(a_k^2 + b_k^2) \right\} . \quad (28)$$

The calculation is simplified by expanding $c_k$ and $d_k$ about the solutions to their "equations of motion". Accordingly, observe that the action (28) is extremized for $c_k$ and $d_k$ that satisfy

$$0 = \frac{\partial S_E}{\partial c_k} = \pi k b_k - \frac{T}{2m} c_k , \quad (29)$$

$$0 = \frac{\partial S_E}{\partial d_k} = -\pi k a_k - \frac{T}{2m} d_k . \quad (30)$$

Thus, define new integration variables $\bar{c}_k$ and $\bar{d}_k$ by

$$c_k = \frac{2\pi m}{T} k b_k + \bar{c}_k , \quad (31)$$

$$d_k = -\frac{2\pi m}{T} k a_k + \bar{d}_k , \quad (32)$$

and the action (28) becomes

$$S_E = ET - \frac{m\omega^2 T}{2} a_0^2 - \frac{T}{2m} c_0^2$$
$$- \frac{1}{2} \sum_{k=1}^{\infty} \left\{ \frac{T}{2m}(\bar{c}_k^2 + \bar{d}_k^2) + \frac{m}{2T}(\omega^2 T^2 - 4\pi^2 k^2)(a_k^2 + b_k^2) \right\} . \quad (33)$$

The integrations over $a_0$, $c_0$, $a_k$, $b_k$, $\bar{c}_k$, and $\bar{d}_k$ are now straightforward since these variables are uncoupled in the action (33). Moreover, for each value of $k$, the integrals over $\bar{c}_k$ are identical to the integrals over $\bar{d}_k$, and the integrals over $a_k$ are identical to the integrals over $b_k$. From these observations it follows that the density of states (27) can be written as

$$\nu(E) = \frac{J}{2\pi\hbar} \int dT\, da_0 dc_0 \exp\left\{ \frac{i}{\hbar} \left[ ET - \frac{m\omega^2 T}{2} a_0^2 - \frac{T}{2m} c_0^2 \right] \right\}$$
$$\times \left( \prod_{k=1}^{\infty} \int da_k d\bar{c}_k \exp\left\{ \frac{i}{\hbar} \left[ \frac{m(4\pi^2 k^2 - \omega^2 T^2)}{4T} a_k^2 - \frac{T}{4m} \bar{c}_k^2 \right] \right\} \right)^2 . \quad (34)$$

Each of these integrals (excluding the integral over $T$) has the form of a Fresnel integral,

$$\int dx \exp(iAx^2) = \sqrt{\frac{\pi}{|A|}} \exp(i\pi \operatorname{sign} A/4) , \quad (35)$$

where the constant $A$ is real. In evaluating (34), it is helpful to note that the square of the Fresnel integral (35) is $i\pi/A$. The result is

$$\nu(E) = -iJ \int dT \frac{1}{\omega T} \prod_{k=1}^{\infty} \left[ \left( \frac{2\hbar}{k} \right)^2 \left( 1 - \frac{\omega^2 T^2}{4\pi^2 k^2} \right)^{-1} \right] \exp\left\{ \frac{i}{\hbar} ET \right\} . \qquad (36)$$

Now use the identity

$$\sin x = x \prod_{k=1}^{\infty} \left( 1 - \frac{x^2}{\pi^2 k^2} \right) \qquad (37)$$

to obtain

$$\nu(E) = \frac{-iJ}{2} \prod_{k=1}^{\infty} \left( \frac{2\hbar}{k} \right)^2 \int dT \frac{1}{\sin(\omega T/2)} \exp\left\{ \frac{i}{\hbar} ET \right\} . \qquad (38)$$

Next, express the inverse of $\sin(\omega T/2)$ as

$$\frac{1}{\sin(\omega T/2)} = \frac{2i}{e^{i\omega T/2} - e^{-i\omega T/2}} = 2i e^{-i\omega T/2} \frac{1}{1 - e^{-i\omega T}}$$

$$= 2i e^{-i\omega T/2} \sum_{n=0}^{\infty} e^{-i\omega Tn} , \qquad (39)$$

and insert this result into (38). Integrating the series term–by–term, we obtain

$$\nu(E) = 2\pi \hbar J \prod_{k=1}^{\infty} \left( \frac{2\hbar}{k} \right)^2 \sum_{n=0}^{\infty} \delta(E - \hbar\omega(n + 1/2)) . \qquad (40)$$

This result shows that the Jacobian $J$ for the change of variables (24), (25) should be identified with the (real, infinite) constant

$$J = \frac{1}{2\pi\hbar} \prod_{k=1}^{\infty} \left( \frac{k}{2\hbar} \right)^2 . \qquad (41)$$

Then the density of states becomes

$$\nu(E) = \sum_{n=0}^{\infty} \delta(E - \hbar\omega(n + 1/2)) , \qquad (42)$$

which is the anticipated result showing that for the harmonic oscillator $\nu(E)$ is a sum of delta functions peaked at half–odd–integer multiples of $\hbar\omega$.

Finally, we note that the various quantum–statistical and thermodynamical properties of a system can be obtained from its density of states. In particular, the canonical partition function $Z(\beta)$ is defined as the Laplace transform of $\nu(E)$, and from $Z(\beta)$ the heat capacity, entropy, and other thermodynamical quantities can be found. For the harmonic oscillator with density of states (42), the partition function is

$$Z(\beta) = \int_0^{\infty} dE \, \nu(E) e^{-\beta E} = \sum_{n=0}^{\infty} e^{-\beta\omega\hbar(n+1/2)} , \qquad (43)$$

which is the well known result.

## ACKNOWLEDGMENTS
This research was supported by National Science Foundation grant PHY-8908741.

## REFERENCES

Braden, H. W., J. D. Brown, B. F. Whiting, and J. W. York (1990). Charged black hole in a grand canonical ensemble. *Physical Review*, **D42**, 3376–3385.

Brown, J. D. and J. W. York (1989). Jacobi's action and the recovery of time in general relativity. *Physical Review*, **D40**, 3312–3318.

Brown, J. D., G. L. Comer, E. A. Martinez, J. Melmed, B. F. Whiting, and J. W. York (1990). Thermodynamic ensembles and gravitation. *Classical and Quantum Gravity*, **7**, 1433–1444.

Brown, J. D., E. A. Martinez, and J. W. York (1991a). Rotating black holes, complex geometry, and thermodynamics. In *Nonlinear Problems in Relativity and Cosmology*, Eds J. R. Buchler, S. L. Detweiler, and J. R. Ipser. New York Academy of Sciences, New York.

Brown, J. D., E. A. Martinez, and J. W. York (1991b). Complex Kerr–Newman geometry and black hole thermodynamics. *Physical Review Letters*, **66**, 2281–2284.

Brown, J. D. and J. W. York (1992). Quasi-local energy in general relativity. In *Mathematical Aspects of Classical Field Theory*, Eds M. J. Gotay, J. E. Marsden, and V. E. Moncrief. American Mathematical Society, Providence.

Brown, J. D. and J. W. York (1993a). Quasilocal energy and conserved charges derived from the gravitational action. To appear in *Physical Review*, **D47**.

Brown, J. D. and J. W. York (1993b). Microcanonical functional integral for the gravitational field. To appear in *Physical Review*, **D47**.

Feynman, R. P. and A. R. Hibbs (1965). *Quantum Mechanics and Path Integrals*. McGraw Hill, New York.

Lanczos, C. (1970). *The Variational Principles of Mechanics*. University of Toronto

Press, Toronto.

Whiting, B. F. and J. W. York (1988). Action principle and partition function for the gravitational field in black–hole topologies. *Physical Review Letters*, **61**, 1336–1339.

York, J. W. (1986). Black-hole thermodynamics and the Euclidean Einstein action. *Physical Review*, **D33**, 2092–2099.

# Decoherence of Correlation Histories

*Esteban Calzetta* *        *B. L. Hu* †

### Abstract

We use a $\lambda\Phi^4$ scalar quantum field theory to illustrate a new approach to
the study of quantum to classical transition. In this approach, the decoherence
functional is employed to assign probabilities to consistent histories defined in
terms of correlations among the fields at separate points, rather than the field
itself. We present expressions for the quantum amplitudes associated with such
histories, as well as for the decoherence functional between two of them. The
dynamics of an individual consistent history may be described by a Langevin-
type equation, which we derive.

*Dedicated to Professor Brill on the occasion of his sixtieth birthday, August 1993*

## 1.   Introduction

## 1.1.   Interpretations of Quantum Mechanics and Paradigms of Statistical Mechanics

This paper attempts to bring together two basic concepts, one from the foundations
of statistical mechanics and the other from the foundations of quantum mechanics,
for the purpose of addressing two basic issues in physics:
1) the quantum to classical transition, and
2) the quantum origin of stochastic dynamics.
Both issues draw in the interlaced effects of dissipation, decoherence, noise, and fluctuation. A central concern is the role played by coarse-graining –the naturalness of its
choice, the effectiveness of its implementation and the relevance of its consequences.

On the fundations of quantum mechanics, a number of alternative interpretations
exists, e.g., the Copenhagen interpretation, the many-world interpretation [1], the
consistent history interpretations [2], to name just a few (see [3] for a recent review).
The one which has attracted much recent attention is the decoherent history approach

---

*IAFE, cc 167, suc 28, (1428) Buenos Aires, Argentina
†Department of Physics, University of Maryland, College Park, MD 20742, USA

of Gell-Mann and Hartle [4]. In this formalism, the evolution of a physical system is described in terms of 'histories': A given history may be either exhaustive (defining a complete set of observables at each instant of time) or coarse-grained. While in classical physics each history is assigned a given probability, in quantum physics a consistent assignment of probabilities is precluded by the overlap between different histories. The decoherence functional gives a quantitative measure of this overlap; thus the quantum to classical transition can be studied as a process of "diagonalization" of the decoherence functional in the space of histories.

On the foundational aspects of statistical mechanics, two major paradigms are often used to describe non-equilibrium processes (see, e.g., [5, 6, 7, 8]): the Boltzmann theory of molecular kinetics, and the Langevin (Einstein-Smoluchowski) theory of Brownian motions. The difference between the two are of both formal and conceptual character.

To begin with, the *setup* of the problem is different: In kinetic theory one studies the overall dynamics of a system of gas molecules, treating each molecule in the system on the same footing, while in Brownian motion one (Brownian) particle which defines the system is distinct from the rest, which is relegated as the environment. The terminology of 'revelant' versus 'irrevelant' variables highlights the discrepancy.

The *object* of interest in kinetic theory is the (one-particle) distribution function (or the nth-order correlation function), while in Brownian motion it is the reduced density matrix. The emphasis in the former is the correlation amongst the particles, while in the latter is the effect of the environment on the system.

The nature of *coarse-graining* is also very different: in kinetic theory coarse-graining resides in the adoption of the molecular chaos assumption corresponding formally to a truncation of the BBGKY hierarchy, while in Brownian motion it is in the integration over the environmental variables. The part that is truncated or 'ignored' is what constitutes the noise, whose effect on the 'system' is to introduce dissipation in its dynamics. Thus the fluctuation-dissipation relation and other features.

Finally the *philosophy* behind these two paradigms are quite different: In Brownian motion problems, the separation of the system from the environment is prescribed: it is usually determined by some clear disparity between the two systems. These models represent "autocratic systems", where some degrees of freedom are more relevant than others. In the lack of such clear distinctions, making a separation 'by hand' may seem rather *ad hoc* and unsatisfactory. By contrast, models subscribing to the kinetic theory paradigm represent "democratic systems": all particles in a gas are equally relevant. Coarse-graining in Boltzmann's kinetic theory appears less contrived, because information about higher correlation orders usually reflects the degree of precision in a measurement, which is objectively definable.

In the last five years we have explored these two basic paradigms of non-equilibrium statistical mechanics in the framework of interacting quantum field theory with the aim of treating dissipative processes in the early universe [9, 10] and decoherence processes in the quantum to classical transition issue [11]. Here we have begun to explore the issues of decoherence with the kinetic model.

Because of the difference in approach and emphasis between these two paradigms and in view of their fundamental character, it is of interest to build a bridge between them. We have recently carried out such a study with quantum fields [12]. By delineating the conditions under which the Boltzmann theory reduces to the Langevin theory, we sought answers to the following questions:

1) What are the factors condusive to the evolution of a 'democratic system' to an 'autocratic system' and vise versa ? A more natural set of criteria for the separation of the system from the environment may arise from the interaction and dynamics of the initial closed system [13, 14].

2) The construction of collective variables from the basic variables, the description of the dynamics of the collective variables, and the depiction of the behavior of a coarser level of structure emergent from the microstructures. [13, 15, 16].

The paradigm of quantum open systems described by quantum Brownian models has been used to analyze the decoherence and dissipation processes, for addressing basic issues like quantum to classical transitions, fluctuation and noise, particle creation and backreaction, which arise in quantum measurement theory [17, 18, 19], macroscopic quantum systems [20], quantum cosmology [21] (for earlier work see references in [22]), semiclassical gravity [23, 24, 25], and inflationary cosmology [26]. The reader is referred to these references and references therein for a description of this line of study.

The aim of this paper is to explore the feasibility for addressing the same set of basic issues using the kinetic theory paradigm. We develop a new approach based on the application of the decoherence functional [2, 4] formalism to histories defined in terms of *correlations* between the fundamental field variables. We shall analyse the decoherence between different histories of an interacting quantum field, a $\lambda\Phi^4$ theory here taken as example, corresponding to different particle spectra and study issues on the physics of quantum to classical transition, the relation of decoherence to dissipation, noise and fluctuation, and the quantum origin of classical stochastic dynamics.

## 1.2.  Quantum to Classical Transition and Coarse-Graining

One basic constraint in the building of quantum theory is that it should reproduce classical mechanics in some limit. (For a schematic discussion of the different criteria of classicality and their relations, see [27]). Classical behavior can be characterized by the existence of strong correlations between position and momentum variables described by the classical equations of motion [28] and by the absense of interference phenomena (decoherence).

Recent research in quantum gravity and cosmology have focussed on the issue of quantum to classical transition. This was highlighted by quantum measurement theory for closed systems (for a general discussion, see, e.g., [29]), the intrinsic incompatibility of quantum physics with general relativity [30], and the quantum origin of classical fluctuations in explaining the large scale structure of the Universe. Indeed, in the inflationary models of the Universe [31], one hopes to trace all cosmic structures to the evolution from quantum perturbations in the inflaton field. More dramatically, in quantum cosmology [32] the whole (classical) Universe where we now live in is regarded as the outcome of a quantum to classical transition on a cosmic scale. In these models, one hopes not only to explain the 'beginning' of the universe as a quantum phenomenon, but also to account for the classical features of the present universe as a consequence of quantum fluctuations. This requires not only a theoretical understanding of the quantum to classical transition issue in quantum mechanics, but also a theoretical derivation of the laws of classical stochastic mechanics from quantum mechanics, the determination of the statistical properties of classical noise (e.g., whether it is white or coloured, local or nonlocal) being an essential step in the formulation of a microscopic theory of the structure of the Universe [26].

Our understanding of the issue of quantum to classical transition has been greatly advanced by the recent development of the decoherent histories approach to quantum mechanics [4]. An essential element of the decoherent histories approach is that the overlap between two exhaustive histories can never vanish. Therefore, the discussion of a quantum to classical transition can only take place in the framework of a coarse grained description of the system, that is, giving up a complete specification of the state of the system at any instant of time.

As a matter of fact, some form of coarse graining underlies most, if not all, successful macroscopic physical theories. This fact has been clearly recognized and exploited at least since the work of Nakajima and Zwanzig [33, 6] on the foundations of nonequilibrium statistical mechanics. Like statistical mechanics, the decoherent histories approach allows a variety of coarse-graining procedures; not all of these, however, are expected to be equally successful in leading to interesting theories. Since the prescription of the coarse graining procedure is an integral part of the implementation of the decoherent histories approach, the development and evaluation of different coarse graining strategies is fundamental to this research program.

When we survey the range of meaningful macroscopic (effective) theories in physics arising from successfully coarse-graining a microscopic (fundamental) theory, one particular class of examples is outstanding; namely, the derivation of the hydrodynamical description of dilute gases from classical mechanics. The crucial step in deriving the Navier-Stokes equation for a dilute gas consists in rewriting the Liouville equation for the classical distribution function as a BBGKY hierarchy, which is then truncated by invoking a 'molecular chaos' assumption. If the truncation is made at the level of the two-particle reduced distribution function, the Boltzmann equation results. In the near-equilibrium limit, this equation leads to the familiar Navier-Stokes theory.

We must stress that in this general class of theories exemplified by Boltzmann's work, coarse-graining is introduced through the truncation of the hierarchy of distribution functions; i.e., by neglecting correlations of some order and above at some singled-out time [5, 6]. This type of coarse graining strategy is qualitatively different from those used in most of the recent work in quantum measurement theory and cosmology, which invoke a system-bath, space-time, or momentum-space separation. In most of these cases, an intrinsically justifiable division of the system from the environment is lacking and one has to rely on case-by-case physical rationales for making such splits. (An example of system-bath split is Zurek's description of the measurement process in quantum mechanics, where a bath is explicitly included to cause decoherence in the system-apparatus complex [17]. Space-time coarse graining has been discussed by Hartle [34] and Halliwell *et al* [35]. An example of coarse-graining in momentum space is stochastic inflation [36], where inflaton modes with wavelenghts shorter than the horizon are treated as an environment for the longer wavelenght modes [37]).

## 1.3.   Coarse-Graining in the Hierarchy of Correlations

In this paper we shall develop a version of the decoherent histories approach where the coarse-graining procedure is patterned after the truncation of the BBGKY hierarchy of distribution functions. For simplicity, we shall refer below to the theory of a single scalar quantum field, with a $\lambda \Phi^4$- type nonlinearity.

The simplest quantum field theoretical analog to the hierarchy of distribution functions in statistical mechanics is the sequence of Green functions (that is, the expectation values of products of $n$ fields) [9]. In this approach, the BBGKY hierarchy of kinetic equations is replaced by the chain of Dyson equations, linking each Green function to other functions of higher order.

The analogy between these two hierarchies is rendered most evident if we introduce "distribution functions" in field theory through suitable partial Fourier transformation of the Green functions. Thus, a "Wigner function" [38] may be introduced as the Fourier transform of the Hadamard function (the symmetric expectation value of the product of two fields) with respect to the difference between its arguments. It obeys both a mass shell constraint and a kinetic equation, and may be regarded as the physical distribution function for a gas of quasi particles, each built out of a cloud of virtual quanta. Similar constructs may be used to introduce higher "distribution functions" [9, 7].

As in statistical mechanics, the part of a given Green function which cannot be reduced to products of lower functions defines the corresponding "correlation function". Thus the chain of Green functions is also a hierarchy of correlations.

To establish contact between the hierarchy of Green functions and the decoherent histories approach, let us recall the well-known fact that the set of expectation values of all field products contains in itself all the information about the statistical state of the field [9]. For a scalar field theory with no symmetry breaking, we can even narrow this set to products of even numbers of fields. This result suggests that a history can be described in terms of the values of suitable composite operators, rather than those of the fundamental field. If products to all orders are specified (binary, quartet, sextet, etc), then the description of the history is exhaustive, and different histories do not decohere. On the other hand, when some products are not specified, or when the information of higher correlations are missing, which is often the case in realistic measurement settings, the description is coarse-grained, which can lead to decoherence.

In this work we shall consider coarse-grained histories where the lower field products (binary, quartic) are specified, and higher products are not. Decoherence will mean that the specified composite operators can be assigned definite values with consistent probabilities. Higher composite operators retain their quantum nature, and therefore cannot be assigned definite values. However, their expectation values can be expressed as functionals of the specified correlations by solving the corresponding Dyson equations with suitable boundary conditions. This situation is exactly analogous to that arising from the truncated BBGKY hierarchy, where the molecular chaos assumption allows the expression of higher distribution functions as functionals of lower ones ( e.g., [5, 6]).

For those products of fields which assume definite values with consistent probabilities, these values can be introduced as stochastic variables in the dynamical equations for the other quantities of interest (usually of lower correlation order). This approach would provide a theoretical basis for the derivation of the equations of classical stochastic dynamics from quantum fields. It can offer a justification (or refutation) for a procedure commonly assumed but never proven in some popular theories like stochastic inflation [36, 37]. Moreover, since in general we shall obtain nontrivial ranges of values for the specified products with nonvanishing probabilities, it can be said that our procedure captures both the average values of the field products and the fluctuations around this average. The statistical nature of these fluctuations is a subject of great interest in itself [26].

There is another conceptual issue that our approach may help to clarify. As we have already noted, in the system-bath split approach to coarse-graining, as well as in related procedures, it is crucial to introduce a hierarchical order among the degrees of freedom of the system, in such a way that some of them may be considered relevant, and others irrelevant. While it is often the case that the application itself suggests which notions of relevance may lead to an interesting theory, in a quantum cosmo-logical model, which purports to be a "first principles" description of our Universe, all these choices are, in greater or lesser degree, arbitrary. Since correlation functions already have a "natural" built-in hierarchical ordering, in this approach the 'arbi-trariness' is reduced to deciding on which level this hierarchy is truncated, and that in turn is determined by the degree of precision one carries out the measurement. In most case one still needs to show the robustness of the macroscopic result against the variance of the extent of coarse-graining, and exceptional situations do exists (an ex-ample is the long time-tail relaxation behavior in multiple particle scattering of dense gas, arising from a failure of the simple molecular chaos assumption). But in general terms correlational coarse-graining seems to us a less *ad hoc* procedure compared to the commonly used system-bath splitting and coarse-graining.

This paper is organized as follows: In Sec. 2 we discuss the implementation of our procedure for the simple case of a $\lambda\Phi^4$ theory in flat space time. We then derive the formulae for the quantum amplitude associated with a set of correlation histories and the decoherence functional between two such histories. In Sec. 3 we discuss the decoherence of correlation histories between binary histories and derive the classical stochastic source describing the effect of higher-order correlations on the lower-order ones, arriving at a Langevin equation for classical stochastic dynamics. In Sec. 4 we summarize our findings.

## 2. Quantum Amplitudes for Correlation Histories and Effective Action

## 2.1. Quantum Mechanical Amplitudes for Correlation Histories

In this section, we shall consider the quantum mechanical amplitudes associated with different histories for a $\lambda\Phi^4$ quantum field theory, defined in terms of the values of time-ordered products of even numbers of fields at various space time points. Let us begin by motivating our ansatz for the amplitudes of these correlation histories.

In the conceptual framework of decoherent histories [4], the "natural" exhaustive specification of a history would be to define the value of the field $\Phi(x)$ at every space time point. These field values are c numbers. The quantum mechanical amplitude for a given history is $\Psi[\Phi] \sim e^{iS[\Phi]}$, where $S$ is the classical action. The decoherence functional between two different specifications is given by $D[\Phi, \Phi'] \sim \Psi[\Phi]\Psi[\Phi']^*$. Since $|D[\Phi, \Phi']| \equiv 1$, there is never decoherence between these histories.

A coarse-grained history would be defined in general through a "filter function" $\alpha$, which is basically a Dirac $\delta$ function concentrated on the set of exhaustive histo-ries matching the specifications of the coarse-grained history. For example, we may have a system with two degrees of freedom $x$ and $y$, and define a coarse-grained history by specifying the values $x_0(t)$ of $x$ at all times. Then the filter function is $\alpha[x, y] = \prod_{t \in R} \delta(x(t) - x_0(t))$. The quantum mechanical amplitude for the coarse-grained history is defined as

$$\Psi[\alpha] = \int D\Phi \, e^{iS} \alpha[\Phi] \tag{1}$$

where the information on the quantum state of the field is assumed to have been included in the measure and/or the boundary conditions for the functional integral. The decoherence functional for two coarse-grained histories is [4]

$$D[\alpha, \alpha'] = \int D\Phi D\Phi' e^{i(S(\Phi) - S(\Phi'))} \alpha[\Phi] \alpha'[\Phi'] \tag{2}$$

In this path integral expression, the two histories $\Phi$ and $\Phi'$ are not independent; they assume identical values on a $t = T = $ constant surface in the far future. Thus, they may be thought of as a single, continuous history defined on a two-branched "closed time-path" [39, 40, 41, 42], the first branch going from $t = -\infty$ to $T$, the second from $T$ back to $-\infty$. Alternatively, we can think of $\Phi = \Phi^1$ and $\Phi' = \Phi^2$ as the two components of a field doublet defined on ordinary space time [9], whose classical action is $S[\Phi^a] = S[\Phi^1] - S[\Phi^2]$. This notation shall be useful later on.

Let us try to generalize this formalism to correlation histories. We begin with the simplest case, where only binary products are specified. In this case a history is defined by identifying a symmetric kernel $G(x, x')$, which purports to be the value of the product $\Phi(x)\Phi(x')$ in the given history, both $x$ and $x'$ defined in Minkowsky space - time. By analogy with the formulation above, one would write the quantum mechanical amplitude for this correlation history as

$$\Psi[G] = \int D\Phi \, e^{iS} \prod_{x \gg x'} \delta(\Phi(x)\Phi(x') - G(x, x')) \tag{3}$$

(In this equation, we have introduced a formal ordering of points in Minkowsky space - time, simply to avoid counting the same pair twice.)

But this straightforward generalization for the correlation history amplitude is unsatisfactory on at least two counts. First, it assumes that the given kernel $G$ can actually be decomposed (maybe not uniquely) as a product of c number real fields at different locations; however, we wish to define amplitudes for kernels (such as the Feynman propagator) which do not have this property. Second, (which is related to the first point,) it is ambiguous, since we do not have a unique way to express higher even products of fields in terms of binary products, and thus of applying the $\delta$ function constraint.

To give an example of this, observe that, should we expand the exponential of the action in powers of the coupling constant $\lambda$, the second order term $\int dx\, dx'\, \Phi(x)^4 \Phi(x')^4$ could become, after integration over the delta function, either

$$\int dx\, dx' \qquad G(x,x)^2 G(x',x')^2,$$

$$\int dx\, dx' \qquad G(x,x')^4,$$

$$(4)$$

or any other combination; of course, if $G$ could be decomposed as a product of fields, this would be unimportant.

Let us improve on these shortcomings. The general idea is to accept Eq. (3) as the definition of the amplitude in the restricted set of kernels where it can be applied, and to define the amplitude for more general kernels through some process of analytical continuation. To this end, we must rewrite the quantum mechanical amplitude in a more transparent form, which we achieve by using an integral representation of the $\delta$ function. Concretely, we redefine

$$\Psi[G] = \int DK \int D\Phi\; e^{iS + \frac{i}{2}\int dx dx'\; K(x,x')(\Phi(x)\Phi(x') - G(x,x'))} \tag{5}$$

where the filter function in the Gell-Mann Hartle scheme is replaced by an integration over "all" symmetric non-local sources $K$. Eq. (5) is not yet a complete definition, since one must still specify both the path and the measure to be used in the $K$ integration. Performing the integration over fields, we obtain

$$\Psi[G] = \int DK\; e^{i(W[K] - (1/2)KG)} \tag{6}$$

where $W[K]$ is the generating functional for connected vacuum graphs with $\lambda\Phi^4$ interaction, and $(\Delta^{-1} - K)^{-1}$ for propagator (see below). Here $\Delta^{-1} = -\nabla^2 + m^2$ is the free propagator for our scalar field theory (our sign convention for the flat space - time metric is $-+++$).

The path integral over kernels can be computed through functional techniques. For example, for a free field, $\lambda = 0$,

$$W[K] = -i \ln \mathrm{Det}[(\Delta^{-1} - K)^{-1/2}] + \text{constant} \tag{7}$$

Through the change of variables

$$(\Delta^{-1} - K) = \kappa G^{-1} \tag{8}$$

we obtain

$$\Psi[G] = \text{constant } [\text{Det } G]^{-1/2} \, e^{(-i/2)\Delta^{-1}G} \tag{9}$$

When the self coupling $\lambda$ is not zero, the evaluation of $\Psi[G]$ is more involved; however, if we are interested in the leading behavior of the amplitude only, we can simply evaluate the functional integral over $K$ by saddle point methods. The saddle lies at the solution to

$$\frac{\partial W[K]}{\partial K} = \frac{1}{2}G \tag{10}$$

We recognize immediately that the exponent, evaluated at the saddle point, is simply the 2 Particle Irreducible (2PI) effective action $\Gamma$, with $G$ as propagator (see below). Including also the integration on gaussian fluctuations around the saddle, we find

$$\Psi[G] \sim [\text{Det}\{\frac{\partial^2\Gamma}{\partial G^2}\}]^{(1/2)} e^{i\Gamma[G]} \tag{11}$$

This is our main result.

As a check, it is interesting to compare the saddle method expression with our exact result for free fields. For a free field $\Gamma[G] = (-i/2)\ln \text{Det}(G) - (1/2)\Delta^{-1}G$, and therefore $\Gamma_{,G} = (-i/2)(G^{-1} - i\Delta^{-1})$, $\Gamma_{,G,G} = (i/2)G^{-2}$, so

$$[\text{Det}\{\frac{\partial^2\Gamma}{\partial G^2}\}]^{(1/2)} e^{i\Gamma[G]} = [\text{Det}\,G]^{-1}[\text{Det}\,G]^{1/2} e^{(-i/2)\Delta^{-1}G} \tag{12}$$

which is exactly the earlier result, Eq. (9).

## 2.2. Quantum Amplitudes and Effective Actions

Eq. (11) is the natural generalization to correlation histories of the quantum mechanical amplitude $e^{iS}$ associated to a field configuration. Let us consider its physical meaning.

The effective action is usually introduced in Field Theory books [43] as a compact device to generate the Feynman graphs of a given theory. Indeed, all Feynman graphs appear in the expansion of the generating functional

$$Z[J] = \int D\Phi e^{i(S+J\Phi)} \tag{13}$$

in powers of the external source $J$ [here, $J\Phi = \int d^4x \, J(x)\Phi(x)$]. $Z$ has the physical meaning of a vacuum persistance amplitude: it is the amplitude for the in vacuum (that is, the vacuum in the distant past) to evolve into the out vacuum (the vacuum in the far future) under the effect of the source $J$. Thus, after proper normalization, $|Z|$ will be unity when the source is unable to create pairs out of the vacuum, and less than unity otherwise.

A more compact representation of the Feynman graphs is provided by the functional $W[J] = -i \ln Z[J]$; the Taylor expansion of $W$ contains only connected Feynman graphs. Thus $W$ developing a (positive) imaginary part signals the instability of the vacuum under the external source $J$.

The external source will generally drive the quantum field $\Phi$ so that its matrix element

$$\phi(x) = \frac{\langle 0out|\Phi(x)|0in\rangle}{\langle 0out|0in\rangle} \tag{14}$$

between the in and out vacuum states will not be zero. Indeed, it is easy to see that

$$\phi = \frac{\partial W}{\partial J} \tag{15}$$

The transformation from $J$ to $\phi$ is generally one to one, and thus it is possible to consider the matrix element, and not the source, as the independent variable. This is achieved by submitting $W$ to a Legendre transformation, yielding the effective action $\Gamma[\phi] = W[J] - J\phi$ [$J$ and $\phi$ being related through Eq. (15)]. This equation can be inverted to yield the dynamic law for $\phi$

$$\frac{\partial \Gamma}{\partial \phi} = -J \tag{16}$$

Eq. (16) shows that $\Gamma$ may be thought of as a generalization of the classical action, now including quantum effects. In the absence of external sources, the in and out vacua agree, so $\phi$ becomes a true expectation value; its particular value is found by extremizing the effective action. Indeed, in this case it can be shown that $\Gamma$ is the energy of the vacuum.

$\Gamma[\phi]$ can be defined independently of the external source through the formula [44]

$$\Gamma[\phi] = S[\phi] + (i/2) \ln \text{Det} \left(\frac{\partial^2 S}{\partial \phi^2}\right) + \Gamma_1[\phi] \tag{17}$$

where $\Gamma_1$ represents the sum of all one particle irreducible (1PI) vacuum graphs of an auxiliary theory whose classical action is obtained from expanding the classical action $S[\phi + \varphi]$ in powers of $\varphi$, and deleting the constant and linear terms. Eq. (17) shows that $\Gamma$ is related to the vacuum persistence amplitude of quantum fluctuations around the matrix element $\phi$. Therefore, an imaginary part in $\Gamma$ also signals a vacuum instability. This situation closely resembles the usual approach to tunneling and phase transitions, where an imaginary part in the free energy signals the onset of instability [45].

Observe that each of the transformations from $Z$ to $W$ to $\Gamma$ entails a drastic simplification of the corresponding Feynman graphs expansions, from all graphs in $Z$ to connected ones in $W$ and to 1PI ones in $\Gamma$. Roughly speaking, it is unneccessary to include non 1PI graphs in the effective action, because the sum of all one-particle insertions is already prescribed to add up to $\phi$. Now the process can be continued: if we could fix in advance the sum of all self energy parts, then we could write down a perturbative expansion where only 2PI Feynman graphs need be considered. This is achieved by the 2PI effective action [46].

Let us return to Eq. (13), and add to the external source a space-time dependent mass term

$$Z[J, K] = \int D\Phi e^{i(S+J\Phi+(1/2)\Phi K \Phi)} \tag{18}$$

where $\Phi K \Phi = \int dx \ dx' \ \Phi(x) K(x, x') \Phi(x')$. Also define $W[J, K] = -i \ln Z[J, K]$. Then the variation of $W$ with respect to $J$ defines the in-out matrix element of the field, as before, but now we also have

$$\frac{\partial W}{\partial K(x, x')} = \frac{1}{2}[\phi(x)\phi(x') + G_F(x, x')] \tag{19}$$

where $G_F$ represents the Feynman propagator of the quantum fluctuations $\varphi$ around the matrix element $\phi$. As before, it is possible to adopt $G$ as the independent variable, instead of $K$. To do this, we define the 2PI effective action (in schematic notation) $\Gamma[\phi, G_F] = W[J, K] - J\phi - (1/2)K[\phi^2 + G_F]$. Variation of this new $\Gamma$ yields the equations of motion $\Gamma_{,\phi} = -J - K\phi$, $\Gamma_{,G_F} = (-1/2)K$.

We can see that the 2PI effective action generates the dynamics of the Feynman propagator, and in this sense it plays for it the role that the classical action plays for the field. In this sense we can say that Eq. (11) generalizes the usual definition of quantum mechanical amplitudes.

The perturbative expansion of the 2PI effective action reads [46]

$$\Gamma[\phi, G_F] = S[\phi] + (i/2) \ln \text{Det} G_F^{-1} + \left(\frac{1}{2}\right) \text{Tr}\left(\frac{\partial^2 S}{\partial \phi^2} G_F\right) + \Gamma_2[\phi, G_F] + \text{constant} \tag{20}$$

where $\Gamma_2$ is the sum of all 2PI vacuum graphs of the auxiliary theory already considered, but with $G_F$ as propagator in the internal lines. As we anticipated, to replace $G_F$ for the perturbative propagator amounts to adding all self energy insertions, and therefore no 2PI graph needs be explicitly included.

Like its 1PI predecessor, the 2PI effective action has the physical meaning of a vacuum persistence amplitude for quantum fluctuations $\varphi$, constrained to have vanishing expectation value and a given Feynman propagator. Therefore, an imaginary part in the 2PI effective action also signals vacuum instability.

The description of the dynamics of a quantum field through both $\phi$ and $G_F$ simultaneously, rather than $\phi$ alone, is appealing not only because it allows one to perform with little effort the resummation of an infinite set of Feynman graphs, but also because for certain quantum states, it is possible to convey statistical information about the field through the nonlocal source $K$. This information is subsequently transferred to the propagator. For this reason, the 2PI effective action formalism is, in our opinion, a most suitable tool to study statistical effects in field theory, particularly for out-of-equilibrium fields [9, 47]. In our earlier studies the object of interest is the on-shell effective action, that is, the effective action for propagators satisfying the equations of motion. Here, in Eq. (11), we find a relationship between the quantum mechanical amplitude for a correlation history and the 2PI effective action which does not assume any restriction on the propagator concerned.

## 2.3.   Quantum Amplitudes for More General Correlation Histories: 2PI CTP Effective Action

One of the peculiarities of the ansatz Eq.(3) for the amplitude of a correlation history is that the kernel $G$ must be interpreted as a time - ordered binary product of fields. This results from the known feature of the path integral, which automatically time orders any monomials occurring within it. Before we proceed to introduce the decoherence functional for correlation histories, it is convenient to discuss how this restriction could be lifted, as well as the restriction to binary products.

The time ordering feature of the path integral is also responsible for the fact that the c-number field $\phi$ in Sec. 2.2 is a matrix element, rather than a true expectation value. As a matter of fact, the Feynman propagator $G_F$ discussed in the previous section is also a matrix element

$$G_F(x, x') = \frac{\langle 0out|T[\varphi(x)\varphi(x')]|0in\rangle}{\langle 0out|0in\rangle} \tag{21}$$

Because $\phi$ and $G_F$ satisfy mixed boundary conditions, the dynamic equations resulting from the 2PI effective action are generally not causal. This drawback has placed limitations in their physical applications.

Schwinger [39] has introduced an extended effective action, whose arguments are true expectation values with respect to some in quantum state. Because the dynamics of these expectation values may be formulated as an initial value problem, the equations of motion resulting from the Schwinger-Keldysh effective action are causal. Schwinger's idea is also the key to solving the restrictions in our definition of quantum amplitudes for correlation histories.

Schwinger's insight was to apply the functional formalism we reviewed in Sec. **2.2** to fields defined on a "closed time-path", composed of a "direct" branch $-T \leq t \leq T$, and a "return" branch $T \geq t \geq -T$ (with $T \to \infty$) [39, 40]. Actually, we have already encountered this kind of path in the discussion of the decoherence functional for coarse - grained histories. Since the path doubles back on itself, the in vacuum is the physical vacuum at both ends; the formalism may be generalized to include more general initial states, but we shall not discuss this possibility[9].

The closed time-path integral time-orders products of fields on the direct branch, anti-time-orders fields on the return branch, and places fields on the return branch always to the left of fields in the direct branch. To define the closed time-path generating functional, we must introduce two local sources $J_a$, and four nonlocal ones $K_{ab}$ (as in Sec. **2.1** an index $a, b = 1$ denotes a point on the first branch, while an index 2 denotes a point on the return part of the path). These sources are conjugated to c number fields $\phi^a$ and propagators $G^{ab}$, which stand for $\langle 0in|\Phi^a(x)|0in \rangle$ and $\langle 0in|\varphi^a(x)\varphi^b(x')]|0in \rangle$. Explicitly, decoding the indices, the propagators are defined as (here and from now on, we assume that the background fields $\phi^a$ vanish):

$$G^{11}(x, x') = \langle 0in|T[\Phi(x)\Phi(x')]|0in \rangle \tag{22}$$

$$G^{12}(x, x') = \langle 0in|\Phi(x')\Phi(x)|0in \rangle \tag{23}$$

$$G^{21}(x, x') = \langle 0in|\Phi(x)\Phi(x')|0in \rangle \tag{24}$$

$$G^{22}(x, x') = \langle 0in|(T[\Phi(x)\Phi(x')])^\dagger|0in \rangle \tag{25}$$

They are, respectively, the Feynman, negative- and positive- frequency Wightman, and Dyson propagators. The definition of the closed time-path (CTP) or in-in 2PI effective action follows the same steps as the ordinary effective action discussed in the previous section, except that now, besides space-time integrations, one must sum over the discrete indexes $a, b$. These indexes can be raised and lowered with the "metric" $h_{ab} = \text{diag}(1, -1)$. Similarly, the "propagator" to be used in Feynman graph expansions is the full matrix $G^{ab}$, and the interaction terms should be read out of the CTP classical action $S[\Phi^1] - S[\Phi^2]$, discussed in Sec. **2.1**.

In the case of vacuum initial conditions, these can be included into the path integral by tilting the branches of the CTP in the complex $t$ plane (the direct branch should acquire an infinitesimal positive slope, and the return branch, a negative one [48]). The CTP boundary condition, that the histories at either branch should fit continuously at the surface $t = T$, may also be explicitly incorporated into the path integral as follows. We first include under the integration sign a term

$$\prod_{x \in R^3} \delta(\Phi^1(x, T) - \Phi^2(x, T)) \tag{26}$$

which enforces this boundary condition; then we rewrite Eq.(26) as

$$exp\{(-1/\alpha^2) \int d^3x \ (\Phi^1(x, T) - \Phi^2(x, T))^2\} \tag{27}$$

where $\alpha \to 0$. This term has the form

$$exp\{i \int d^4x \ d^4x' K_{ab}(x, x')\Phi^a(x)\Phi^b(x')\}, \tag{28}$$

where

$$K_{ab}(x, x') = (i/\alpha^2)\delta(x - x')\delta(t - T)[2\delta_{ab} - 1]. \tag{29}$$

In this way, we have traded the boundary condition by an explicit coupling to a non local external source.

As before, variation of the CTP 2PI effective action yields the equations of motion for background fields and propagators. The big difference is that now these equations are real and causal [42, 9].

We can now see how the CTP technique solves the ordering problem in the definition of quantum amplitudes for correlation histories. One simply considers the specified kernels as products of fields defined on a closed time - path. In this way, we may define up to four different kernels $G^{ab}$ independently, to be identified with the four different possible orderings of the fields (for simplicity, we assume the background fields are kept equal to zero). If the kernels $G^{ab}$ can actually be decomposed as products of c-number fields on the CTP, then we associate to them the quantum amplitude

$$\Psi[G^{ab}] = \int D\Phi^a \ e^{iS} \prod_{x \gg x', ab} \delta(\Phi^a(x)\Phi^b(x') - G^{ab}(x, x')) \tag{30}$$

(where $S$ stands for the CTP classical action) The path integral can be manipulated as in Sec. 2.1 to yield

$$\Psi[G^{ab}] \sim [Det\{\frac{\partial^2 \Gamma}{\partial G^{ab}\partial G^{cd}}\}]^{(1/2)}e^{i\Gamma[G^{ab}]} \tag{31}$$

where $\Gamma$ stands now for the CTP 2PI effective action. This last expression can be analytically extended to more general propagator quartets, and, indeed, even to kernels which do not satisfy the relationships $G^{11}(x,x') = G^{21}(x,x') = G^{12*}(x,x') = G^{22*}(x,x')$ for $t \geq t'$, which follow from their interpretation as field products.

Quantum amplitudes for correlation histories including higher order products are defined following a similar procedure. For example, four particle correlations are specified by introducing 16 kernels [9]

$$G^{abcd} \sim \Phi^a \Phi^b \Phi^c \Phi^d - G^{ab} G^{cd} - G^{ac} G^{bd} - G^{ad} G^{bc} \tag{32}$$

If the new kernels are simply products of the binary ones, then the amplitude is given by

$$
\begin{aligned}
\Psi[G^{ab}, G^{abcd}] &= \int D\Phi^a \; e^{iS} \prod_{ab} \delta(\Phi^a \Phi^b - G^{ab}) \\
&\quad \prod_{abcd} \delta(\Phi^a \Phi^b \Phi^c \Phi^d - G^{ab} G^{cd} - G^{ac} G^{bd} - G^{ad} G^{bc} - G^{abcd})
\end{aligned}
\tag{33}
$$

(In the last two equations, we have included the space - time index $x$ and the branch index $a$ into a single multi index). Here, each pair appears only once in the product, as well as each quartet $abcd$. Exponentiating the $\delta$ functions we obtain

$$
\begin{aligned}
\Psi[G^{ab}, G^{abcd}] = \quad & \int DK_{abcd} \int DK_{ab} \int D\Phi \; exp\{i[S + \frac{1}{2}K_{ab}(\Phi^a \Phi^b - G^{ab}) \\
& + \frac{1}{24}K_{abcd}(\Phi^a \Phi^b \Phi^c \Phi^d - G^{ab} G^{cd} - G^{ac} G^{bd} - G^{ad} G^{bc} - G^{abcd})]\}
\end{aligned}
\tag{34}
$$

Now the integral over fields yields the CTP generating functional for connected graphs, for a theory with a non local interaction term. Thus

$$
\begin{aligned}
\Psi[G^{ab}, G^{abcd}] \quad &= \int DK_{abcd} \int DK_{ab} \; exp\{i[W[K_{ab}, K_{abcd}] - \frac{1}{2}K_{ab}G^{ab} \\
& - \frac{1}{24}K_{abcd}(G^{ab} G^{cd} + G^{ac} G^{bd} + G^{ad} G^{bc} + G^{abcd})]\}
\end{aligned}
\tag{35}
$$

The integral may be evaluated by saddle point methods, the saddle being the solution to $W_{,K_{ab}} = (1/2)G^{ab}$, $W_{,K_{abcd}} = \frac{1}{24}(G^{ab} G^{cd} + G^{ac} G^{bd} + G^{ad} G^{bc} + G^{abcd})$. To evaluate the exponential at the saddle is the same as to perform a Legendre transform on $W$ –it yields the higher order CTP effective action $\Gamma[G^{ab}, G^{abcd}]$. Variation of $\Gamma$ yields the equation of motion for its arguments, which are also the inversion of the saddle point conditions

$$\begin{aligned}
\Gamma_{,G^{ab}} &= (-1/2)K_{ab} - (1/4)K_{abcd}G^{cd} \\
\Gamma_{,G^{abcd}} &= (-1/24)K_{abcd}
\end{aligned}$$

(36)

Thus up to quartic correlations, the quantum mechanical amplitude is given by

$$\Psi[G^{ab}, G^{abcd}] \sim e^{i\Gamma[G^{ab}, G^{abcd}]}$$

(37)

This expression can likewise be extended to more general kernels.

As a check on the plausibility of this result, let us note the following point. Since quantum mechanical amplitudes are additive, it should be possible to recover our earlier ansatz Eq. (11) for binary correlation histories from the more general result Eq. (37), by integration over the fourth order kernels. Within the saddle point approximation, integration amounts to substituting these kernels by the solution to the second Eq. (36) for the given $G^{ab}$, with $K_{abcd} = 0$, and with null initial conditions. (Indeed, since initial conditions can always be included as delta function - like singularities in the external sources, the third condition is already included in the second.) This procedure effectively reduces the fourth order effective action to the 2PI CTP one [9], as we expected.

A basic point which emerges here relevant to our study of decoherence is that, while quantum field theory is unitary and thus time reversal invariant, the evolution of the propagators derived from the 2PI CTP effective action is manifestly irreversible [9, 49]. The key to this apparent paradox is that, while the evolution equations are indeed time reversal invariant, when higher order kernels are retained as independent variables, their reduction to those generated by the 2PI effective action involves the imposition of trivial boundary conditions in the past. Thus the origin of irreversibility in the two point functions is the same as in the BBGKY formulation in statistical mechanics [5]. The lesson for us in the present context is that there is an intrinsic connection between dissipation and decoherence [26, 23]. Knowledge that the evolution of the propagators generated by the 2PI effective action is generally dissipative leads us to expect that histories defined through binary correlations will usually decohere. We proceed now to a detailed study of this point.

## 3.   Decoherence of Correlation Histories

## 3.1.   Decoherence Functional for Correlation Histories

Having found an acceptable ansatz for the quantum mechanical amplitude associated with a correlation history, we are in a position to study the decoherence functional between two such histories. As was discussed in the Introduction, if the decoherence functional is diagonal, then correlation histories support a consistent probability assignment, and may thus be viewed as classical (stochastic) histories.

For concreteness, we shall consider the simplest case of decoherence among histories defined through (time-ordered) binary products. Let us start by considering two histories, associated with kernels $G(x, x')$ and $G'(x, x')$, which can in turn be written as products of fields. Taking notice of the similarity between the quantum amplitudes Eqs. (1) and (3), we can by analogy to Eq. (2) define the decoherence functional for second correlation order as

$$
\begin{aligned}
D[G, G'] \quad &= \int d\Phi d\Phi' \; e^{i(S[\Phi] - S[\Phi'])} \\
&\prod_{x \gg x'} \delta((\Phi(x)\Phi(x') - G(x, x'))\delta((\Phi'(x)\Phi'(x') - G'^*(x, x'))
\end{aligned}
$$

(38)

Recalling the expression Eq. (30) for the quantum amplitude associated with the most general binary correlation history, we can rewrite Eq. (38) as

$$
D[G, G'] = \int DG^{12} \, DG^{21} \; \Psi[G^{11} = G, G^{22} = G'^*, G^{12}, G^{21}]
$$

(39)

This expression for the decoherence functional can be extended to arbitrary kernels.

In the spirit of our earlier remarks, we use the ansatz Eq. (31) for the CTP quantum amplitude and perform the integration by saddle point methods to obtain

$$
D[G, G'] \sim e^{i\Gamma[G^{11}=G, G^{22}=G'^*, G_0^{12}, G_0^{21}]}
$$

(40)

where the Wightman functions are chosen such that

$$
\frac{\partial \Gamma}{\partial G_0^{12}} = \frac{\partial \Gamma}{\partial G_0^{21}} = 0
$$

(41)

for the given values of the Feynman and Dyson functions. These last two equations are the sought-for expression for the decoherence functional.

As an application, let us study the decoherence functional for Gaussian fluctuations around the vacuum expectation value (VEV) of the propagators for a $\lambda\Phi^4$ theory, carrying the calculations to two-loop accuracy. Gaussian fluctuations means that we only need the closed time-path 2PI effective action to second order in the fluctuations $\delta G^{ab} = G^{ab} - \Delta_0^{ab}$, where $\Delta_0^{ab}$ stands for the VEVs. Since the effective action is stationary at the VEV, there is no linear term. Formally

$$\Gamma[\delta G^{ab}] = (1/2)\{\Gamma_{,(aa),(bb)}\delta G^{aa}\delta G^{bb} + 2\Gamma_{,(a\neq b),(cc)}\delta G^{a\neq b}\delta G^{cc} + \Gamma_{,(a\neq b),(c\neq d)}\delta G^{a\neq b}\delta G^{c\neq d}\}$$
(42)

so the saddle point equations (41) become

$$\{\Gamma_{,(a\neq b),(c\neq d)}\}\delta G_0^{c\neq d} = -\Gamma_{,(a\neq b),(ee)}\delta G^{ee}$$
(43)

The formal Feynman graph expansion of the 2PI effective action is given in Eq. (20). To two-loop accuracy, we find [9]

$$\begin{aligned}
\Gamma_2[G^{ab}] &= -\frac{\lambda}{8}h_{abcd}\int d^4x\, G^{ab}(x,x)G^{cd}(x,x) \\
&+\frac{i\lambda^2}{48}h_{abcd}h_{efgh}\int d^4x\, d^4x'\, G^{ae}(x,x')G^{bf}(x,x')G^{cg}(x,x')G^{dh}(x,x')
\end{aligned}$$
(44)

where $h_{ab}, h_{abcd} = 1$ if $a = b = c = d = 1$, $-1$ if $a = b = c = d = 2$, and vanish otherwise.

Computing the necessary derivatives, we find

$$\frac{\partial^2\Gamma}{\partial G^{ab}(x,x')\partial G^{cd}(x'',x''')} =$$

$$\begin{aligned}
&(\frac{-1}{2})[-i(G^{-1})_{ac}(x,x'')(G^{-1})_{db}(x''',x') \\
&+(1/2)\lambda h_{abcd}\delta(x'-x)\delta(x''-x)\delta(x'''-x) \\
&-(i/2)\lambda^2 h_{aceg}h_{bdfj}\delta(x''-x)\delta(x'''-x')G^{ef}(x,x')G^{gj}(x,x')]
\end{aligned}$$
(45)

These derivatives are evaluated at $G^{ab} = \Delta_0^{ab}$, where

$$\begin{aligned}
(\Delta_0^{-1})_{ab}(x,x') &= i[h_{ab}(-\nabla^2 + m^2 - ih_{ab}\epsilon)\delta(x'-x) \\
&+(\lambda/2)h_{abcd}\delta(x'-x)\Delta_0^{cd}(x,x) \\
&-(i/6)\lambda^2 h_{aecd}h_{bfgh}\Delta_0^{ef}(x,x')\Delta_0^{cg}(x,x')\Delta_0^{dh}(x,x')] \\
&+\frac{1}{2\alpha^2}\delta(x'-x)\delta(t-T)[2\delta_{ab}-1]
\end{aligned}$$
(46)

where it is understood that the limits $\epsilon, \alpha \to 0$, $T \to \infty$ are taken. The first infinitesimal is included to enforce appropiate Feynman/Dyson orderings, the second to carry the CTP boundary conditions in the far future.

In computing the Feynman graphs in these expressions, the usual divergences crop up. They may be regularized and renormalized by standard methods, which we will not discuss here. The "tadpole" graph $\Delta_0^{cd}(x,x)$ can be made to vanish by a suitable choice of the renormalization point, which we shall assume.

Let us narrow our scope to a physically meaningful set of histories, namely, those describing ensembles of real particles distributed with a position-independent spectrum $f(k)$, $k$ being the four momentum vector. Such ensembles are described by propagators [9]

$$\delta G(x,x') = 2\pi \int \left(\frac{d^4k}{(2\pi)^4}\right) e^{ik(x-x')} \delta(k^2 + m^2) f(k) \tag{47}$$

The distribution functions $f$ are real, positive, and even in $k$. We wish to analyze under what conditions it is possible to assign consistent probabilities to different spectra $f$. To this end we must compute the decoherence functional between the propagator in Eq. (47) and another, say, associated with a function $f'$.

Let us begin by investigating Eqs (43) for the missing propagators $G^{12}$ and $G^{21}$. We shall first disregard the boundary condition enforcing terms in these equations, introducing them at a later stage. When this is done, the right hand side of Eqs. (43) vanishes, since $(-\nabla^2 + m^2)G^{aa}(x,x') \equiv 0$ in the present case.

On the other hand, we only need the left hand side to zeroth order in $\lambda$, since any other term would be of too high an order to contribute to the decoherence functional at the desired accuracy. With this in mind, Eq. (43) reduces to the requirement that the unknown propagators should be homogeneous solutions to the Klein-Gordon equation on both of their arguments.

To determine the proper boundary conditions for these propagators, we may consider the boundary terms in Eq. (46), or else appeal to their physical interpretation. We shall choose the second approach.

To this end, we observe that the physical meaning of the propagators as (non standard) products of fields, Eqs. (22) to (25), entails the identity $G^{12} + G^{21} = G^{11} + G^{22}$, which is consistent in this case, since both sides solve the Klein - Gordon equation. Actually, this identity is satisfied by the VEV propagators, so it can be imposed directly on their variations.

Physically, a change in the propagators reflects a corresponding change in the statistical state of the field. To zeroth order in the coupling constant, however, the commutator of two fields is a c-number , and does not depend on the state. Therefore, to this accuracy, $G^{12} - G^{21}$ should not change; that is, $\delta G^{12}$ should be equal to $\delta G^{21}$. We thus conclude that the correct solution to Eq. (43) is

$$\delta G^{12} = \delta G^{21} = (\frac{1}{2})\{\delta G^{11} + \delta G^{22}\} \tag{48}$$

Consideration of the CTP boundary conditions would have led to the same result.

We may now evaluate the second variation of the 2PI CTP effective action, Eq. (42). We should stress that the Klein-Gordon operator annihilates all propagators involved, and that the $O(\lambda)$ term in $\Delta_0^{-1}$ vanishes because of our choice of renormalization point. Therefore the second (mixed) term in Eq. (42) is of higher than second order and may be disregarded. The same holds for terms of the form $(\Delta_0)_{ac}^{-1}\delta G^{cd}(\Delta_0)_{db}^{-1}\delta G^{ab}$, disregarding boundary terms.

The remaining terms can be read out of Eq. (45), with the input of the "fish" graph [43, 9]

$$
\begin{aligned}
\Sigma(x, x') \quad &= (\Delta_0^{11})^2(x, x') = \frac{i\mu^\epsilon}{(4\pi)^2} \int \frac{d^4k}{(2\pi)^4} e^{ik(x-x')} [\frac{2}{\epsilon} + \ln \frac{m^2}{4\pi\mu^2} - \psi(1) \\
&- k^2 \int_{4m^2}^\infty \frac{d\sigma^2}{\sigma^2(\sigma^2 + k^2 + i\epsilon)} \sqrt{1 - \frac{4m^2}{\sigma^2}}]
\end{aligned}
\tag{49}
$$

where $\epsilon = d - 4$ and $\mu$ is the renormalization scale. Clearly, the local terms in $\Sigma$ can be absorbed into a coupling-constant renormalization.

The important thing for us to realize is that the $O(\lambda)$ terms in Eq. (45), as well as the imaginary part of $\Sigma$, contribute only to the phase of the decoherence functional, and thus are totally unrelated to decoherence. The only contribution to a decoherence effect comes from the real part of $\Sigma$. Reading it out of Eq. (49), we obtain the sought for result

$$
\begin{aligned}
|D[f, f']| \quad &\sim exp\{(\frac{-\pi\lambda^2}{8}) \int \frac{d^4p \, d^4q}{(2\pi)^8} \delta(p^2 + m^2)\delta(q^2 + m^2) \\
&(f(p) - f'(p))(f(q) - f'(q))\theta[-((p + q)^2 + 4m^2)]\sqrt{1 + \frac{4m^2}{(p + q)^2}}\}
\end{aligned}
\tag{50}
$$

where $\theta$ is the usual step function. As expected, we do find decoherence between different correlation histories. Moreover, decoherence is related to dissipative processes, which in this case arise from pair production [49]. Indeed, the real part of the kernel $\Sigma$ is essentially the probability of a real pair being produced out of quanta with momenta $p$ and $q$, with $p + q = k$ [43].

Let us mention two obvious consequences of our result for the decoherence functional. The first point is that decoherence is associated with instability of the vacuum: the distribution functions whose overlap is suppressed represent ensembles which are unstable against non trivial scattering of the constituent particles. This scattering produces correlations between particles. Therefore, truncation of the correlation hierarchy leads to an explicitly dissipative evolution. This would not be the case if there were no scattering.

The second point is that $|D|$ remains unity on the diagonal. Thus, at least for Gaussian fluctuations, and to two-loop accuracy, all histories are equally likely. What this means physically is that the two-point functions to be perceived by an observer after the quantum to classical transition need not be close to their VEV in any stringent sense. Indeed, what is observed will not even be "vacuum fluctuations" in the proper sense of the word; they are real physical particles whose momenta are on shell, and may propagate to the asymptotic region, if they manage not to collide with other particles.

## 3.2.   Beyond Coarse Graining

For the observer confined to a single consistent history, as is the case for the quantum cosmologist, questioning the probability distribution of histories is somewhat academic. What would be relevant is one's ability to predict the future behavior of one's particular history. This ability is impaired by the lack of knowledge about the coarse-grained elements of the theory, which, in our case, are the higher correlations of the field.

As we have already seen, variation of the 2PI CTP effective action, id est, of the phase of the decoherence functional, yields the evolution equations for the VEVs of the two-point functions. These equations should be regarded as the Hartree-Fock approximation to the actual evolution, since in them the effect of higher correlations is represented only in the average. Deviations of the actual evolution from this ideal average may be represented by adding a source term to the Hartree-Fock equation. As the detailed state of the higher correlations is unknown, this right hand side should take the form of a stochastic binary external source.

The non-diagonal terms of the decoherence functional represented in Eq. (50), while not contributing to the Hartree-Fock equations, contain the necessary information to build a phenomenological model of the back reaction of the higher correlations on the relevant sector. To build this model, we compare the actual form of the decoherence functional against that resulting from the coupling of the propagators to an actual gaussian random external source [50].

The result of this comparison is that higher correlations react on the propagators as if these obey a Langevin- type equation

$$\frac{\partial \Gamma[\delta G^{11} = \delta G, \delta G^{22} = \delta G'^*, \delta G_0^{12}, \delta G_0^{21}]}{\partial(\delta G(x, x'))} = \frac{-1}{2v} F(x - x') J(x - x') \qquad (51)$$

where, after the variational derivative is taken, we must take the limit $\delta G' \to \delta G$. In Eq. (51) $v$ is (formally) "the space - time volume", the gaussian stochastic source $J$ has autocorrelation $\langle J(u)J(u') \rangle = \delta(u - u')$, and

$$F^2(u) = \lambda^2 \int_0^\infty \frac{ds}{(4\pi s)^2} \int_{4m^2}^\infty d\sigma^2 \sin(s\sigma^2 - \frac{u^2}{4s}) \sqrt{1 - \frac{4m^2}{\sigma^2}} \qquad (52)$$

Because the limit $\delta G' \to \delta G$ is taken, the imaginary terms of the CTP effective action reproduced in Eq. (50) do not contribute to the left hand side of Eq. (51); as far as the "Hartree - Fock" equations are concerned, they could as well be deleted from the effective action.

However, the stochastic source in the right hand side of Eq. (51) modifies the quantum amplitude associated with the correlation history by a factor

$$\exp\{(i/2) \int d^4u \; F(u)J(u)\delta G(u)\}, \qquad (53)$$

where $G(u) = (1/v) \int d^4X \; G(X + (u/2), X - (u/2))$. Correspondingly, the decoherence functional gains a factor

$$\exp\{(i/2) \int d^4u \; F(u)J(u)(\delta G(u) - \delta G'(u))\}. \qquad (54)$$

Upon averaging over all possible external sources, each having a probability

$$\exp\{(-1/2) \int d^4u \; J^2(u)\}, \qquad (55)$$

the new factor in the decoherence functional becomes

$$\exp\{(-1/8) \int d^4u \; F^2(u)(\delta G(u) - \delta G'(u))^2\}, \qquad (56)$$

which exactly reproduces Eq. (50). Observe that the assumed form for the right hand side of Eq. (51), and the requirement of recovering Eq. (50) upon averaging, uniquely determines the function $F$.

In this way, Eq.(51) yields the correct, if only a phenomenological, description of the dynamics of classical fluctuations in the aftermath of the quantum to classical transition. It should be obvious that nonlinearity is essential to the generation of these fluctuations.

## 4. Discussion

This paper presents three main results. The first is the ansatz Eq. (11) for the quantum amplitude associated with a correlation history. The second is the ansatz Eq. (40) for the decoherence functional between two such histories. On the basis of this ansatz, we have shown in Eq. (50) that the quantum interference between histories corresponding to different particle spectra is suppressed whenever these spectra differ by particles whose added momenta go above the two particle treshold $4m^2$, $m^2$ being the one-loop radiative-corrected physical mass. The third result is the phenomenological description in Eq. (51) of the dynamics of an individual consistent correlation history.

What we have presented in the above, despite its embryonic form, is a framework for bringing together the correlational-hierarchy idea in non-equilibrium statistical mechanics and the consistent-history interpretation of quantum mechanics. This framework puts decoherence and dissipation due to fluctuations and noise (manifested here through particle creation) on the same footing. It suggests a natural (intrinsic) measure of coarse-graining which is commensurate with ordinary accounts of dissipative phenomena, and with it addresses the issue of quantum to classical transition. It also provides a theoretical basis for the derivation of classical stochastic equations from quantum fluctuations, and identifies the nature of noise in these equations.

It should be noticed that a formal identity exists between the present results and those previously obtained from the influence functional formalism [51, 20, 11]. Indeed, our decoherence functional has the same structure as the influence functional, with the non diagonal terms in Eq. (50) playing the role of the "noise kernel". This is more than an analogy, as it should be clear from the discussions above and elsewhere.

While for reasons of clarity and economy of space, we have focused on a simple application from quantum field theory to develop our arguments, the implications on quantum mechanics and statistical mechanics go beyond what this example can show. The theoretical issues raised here in the context of quantum mechanics and statistical mechanics, as well as the consequences of problems raised in the context of quantum and semiclassical (especially the inflationary universe) cosmology, which motivated us to make these inquiries in the first place, will be explored in greater detail elsewhere.

This work is part of an on-going program which draws on many year's worth of pondering on the role of statistical mechanics ideas in quantum cosmology, using quantum field theoretical methods while placing the issues in the larger context of general physics. The project began in 1985, when one of us (EC) was invited by Dieter Brill to join the General Relativity Group at Maryland. It is therefore an honour and a pleasure for us to dedicate this paper to him on this happy occasion.

**Acknowledgments**

E. C. is partially supported by the Directorate General for Science Research and Development of the Commission of the European Communities under Contract N° C11 - 0540 - M(TT), and by CONICET, UBA and Fundación Antorchas (Argentina). B. L. H's research is supported in part by the US NSF under grant PHY91-19726 This collaboration is partially supported by NSF and CONICET as part of the Scientific and Technological Exchange Program between Argentina and the USA.

# References

[1] H. Everitt, III, Rev. Mod. Phys. **29**, 454 (1957); B. S. DeWitt and N. Graham, eds., *The Many-Worlds Interpretation of Quantum Mechanics* (Princeton Univ., Princeton, 1973).

[2] R. B. Griffiths, J. Stat. Phys. **36**, 219 (1984); R. Omnés, J. Stat Phys. **53**, 893, 933, 957 (1988); Ann. Phys. (NY) **201**, 354 (1990); Rev. Mod. Phys. **64**, 339 (1992)

[3] J. B. Hartle, "Quantum Mechanics of Closed Systems" in *Directions in General Relativity* Vol. 1, eds B. L. Hu, M. P. Ryan and C. V. Vishveswara (Cambridge Univ., Cambridge, 1993)

[4] M. Gell-Mann and J. B. Hartle, in *Complexity, Entropy and the Physics of Information*, ed. by W. H. Zurek (Addison-Wesley, Reading, 1990); Phys. Rev. **D47**, (1993) H. F. Dowker and J. J. Halliwell, Phys. Rev. **D46**, 1580 (1992); Brun, Phys. Rev. **D47**, (1993)

[5] A. I. Akhiezer and S. V. Peletminsky, *Methods of Statistical Physics* (Pergamon, London, 1981).

[6] I. Prigogine, *Non Equilibrium Statistical Mechanics* (John Wiley, New York, 1962); R. Balescu, *Equilibrium and Non Equilibrium Statistical Mechanics* (John Wiley, New York, 1975).

[7] L. Kadanoff and G. Baym, *Quantum Statistical Mechanics* (Benjamin, New York, 1962).

[8] R. Kubo, M. Toda and N. Hashitsume *Statistical Physics II*, (Springer-Verlag, Berlin, 1978); J. A. McLennan, *Introduction to Non-Equilibrium Statistical Mechanics* (Prentice-Hall, New Jersey, 1989); N. G. van Kampen, *Stochastic Processes in Physics and Chemistry* (North Holland, Amsterdam, 1981).

[9] E. Calzetta and B. L. Hu, Phys. Rev. **D37**, 2878 (1988).

[10] E. Calzetta, S. Habib and B. L. Hu, Phys. Rev. **D37**, 2901 (1988); S. Habib, Ph. D. Thesis, University of Maryland, 1988 (unpublished).

[11] Yuhong Zhang, Ph. D. Thesis, University of Maryland, 1990 (unpublished); B. L. Hu, J. P. Paz and Y. Zhang, Phys. Rev. **D45**, 2843 (1992); "Quantum Brownian Motion in a General Environment II. Nonlinear coupling and perturbative approach" Phys. Rev. **D47**, (1993); "Stochastic Dynamics of Interacting Quantum Fields" (paper III).

[12] E. Calzetta and B. L. Hu, "From Kinetic Theory to Brownian Motion" (1993); "Quantum and Classical Fluctuations" (1993); "On Correlational Noise" (1993)

[13] B. L. Hu, "Fluctuation, Dissipation and Irreversibility in Cosmology" in *The Physical Origin of Time-Asymmetry* Huelva, Spain, 1991 eds. J. J. Halliwell, J. Perez-Mercader and W. H. Zurek (Cambridge University Press, 1993).

[14] B. L. Hu, "Quantum Statistical Processes in the Early Universe" in *Quantum Physics and the Universe*, Proc. Waseda Conference, Aug. 1992 ed. Namiki, K. Maeda, et al (Pergamon Press, Tokyo, 1993).

[15] B. L. Hu "Quantum and Statistical Effects in Superspace Cosmology" in *Quantum Mechanics in Curved Spacetime*, ed. J. Audretsch and V. de Sabbata (Plenum, London 1990).

[16] R. Balian and M. Veneroni, Ann. Phys. (N. Y.) **174**, 229 (1987)

[17] W. H. Zurek, Phys. Rev. **D24**, 1516 (1981); **D26**, 1862 (1982); in *Frontiers of Nonequilibrium Statistical Physics*, ed. G. T. Moore and M. O. Scully (Plenum, N. Y., 1986); W. G. Unruh and W. H. Zurek, Phys. Rev. **D40**, 1071 (1989); Physics Today **44**, 36 (1991); W. H. Zurek, J. P. Paz and S. Habib, Phys. Rev. **47**, 488 (1993).

[18] E. Joos and H. D. Zeh, Z. Phys. **B59**, 223 (1985); H. D. Zeh, Phys. Lett. A **116**, 9 (1986).

[19] J. A. Wheeler and W. H. Zurek, *Quantum Theory and Measurement* (Princeton Univ., Princeton, 1983)

[20] A. O. Caldeira and A. J. Leggett, Physica **121A**, 587 (1983); Phys. Rev. **A31**, 1059 (1985); A. J. Leggett, S. Chakravrty, A. T. Dorsey, M. P. A. Fisher, A. Garg and W. Zwerger, Rev. Mod. Phys. **59**, 1 (1987).

[21] B. L. Hu, "Statistical Mechanics and Quantum Cosmology", in *Proc. Second International Workshop on Thermal Fields and Their Applications*, eds. H. Ezawa et al (North-Holland, Amsterdam, 1991); E. Calzetta, Phys. Rev. **D43**, 2498 (1991); S. Sinha, Ph. D. Thesis, University of Maryland, 1991 (unpublished); S. Sinha and B. L. Hu, Phys. Rev. **D44**, 1028 (1991); B. L. Hu, J. P. Paz and S. Sinha, "Minisuperspace as a Quantum Open System" in *Directions in General Relativity* Vol. 1, (Misner Festschrift) eds B. L. Hu, M. P. Ryan and C. V. Vishveswara (Cambridge Univ., Cambridge, 1993).

[22] J. J. Halliwell, in Proc. 13th GRG Meeting, Cordoba, Argentina, July 1992

[23] J. P. Paz and S. Sinha, Phys. Rev. **D44**, 1038 (1991); *ibid* **D45**, 2823 (1992); E. Calzetta and D. Mazzitelli, Phys. Rev. **D42**, 4066 (1990).

[24] B. L. Hu, Physica A **158**, 399 (1989).

[25] E. Calzetta, Class. Quan. Grav. **6**, L227 (1989)

[26] B. L. Hu, J. P. Paz and Y. Zhang, "Quantum Origin of Noise and Fluctuation in Cosmology" in *Proc. Conference on the Origin of Structure in the Universe* Chateau du Pont d'Oye, Belgium, April, 1992, ed. E. Gunzig and P. Nardone (NATO ASI Series) (Plenum Press, New York, 1993); S. Habib and H. E. Kandrup, Phys. Rev. **D46**, 5303 (1992).

[27] B. L. Hu and Y. Zhang, in *Proc. Third International Workshop on Quantum Nonintegrability*, Drexel University, Philadelphia, May 1992, ed. D. H. Feng, J. Yuan (Gordon and Breach, New York, 1993)

[28] R. Geroch, Noûs **18**, 617 (1984); J.B Hartle, in *Gravitation in Astrophysics*, 1986 NATO Advanced Summer Institute, Cargese, ed. B.Carter and J. Hartle (NATO ASI Series B: Physics Vol. 156, Plenum, N. Y., 1987)

[29] J. B. Hartle, in *Directions in General Relativity* Vol 2 (Brill Festschrift) eds. B. L. Hu and T. A. Jacobson (Cambridge Univ., Cambridge, 1993)

[30] R. Penrose, in Proc. 13th GRG Meeting, Cordoba, Argentina, July, 1992.

[31] A. H. Guth, Phys. Rev. **D23**, 347 (1981); K. Sato, Phys. Lett. **99B**, 66 (1981); A. D. Linde, Phys. Lett. **108B**, 389 (1982); A. Albrecht and P. J. Steinhardt, Phys. Rev. Lett. **48**, 1220 (1982); Also see E. Kolb and M. Turner *the Early Universe* (Addison-Wesley, Menlo Park, 1990), and A. Linde, *Inflationary and Quantum Cosmology* (Academic Press, San Diego, 1991)

[32] J. B. Hartle and S. W. Hawking, Phys. Rev. **D28**, 1960 (1983); A. Vilenkin, Phys. Lett. **117B**, 25 (1985); Phys. Rev. **D27**, 2848 (1983), **D30**, 509 (1984);

[33] S. Nakajima, Progr. Theor. Phys. **20**, 948 (1958); R. Zwanzig, J. Chem. Phys. **33**, 1338 (1960), and in *Lectures in Theoretical Physics III*, edited by W. Britten *et al.*(Wiley, New York, 1961), p. 106; H. Mori, Prog. Theor. Phys. **34**, 423 (1965)

[34] J. B. Hartle, Phys. Rev. **D37**, 2818 (1988),

[35] J. J. Halliwell and M. Ortiz, "Sum over Histories Origin of the Composition Laws of Relativistic Quantum Mechanics" Phys. Rev. **D47** (1993).

[36] A. A. Starobinsky, in *Field Theory, Quantum Gravity and Strings*, ed. H. J. de Vega and N. Sanchez (Springer, Berlin 1986); J. M. Bardeen and G. J. Bublik, Class. Quan. Grav. **4**, 473 (1987).

[37] B. L. Hu and Y. Zhang, "Coarse-Graining, Scaling, and Inflation" Univ. Maryland Preprint 90-186; B. L. Hu, in *Relativity and Gravitation: Classical and Quantum* Proc. SILARG VII, Cocoyoc, Mexico 1990. eds. J. C. D' Olivo et al (World Scientific, Singapore 1991); S. Habib, Phys. Rev. **D46**, 2408 (1992).

[38] E. P. Wigner, Phys. Rev. **40**, 749 (1932)

[39] J. Schwinger, J. Math. Phys. **2** (1961) 407; L. V. Keldysh, Zh. Eksp. Teor. Fiz. **47** , 1515 (1964) [Engl. trans. Sov. Phys. JEPT **20**, 1018 (1965)].

[40] G. Zhou, Z. Su, B. Hao and L. Yu, Phys. Rep. **118**, 1 (1985); Z. Su, L. Y. Chen, X. Yu and K. Chou, Phys. Rev. **B37**, 9810 (1988).

[41] B. S. DeWitt, in *Quantum Concepts in Space and Time* ed. R. Penrose and C. J. Isham (Claredon Press, Oxford, 1986); R. D. Jordan, Phys. Rev. **D33**, 44 (1986).

[42] E. Calzetta and B. L. Hu, Phys. Rev. **D35**, 495 (1987).

[43] P. Ramond, *Field Theory, a Modern Primer* (Benjamin, New York, 1981).

[44] R. Jackiw, Phys. Rev. **D9**, 1686 (1974); J. Iliopoulos, C. Itzykson and A. Martin, Rev. Mod. Phys. **47**, 165 (1975).

[45] J. S. Langer, Ann. Phys. (NY) **41**, 108 (1967); **54**, 258 (1969).

[46] H. D. Dahmen and G. Jona - Lasinio, Nuovo Cimento **52A**, 807 (1962); C. de Dominicis and P. Martin, J. Math. Phys. **5**, 14 (1964); J. M. Cornwall, R. Jackiw and E. Tomboulis, Phys. Rev. **D10**, 2428 (1974); R. E. Norton and J. M. Cornwall, Ann. Phys. (NY) **91**, 106 (1975).

[47] E. Calzetta, Ann. Phys. (NY), **190**, 32 (1989).

[48] R. Mills, *Propagators for Many Particle Systems* (Gordon and Breach, New York, 1969).

[49] E. Calzetta and B. L. Hu, Phys. Rev. **D40**, 656 (1989).

[50] Our argument is adapted from a similar problem in R. Feynman and A. Hibbs, *Quantum Mechanics and Path Integrals*, (McGraw - Hill, New York, 1965).

[51] R. Feynman and F. Vernon, Ann. Phys. (NY) **24**, 118 (1963).

# The Initial Value Problem in Light of Ashtekar's Variables

*Riccardo Capovilla* *      *John Dell* [t]      *Ted Jacobson* [t]

### Abstract

The form of the initial value constraints in Ashtekar's hamiltonian formulation of general relativity is recalled, and the problem of solving them is compared with that in the traditional metric variables. It is shown how the general solution of the four diffeomorphism constraints can be obtained algebraically provided the curvature is non-degenerate, and the form of the remaining (Gauss law) constraints is discussed. The method is extended to cover the case when matter is included, using an approach due to Thiemann. The application of the method to vacuum Bianchi models is given. The paper concludes with a brief discussion of alternative approaches to the initial value problem in the Ashtekar formulation.

## 1. Introduction

It is with great pleasure that we dedicate this paper to Dieter Brill, our teacher, advisor, and colleague, on the occasion of his 60th birthday. Our contribution concerns the initial value problem for general relativity, which is amongst Dieter's many areas of expertise. As is the case with almost all research activity developed around the general relativity group at the University of Maryland, the ideas we will present have benefitted from Dieter's always kind and sometimes maddening insightful questioning. Of course it is our wish that this paper will prompt some more such questioning.

General relativity is invariant under four dimensional diffeomorphisms. In the hamiltonian formulation, this invariance manifests itself in the presence of constraints on

*Centro de Investigacion y de Estudios Avanzados, I.P.N., Apdo. Postal 14-740, Mexico 14, D.F., Mexico. This author was partially supported by CONACyT Grant No. 1155-E9209.

[t]Thomas Jefferson High School for Science and Technology, 6560 Braddock Road, Alexandria, VA 22312, USA.

[t]Department of Physics, University of Maryland, College Park, MD 20742, USA. This author's research was supported in part by the National Science Foundation under Grants No. PHY91-12240 at the University of Maryland and PHY89-04035 at the Institute for Theoretical Physics, Santa Barbara, California.

the canonical variables. If the constraints are satisfied at one instant of time, they continue to hold at all times. The initial value problem is the problem of giving a construction for a parametrization of the general solution of the constraints.

The most well-developed approach to the initial value problem is based on the "conformal technique" (see for example [1] and references therein). In this approach, the phase space variables for general relativity are the 3-metric $q_{ab}$ and its conjugate momentum $p^{ab}$ which is a tensor density of weight 1, closely related to the extrinsic curvature of the spatial hypersurface in a solution to the field equations. There are four constraints, consisting of one Hamiltonian constraint, which generates diffeomorphisms normal to the spacelike surface on which the initial data are given, and three momentum constraints, which generate spatial diffeomorphisms. The Hamiltonian constraint is viewed as a (quasi-linear, elliptic) equation determining the conformal factor of the 3-metric. The momentum constraints determine the "longitudinal" piece of $p^{ab}$. In this approach, the freely specified data are, in principle, the conformal equivalence class of the 3-metric, and the transverse traceless part of the momentum $p^{ab}$.

The conformal technique is well suited to addressing questions of existence and uniqueness of solutions to the field equations. However, for two reasons it is not suited to reduction of the phase space to the unconstrained degrees of freedom. First, the "freely specifiable" transverse traceless tensor densities are not known for an arbitrary conformal 3-metric. Second, the solutions to the constraint equations determining the conformal factor and longitudinal part of the momentum are not given explicitly.

An alternative representation for hamiltonian general relativity was introduced in the mid-eighties by Ashtekar [2]. The Ashtekar representation may be understood (and was discovered originally) as resulting from a complex canonical transformation that goes from the (triad-extended) ADM gravitational phase space variables to a new set of variables. The new variables are a spatial SO(3,C) connection, and an $so(3,C)$-valued vector density as its conjugate momentum. The major benefit of this representation is that the constraints take a simpler form in terms of these variables than in the ADM representation. First, the constraints turn out naturally to be polynomial of low order in the Ashtekar phase space variables. Moreover, for the vacuum case, the scalar constraint is homogenous in the Ashtekar canonical momenta. This simplification has allowed considerable progress in many questions of gravitational physics. Perhaps the most impressive results have been obtained in the quantum theory, where this formulation has made it possible to perform the first few steps in the Dirac quantization of general relativity [3, 4, 5].

Given that the Ashtekar representation simplifies the form of the constraints, it is natural to reconsider the classical initial value problem in this context. As it turns out, generically in phase space, it is possible to obtain *algebraically* the general solution to the Ashtekar version of the diffeomorphism constraints of general relativity! [6, 7]

However, this does not give an easy solution of the initial value problem. The reason is that, due to the covariance of the Ashtekar formalism under local SO(3,C) rotations, there are three additional constraints whose form is identical to the non-abelian form of Gauss' law familiar from Yang-Mills theories. Once the diffeomorphism contraints are solved by the method to be described below, one still has to face the Gauss law, which has now become a non-linear first order partial differential equation on the remaining variables. In addition, the reality conditions restricting to the phase space of real, Lorentzian GR need to be enforced.

Still, one can hope that the relocation of the difficulties afforded by the Ashtekar approach may prove to be fruitful in some applications. This already seems to be the case in the context of GR reduced by symmetry conditions. In addition, the interplay between "gauge-fixing" and solving the constraints is different in the Ashtekar approach, which is a fact that remains to be fully explored.

The rest of this paper is organized as follows. The form of the initial value constraints in Ashtekar's hamiltonian formulation of general relativity is recalled in section 2, and in section 3 it is shown how the general solution of the four diffeomorphism constraints in vacuum (or with a cosmological constant) can be obtained algebraically provided the curvature is non-degenerate. The form of the remaining (Gauss law) constraints is also discussed, as is the relation between the solution given and the structure of the 4-dimensional curvature. In section 4 the method is extended to cover the case when matter is included, using an approach due to Thiemann. The application of the method to vacuum Bianchi models is given in section 5, and the paper concludes in section 6 with a brief discussion of alternative approaches to the initial value problem in the Ashtekar formulation.

## 2.   Ashtekar's variables

Ashtekar's representation of hamiltonian general relativity has been the subject of extensive reviews (cf. [5, 8]), so we shall just recall its main features.

The Ashtekar canonical coordinates are an SO(3,C) spatial connection, $A_a^i$, and an $so(3,C)$-valued vector density of weight 1, $E_i^a$. The fundamental Poisson bracket is given by $\{A_a^i(x), E_j^b(y)\} = i\delta_j^i \delta_a^b \delta^3(x,y)$. (Our notation is as follows. Latin letters from the beginning of the alphabet denote spatial indices, e.g. $a, b, \ldots = 1, 2, 3$. Latin letters from the middle of the alphabet are SO(3,C) indices, $i, j, \ldots = 1, 2, 3$. They are raised and lowered with the Kronecker delta $\delta^{ij}$ and $\delta_{ij}$. We will also use the totally antisymmetric symbol $\epsilon_{ijk}$, with $\epsilon_{123} = 1$. We use units with $c = G = 1$.)

The vector density $E_i^a$ may be identified with the densitized spatial triad $\sqrt{q}e_i^a$, with the contravariant spatial metric given by $q^{ab} = e_i^a e^{bi}$. In turn, the SO(3,C) connection may be identified in a solution with the spatial pullback of the self-dual part of the spin-connection.

These variables parametrize the phase space of *complex* general relativity. A real metric with Lorentzian signature may be recovered by imposing appropriate reality conditions (cf. [5]). These conditions amount to the requirement that $E_i^a E^{bi}$ be real, and that its time derivative be real. If these conditions are satisfied initially then the dynamical evolution will preserve them in time. A metric of Euclidean signature is obtained by simply taking $A_a^i$ and $E_i^a$ real.

In terms of these variables, the constraints of (complex) general relativity take the form

$$\varepsilon_{abc}\epsilon^{ijk}E_i^a E_j^b B_k^c = -\rho \tag{1}$$

$$\varepsilon_{abc}E_i^b B^{ci} = -iJ_a \tag{2}$$

$$D_a E_i^a = K_i . \tag{3}$$

Here $\varepsilon_{abc}$ is the standard Levi-Civita tensor density. $B_i^a$ is the "magnetic field" of the connection $A_a^i$, defined by $B_i^a := \varepsilon^{abc} F_{bci}$, and $F_{ab}^i := \partial_a A_b^i - \partial_b A_a^i + \epsilon^i_{jk} A_a^j A_b^k$. $D_a$ is the SO(3)-covariant derivative determined by $A_a^i$. $\rho/\sqrt{q}$ and $J_a$ are (up to coefficients) the matter energy and momentum densities respectively. $K_i$ is a spin density, present only for half-integer spin matter fields.

In the following, we will call the constraints (1) and (2) "diffeomorphism constraints". The first generates diffeomorphisms normal to the spatial hypersurface $\Sigma$ together with some SO(3) rotation, so it takes the place in Ashtekar's formalism of the ADM hamiltonian constraint. It will be called here the "scalar constraint". The second generates diffeomorphisms tangential to $\Sigma$ together with some SO(3) rotation, so it takes the place of the ADM momentum constraint. It will be called here the "vector constraint".

There are 3 additional constraints, (3), of the Gauss-law type familiar from Yang-Mills theories. They generate local SO(3,C) rotations, under which the Ashtekar formalism is covariant.

The constraints are polynomial in the gravitational phase space variables and for scalar and spin-1/2 matter. For Yang-Mills type fields, polynomiality is mantained only if one multiplies the scalar constraint (1) through by $detq = detE_i^a$, which is cubic in $E_i^a$. Moreover, it is remarkable that the gravitational contribution to the scalar constraint (1) is *homogenous* in the momenta $E_i^a$. This should be compared with the ADM Hamiltonian constraint, where the term quadratic in the canonical momenta must be balanced by a "potential" term given by the 3D Ricci scalar times the determinant of the 3-metric. Note that while the vector and Gauss constraints (2) and (3) are densities of weight 1, the scalar constraint (1) is of weight 2.

# 3.   Algebraic solution of the vacuum diffeomorphism constraints

We now proceed to show how one can use the simplification of the constraints for general relativity provided by the Ashtekar formalism to tackle the initial value problem. In particular, we will show how the general solution of the the scalar and vector diffeomorphism constraints can be obtained by algebraic methods. We originally discovered this solution in the context of a Lagrangian pure spin-connection formulation of GR, in which one solves for the metric variables (self-dual 2-forms) in terms of the connection [6]. However, it is unnecessary to view the technique in that context.

The first step is to assume that the magnetic field $B_i^a$ is non-degenerate as a $3 \times 3$ matrix. In a generic real, Lorentzian spacetime, the real and imaginary parts of the complex equation $det B = 0$ will define two surfaces, and their intersection will give a one-dimensional submanifold on which $B_i^a$ is degenerate. That is, generically $B_i^a$ is non-degenerate except on a set of measure zero. (The points in phase space with $det B = 0$ *everywhere* form a set of measure zero in phase space. The identity of these points is best seen in the covariant formalism (cf. [6, 9]). It turns out that they correspond to space-time metrics of Petrov type 0, (4), (3,1), when $E_i^a$ is non-degenerate.) We will ignore here any problems that might arise as a consequence of degeneracy of $B_i^a$.

Now, since $B_i^a$ is assumed to be non degenerate, one may use it as a basis for the space of SO(3,C) vector densities. In particular, one can expand the momentum $E_i^a$ with respect to it,

$$E_i^a = P_{ij} B^{aj} , \qquad (4)$$

for some (non-degenerate) $3 \times 3$ matrix $P_{ij}$.

We first consider the vacuum case, $\rho = 0$, $J_a = 0$, $K_i = 0$. It is easy to see that the vector constraint (2) implies that $P_{ij}$ must be symmetric. The scalar constraint (1) implies that $P_{ij}$ must satisfy a quadratic algebraic condition,

$$(P_i^i)^2 - P_{ij} P^{ij} = 0 . \qquad (5)$$

This condition fixes one of the six components of $P_{ij}$ with respect to the others.

We know of two methods for solving the scalar constraint (5). In the first method, one solves (5) for the trace of $P$. Decomposing $P_{ij}$ into the trace $TrP$ and tracefree part $\hat{P}_{ij} := P_{ij} - \frac{1}{3} TrP \delta_{ij}$, (5) becomes the condition

$$TrP = \pm(\tfrac{3}{2} Tr \hat{P}^2)^{1/2} . \qquad (6)$$

In the second method for solving (5), one notes that the characteristic equation for $P$ implies that $TrP^{-1} = [TrP^2 - (TrP)^2]/2det P$. Thus (5) is equivalent to the tracelessness of $P^{-1}$, provided $P$ is invertible (which it must be if both $B_i^a$ and the 3-metric are nondegenerate). Let $\psi$ denote $P^{-1}$. Then the characteristic equation for $\psi$, assuming $Tr\psi = 0$, reads $\psi^3 - \frac{1}{2}(Tr\psi^2)\psi - det\psi I = 0$, giving $\psi^{-1} = [\psi^2 - \frac{1}{2}(Tr\psi^2)I]/det\psi$. Thus the general solution to (5) for invertible $P$ can be explicitly written in terms of the 5 independent components of a tracefree $3 \times 3$ matrix $\phi$ in the form

$$P = \phi^2 - \frac{1}{2}(Tr\phi^2)I . \qquad (7)$$

One is left with the Gauss constraint (3), which becomes

$$B_i^a D_a P^{ij} = 0 , \qquad (8)$$

where we have used the Bianchi identity $D_a B_i^a = 0$. In view of (6) or (7), the Gauss constraint becomes a *non-linear* equation in $P_{ij}$ or $\phi$ respectively. It is here, and in the reality conditions, that the initial value problem now resides.

We can consider the Gauss law constraint as 3 conditions on the 5 independent components of $P^{ij}$, for a fixed $A_a^i$. Then there are presumably 2 free functions worth of solutions for this equation, yielding the two indepenent metric degrees of freedom of GR. Up to this stage the connection $A_a^i$ has remained entirely arbitrary. By use of the 4 parameter diffeomorphism and 3 parameter SO(3) transformations, one could now fix all but 2 of the 9 components of the connection, yielding the other half of the coordinates on the reduced phase space.

If a spherically symmetric ansatz is assumed, then the above method of solving the constraints can be carried out entirely, and the Gauss constraint reduces to an ordinary differential equation that can be solved [10].

The existence in general of solutions to the Gauss law constraint (8) with $P^{ij}$ restricted by (6) or (7) is a problem that has not been addressed. To indicate just one complication that can arise, consider the asymptotically flat case. If $A_a^i$ is asymptotically the spin-connection corresponding to a negative-mass spacetime, then we know by the positive energy theorem that it must be impossible to find a regular solution to (8) for $P^{ij}$, subject to the reality conditions.

The geometrical interpretation of the matrix $P^{ij}$ is easily seen from the covariant point of view [11]. The densitized triad $E^{ai}$ of the Ashtekar formalism is related to the triad of anti-self-dual 2-forms by $E^{ai} = \epsilon^{abc} \Sigma_{bc}^i$ [12]. Now the curvature 2-form $R^i$ of the spin-connection can be expanded in terms of the self-dual and anti-self-dual 2-forms as as $R_i = \psi_{ij} \Sigma^j + \frac{1}{3}\Lambda \Sigma_i + \Phi_{ij}\bar{\Sigma}^j$. $\psi_{ij}$ is symmetric and tracefree and is just the Weyl spinor in SO(3) notation. $\Lambda$ is proportional to the Ricci scalar, and $\Phi_{ij}$ is equivalent to the tracefree part of the Ricci tensor. Thus in vacuum, $\Lambda$ and $\Phi_{ij}$ vanish, and one can solve for the anti-self-dual 2-forms in terms of the curvature as $\Sigma^i = (\psi^{-1})^{ij} R_j$. The dual of the spatial pullback of this equation yields immediately the general solution given above for the four diffeomorphism constraints in Ashtekar's formalism, with $P^{ij}$ identified with $(\psi^{-1})^{ij}$.

Let us now consider the addition of a cosmological constant $\Lambda$. This modifies only the scalar constraint, which becomes

$$\varepsilon_{abc}\epsilon^{ijk}E_i^a E_j^b B_k^c - (1/3)\Lambda\varepsilon_{abc}\epsilon^{ijk}E_i^a E_j^b E_k^c = 0 . \tag{9}$$

One may now proceed as in the vacuum case. The only difference is that the algebraic condition (5) is replaced by

$$(P_i^i)^2 - P_{ij}P^{ij} = 2\Lambda(det P) . \tag{10}$$

This can be regarded as a cubic equation for $TrP$. It is equivalent to the statement that the inverse of $P$ has trace equal to $\Lambda$. In this case, P corresponds to the inverse of the matrix $(\psi_{ij} + \frac{1}{3}\Lambda\,\delta_{ij})$.

An interesting special solution of the constraints in the case of a nonvanishing $\Lambda$ is given by the Ansatz $P_{ij} = (3/\Lambda)\delta_{ij}$, or

$$E_i^a = (3/\Lambda)B_i^a , \tag{11}$$

which solves automatically all of the constraints. (The Gauss constraint is satisfied as a consequence of the Bianchi identity $D_a B_i^a = 0$.) This Ansatz was first introduced by Ashtekar and Renteln [13], who observed that it gives rise to self-dual solutions of the Einstein equation with a non-vanishing $\Lambda$.

## 4.   Matter couplings

It is possible to extend the method just described to solve the diffeomorphism constraints in the presence of matter using an approach that has recently been brought to our attention by Thomas Thiemann [7]. Thiemann's method proceeds as follows.

One begins with the constraints in the form (1), (2), (3)[14]. Then expanding $E_i^a = P_{ij}B^{aj}$ as in (4), one finds that the vector constraint (2) implies that the anti-symmetric part of $P_{ij}$ no longer vanishes, but is given instead by

$$A_{ij} := P_{[ij]} = \frac{i}{2B}\, \epsilon_{ijk}B_k^a J_a , \tag{12}$$

where $B := det B_i^a$. This is an explicit expression for $P_{[ij]}$ provided $J_a$ does not depend upon $P_{[ij]}$ as well. Since $J_a$ is just the generator of spatial diffeomorphisms for the matter variables, it does not depend on the gravitational field variables for the case of integer spin matter. In the spin-1/2 case, the diffeomorphism is accompanied by an SO(3) rotation, which involves the spin connection but not $E_i^a$. In all cases therefore, $J_a$ is independent of $P_{ij}$. (In particular, in the scalar, spin-1/2, and Yang-Mills cases, the structure of $J_a$ is $\pi\partial_a\phi$, $\pi^A D_a\psi_A$, and $e^{bI}f_{abI}$ respectively, where $e^{bI}$ and $f_{abI}$ are the Yang-Mills-electric and (dual of) magnetic fields.)

Now turning to the scalar constraint, we begin by noting that when $P_{ij}$ is decomposed into its symmetric and antisymmetric parts $P_{ij} = S_{ij} + A_{ij}$, the left hand side of the scalar constraint (1) is given by

$$\epsilon_{ijk}\epsilon_{abc}E^{ai}E^{bj}B^{ck} = det B \left[ (TrS)^2 - TrS^2 - TrA^2 \right]. \qquad (13)$$

Let us first consider the case of a single scalar field. Then we have

$$\rho = \pi^2 + E^{ai}E_i^b \partial_a\phi\partial_b\phi + \epsilon_{ijk}\epsilon_{abc}E^{ai}E^{bj}E^{ck}\, V(\phi). \qquad (14)$$

Now as in the vacuum case there are two approaches to solving the scalar constraint. In the first method, one regards the constraint as a cubic equation on the trace of $S_{ij}$, which can be solved (albeit messily) in closed form. In the second approach, due to Thiemann [7], one notes that every term of the constraint is either independent of the scalar field momentum $\pi$ or depends on $\pi$ quadratically! To see why, observe that $J_a$ is linear in $\pi$, and therefore according to (12), so too is $A_{ij}$. Thus the gravitational part of the constraint (13) is quadratic in $\pi$. As for $\rho$ (14), the second (gradient) term is independent of $\pi$ because, as one easily sees using (12), $A_{ij}$ drops out of the expression $E^{ai}\partial_a\phi = S^{ij}B_j^a\partial_a\phi$. Moreover, in the third term one can show that $detE$ involves $A_{ij}$, and therefore $\pi$, also only quadratically.

Thus one can simply solve the constraint for $\pi$ in closed form. Even for the case $V(\phi) = 0$, the resulting expression is fairly complicated:

$$\pi = \pm\left[\left((TrS)^2 - TrS^2 + (S\xi)^2\right)/\left(\tfrac{\xi^2}{2B} - 1\right)\right]^{1/2}, \qquad (15)$$

where $\xi^i := B^{ai}\partial_a\phi$.

For the real theory, there is also the constraint that the argument of the square root be a positive real number. When $V(\phi) \neq 0$, the solution for $\pi$ becomes significantly more complicated. It is unlikely that this is of any practical use in general. However, Thiemann's intended application is quantization of the spherically symmetric scalar-gravitational system, for which the resulting expressions may be more useful.

Note that even when the argument of the square root in (15) is real, the solution for $\pi$ has a sign ambiguity. This can lead to problems if one wishes to pass to the reduced phase space for the purpose of quantization, since it is difficult eliminate the momentum and not lose part of the reduced phase space[15].

For other types of matter coupling the situation becomes yet more complicated. For a massless spin-1/2 field, the scalar constraint remains quadratic (but not homogeneous) in the two components $\pi^A$ of the spinor field momentum, so one can in principle solve for one of these components. In addition, the Gauss constraint is augmented by a spinor field term in this case. When Yang-Mills fields are included, the scalar constraint remains polynomial only if it is multiplied through by $detE$, resulting in a 4th order polynomial in the scalar field momentum.

## 5.   Bianchi models

In this section, we apply to spatially homogenous vacuum space-times, *i.e.* Bianchi models, the method for the solution of the constraints illustrated in section 3 for the vacuum case. The hope is to gain some insight on how to deal with the remaining, Gauss constraint in the form (8). In any case, this analysis may be of use in a minisuperspace approximation of vacuum general relativity.

For Bianchi models, as follows from the assumption of spatial homogeneity, this last constraint reduces to an algebraic condition. There are some simplifications, but the condition is still rather complicated. We record it here for the benefit of inventory. One interesting feature is that only the traceless part of $P_{ij}$ enters in it.

The formulation of Bianchi models in terms of Ashtekar variables was first worked out by Kodama [16]. Here we follow the approach of Ashtekar and Pullin [17].

As is familiar from the ADM treatment of Bianchi models (see *e.g.* [18, 19]), we consider a kinematical triad of vectors, $X_I^a$, which commute with the three Killing vectors on the spatial hypersurface $\Sigma$. (Capital latin letters $I, J, K, ...$ will be used to label the triad vectors.) The triad satisfies

$$[X_I, X_J]^a = C_{IJ}{}^K X_K^a ,  \tag{16}$$

where $C_{IJ}{}^K$ denote the structure constants of the Bianchi type under consideration. The basis dual to $X_I^a$, $\chi_a^I$, satisfies,

$$2\partial_{[a}\chi_{b]}^I = -C_{JK}{}^I \chi_a^J \chi_b^K .  \tag{17}$$

Without loss of generality, one may set

$$C_{IJ}{}^K = \epsilon_{IJL}S^{LK} + 2\delta_{[I}^K V_{J]}  \tag{18}$$

with $S^{IJ}$ symmetric. (This $S^{IJ}$ has nothing to do with the symmetric part $S^{ij}$ of $P^{ij}$ referred to earlier.) From the Jacobi identities, it follows that $S^{IJ}V_J = 0$. The Bianchi classification may be described in terms of the vanishing or not of $V_I$, and the signature of $S^{IJ}$, subject to this condition. The most popular models are Bianchi I, selected by $C_{IJ}{}^K = 0$, and Bianchi IX, selected by $V_I = 0$ and $S^{IJ} = \delta^{IJ}$.

The Ashtekar gravitational phase space variables may be expanded with respect to the kinematical triad $X_I^a$, and $\chi_a^I$, as

$$A_a^i = A_M^i \chi_a^M  \tag{19}$$
$$E_i^a = det\chi \, E_i^M X_M^a  \tag{20}$$

where $det\chi$ denotes the determinant of $\chi_a^I$, which is introduced in order to de-densitize $E_i^a$. Similarly, the magnetic field may be expanded as

$$B_i^a = det\chi \, B_i^M X_M^a ,  \tag{21}$$

and $B_i^M$ is given by

$$B_i^M = -\epsilon^{MNP}C_{NP}{}^Q A_{Qi} + \epsilon^{MNP}\epsilon_{ijk}A_N^j A_P^k \,. \tag{22}$$

The gravitational phase space has thus been reduced to the matrices $\{A_M^i, E_i^M\}$, i.e. from 18 degrees of freedom per space point to only 18 in total.

The constraints for vacuum reduce to

$$\epsilon_{MNP}\epsilon^{ijk}E_i^M E_j^N B_k^P \;=\; 0 \tag{23}$$

$$\epsilon_{MNP}E_i^N B^{Pi} \;=\; 0 \tag{24}$$

$$C_{KM}{}^K E_i^M + \epsilon_{ijk}A_M^j E^{Mk} \;=\; 0 \tag{25}$$

where $B_i^M$ is given by (22).

We can follow now the footsteps of section 3, for the solution of these constraints. Assuming that $B_i^M$ is non degenerate, we expand $E_i^M$ in terms of it:

$$E_i^M = P_{ij}B^{Mj} \,. \tag{26}$$

From (24), we find that $P_{ij}$ must be symmetric, and from (23) that it must satisfy the algebraic condition (5). We arrive at the last constraint, which now takes the form

$$C_{KM}{}^K B^{Mj}P_{ij} + \epsilon_{ijk}A_M^j B_l^M P^{lk} = 0 \,. \tag{27}$$

Using (22) and the Jacobi identities, after some algebraic manipulations, this may be written in the form

$$W^j \hat{P}_{ij} = Z_{ijk}\hat{P}^{jk} \,, \tag{28}$$

where $\hat{P}_{ij}$ corresponds to the traceless part of $P_{ij}$, and

$$W^j \;:=\; 3V_Q\epsilon^{QMN}\epsilon^{jkl}A_{Mk}A_{Nl} \tag{29}$$

$$Z_{ijk} \;:=\; -2\epsilon_{ijl}A_M^l S^{MN}A_{Nk} \,. \tag{30}$$

Note that this condition involves only the traceless part of $P^{ij}$.

We can now go through the Bianchi classification for some understanding of this condition. For a Bianchi model of type I, defined with $C_{IJ}{}^K = 0$, this condition is trivially satisfied. If $S^{IJ} = 0$, which defines a model of type V, the condition reduces to the requirement that $W^i$ be a null eigenvector for $\hat{P}_{ij}$. If $V_Q = 0$, one can look at the simple case in which the connection matrix $A_{Mi}$ is assumed to be diagonal (see Ref. [17]). Then, recalling that without loss of generality $S^{IJ}$ can be put in diagonal form with $\pm 1$ or 0 entries, and letting $S^{IJ} = diag(s_1, s_2, s_3)$, the condition takes the form

$$[s_1(A_{11})^2 - s_2(A_{22})^2]\hat{P}_{12} = 0 \,, \tag{31}$$

together with its cyclic permutations. An obvious solution is obtained if $\hat{P}_{ij}$ itself is diagonal.

## 6.    Other Approaches

The introduction of Ashtekar's variables has generated other approaches to the initial
value problem in general relativity. One is the recent proposal by Newman and Rovelli
[20]. For a Yang-Mills theory they use a Hamilton-Jacobi technique to solve Gauss'
law.  The Yang-Mills physical degrees of freedom are coded in one pair of scalar
functions per dimension of the gauge group, together with conjugate momenta. Each
pair of scalar fields defines a congruence of lines ("generalized lines of force") by
the intersections of their level surfaces. The method can also be applied to gravity
in the Ashtekar formulation, where Newman and Rovelli solve not only the Gauss
constraint, but also the vector constraint, by the device of using three of the scalar
fields as spatial coordinates and letting the remaining fields be given as functions of
these. It is not known at present if also the scalar constraint can be solved using this
technique.

Another potential avenue to the initial value problem is the application of the Goldstone-
Jackiw solution [21] of the Gauss constraint in the SU(2) Yang-Mills theory to gravity
in the Ashtekar formulation [22]. In this approach, one would start by solving the
Gauss law. The solution could then be used in the diffeomorphism constraints. For
Yang-Mills, this procedure is not particularly useful, because it results in a compli-
cated hamiltonian. For gravity, the jury is still out.

Finally, Thiemann [7] has just introduced a new approach that enables him to solve
*all* of the constraints. The key idea is to start with the ansatz $E^{ai} = \epsilon^{abc} D_{[b} v^i_{c]}$. This
ansatz involves no loss of generality provided the curvature is non-degenerate, or
provided the spin density $K_i$ vanishes (so that the Gauss constraint implies $D_a E^{ai} =$
0). Using this ansatz, the Gauss constraint becomes a linear condition on $v^i_a$. The
vector and scalar constraints are then solved in the presence of matter by solving for
some of the matter momenta.

## References

[1] Choquet-Bruhat Y. and York J.W. Jr., in *General Relativity and Gravita-
    tion, vol. 1*, ed. A. Held, *Plenum Press, 1980.*

[2] Ashtekar A., *Phys. Rev. Lett.* 57, 2244 (1986) ; *Phys. Rev D* 36, 1587 (1987)

[3] Jacobson T., and Smolin L. *Nucl. Phys.* B299, 295 (1988).

[4] Rovelli C., and Smolin L. *Nucl. Phys.* B331, 80 (1990).

[5] Ashtekar A., (in collaboration with R.S. Tate) *Lectures on Non-Perturbative
    Canonical Gravity, World Scientific, 1991.*

[6] Capovilla R., Dell J., and Jacobson T., *Phys. Rev. Lett.* 63, 2325 (1989).

[7] personal communication, October, 1992; Thiemann T., "On the initial value problem for general relativity coupled to matter in terms of Ashtekar's constraints", preprint PITHA 93-1, ITP, RWTH Aachen, Germany, January 1993.

[8] Rovelli C., *Class. Quant. Grav.* 8, 1613 (1991).

[9] Capovilla R., Dell J., and Jacobson T., *Class. Quant. Grav.* 8, 59 (1991); Erratum 9, 1839 (1992).

[10] Bengtsson I., *Class. Quant. Grav.* 8, 1847 (1991).

[11] Capovilla R., Dell J., Jacobson T., and Mason L., *Class. Quant. Grav.* 8, 41 (1991).

[12] Mason L., and Frauendiener J., in *Twistors in Mathematics and Physics*, eds. Baston R., and Bailey T., *Cambridge University Press*, 1990.

[13] Ashtekar A. and Renteln P., in *Mathematics and General Relativity*, ed. J. Isenberg, *American Mathematical Society*, 1988.

[14] Ashtekar A., Romano J.D., and Tate R.S., *Phys. Rev. D* 40, 2572 (1989); Jacobson T., *Class. Quant. Grav.* 5, L143 (1988).

[15] The importance of this point was stressed to us by R. Tate in a private communication, January 1993.

[16] Kodama H., *Prog. Theor. Phys.* 80, 1024 (1988).

[17] Ashtekar A. and Pullin J., *Proc. Roy. Soc. Israel* 9, 65 (1990).

[18] MacCallum M.A.H., in *General Relativity. An Einstein Centenary Survey*, eds. S.W. Hawking and W. Israel, *Cambridge University Press, 1979*.

[19] Ryan M.P. Jr., and Shepley L.C., *Homogeneous Relativistic Cosmologies*, *Princeton University Press, 1975*.

[20] Newman E.T. and Rovelli C., *Phys. Rev. Lett.* 69, 1300 (1992).

[21] Goldstone J. and Jackiw R., *Phys. Lett. B* 74, 81 (1978).

[22] Manojlovic N., Syracuse University Preprint, to appear.

# Status Report on an Axiomatic Basis
# for Functional Integration

*Pierre Cartier* *        *Cécile DeWitt-Morette* †

*The early stage of the work was reported in a note in the Comptes Rendus de l'Académie des Sciences [1] and in a contribution to the Proceedings of the International Colloquium in Honour of Yvonne Choquet-Bruhat [2].*

*The next stage involves testing the proposed framework on nontrivial problems; some of it will be ready in time for the Symposium, but is not ready for the Proceedings. We are happy that we shall nevertheless be included with the colleagues and friends of Dieter Brill and Charles Misner who are celebrating their contributions to physics, in these volumes, and we thank the editors for accepting a status report.*

The goal of our project is to extract from work done by physicists during the last fifty years an axiomatic basis for functional integration which will provide simple and robust methods of calculation, in particular for integration by parts, change of variable of integration, sequential integrations.

For instance, physicists write formally

$$\int \exp\left(\frac{i}{\hbar}\left(S(\varphi) - \hbar\langle J, \varphi\rangle\right)\right) \mathscr{D}\varphi = Z(J) \tag{1}$$

where $S$ is the (bare) action of a system, $\varphi$ is a field, $J$ is a source, $\mathscr{D}\varphi$ an integrator left undefined. The discomfort of physicists and mathematicians is due to the mythical nature of $\mathscr{D}\varphi$—but no more mythical than $dx$ was in the seventeenth century. We can define $dx$ by

$$\int_a^b dx = b - a\,,$$

or by

$$\int_{\mathbb{R}} \exp\left(-\pi x^2 - 2\pi i x y\right) dx = \exp\left(-\pi y^2\right)\,;$$

we can similarly define an integrator $\mathscr{D}_Q\,\varphi$ on a Banach space $\Phi$ with respect to a quadratic form $Q$ on $\Phi$ by the equation

*Département de Mathématique et Informatique, Ecole Normale Superieure, F-75005 Paris, France.

†Department of Physics and Center for Relativity, The University of Texas, Austin, Texas 78712-1081, USA.

$$\int_\Phi \exp\left(-\pi Q(\varphi) - 2\pi i \left\langle J, \varphi \right\rangle\right) \mathscr{D}_Q \, \varphi \; = \; \exp\left(-\pi W(J)\right) \tag{2}$$

where $J$ is an element of the dual $\Phi'$ of $\Phi$, and $W$ is a quadratic form on $\Phi'$ "inverse" to $Q$, in the sense that

$$\frac{\delta^2 W(J)}{\delta J_a \, \delta J_b} \quad \text{is the inverse of} \quad \frac{\delta^2 Q(\varphi)}{\delta \varphi^c \delta \varphi^d}. \tag{3}$$

An integrator $\mathscr{D}_Q \, \varphi$ defined by equation (2) will be called a gaussian integrator, whether $Q$ is real or complex. In section 1 we investigate gaussian integrators: we identify a space $\mathscr{F}_Q$ of functionals on $\Phi$ integrable by $\mathscr{D}_Q$; we choose a norm on $\mathscr{F}_Q$ such that we can generalize the Fubini theorem to iterated integrations on function spaces; we give the rule for linear change of variable of integrations. In section 2 we investigate the possibility of generalizing section 1 to integrators $\mathscr{D}_{\Theta, Z}$ such that

$$\int_\Phi \Theta(\varphi, J) \mathscr{D}_{\Theta, Z} \, \varphi = Z(J) \tag{4}$$

where $\Theta$ and $J$ are two given continuous bounded maps

$$\begin{aligned} \Theta &: \Phi \times \Phi' \to \mathbb{C} \\ Z &: \Phi' \quad\;\; \to \mathbb{C}. \end{aligned} \tag{5}$$

In section 3 we consider the case of Quantum Field theory where $\Theta$ is of the type which appears in equation (1) and where $Z(J)$ is an experimental function, in the sense that it is related to the effective action, and therefore not known a priori.

The key idea is that an integrator on an infinite dimensional space cannot, in general, be expected to be a measure, therefore there is no universal $\mathscr{D}$ on infinite dimensional spaces.

## 1.  Gaussian Integrators

Let

$$\Theta(\varphi, J) = \exp\left(-i\pi Q(\varphi) - 2\pi i \left\langle J, \varphi \right\rangle\right), \qquad Z(J) = \exp\left(i\pi W(J)\right) \tag{6}$$

where $Q$ is a quadratic form on $\Phi$ and $W$ is a quadratic form on the dual $\Phi'$ of $\Phi$, "inverse" of $Q$ in the sense of (3). We extend the work of Albeverio and Hoegh-Krohn to the case where $Q$ is not necessarily positive definite, nor necessarily real.

Let $\mathscr{D}_Q$ be the gaussian integrator defined by (2), namely

$$\int_\Phi \Theta(\varphi, J) \mathscr{D}_Q \, \varphi = Z(J). \tag{7}$$

A possible space $\mathscr{F}_Q$ of integrands is the Albeverio and Hoegh-Krohn space of complex valued functionals $F$ on $\Phi$ such that

$$F(\varphi) = \int_{\Phi'} \Theta(\varphi, J) \, d\mu(J) \tag{8}$$

where $\mu$ is a bounded measure on $\Phi'$, possibly complex. There may be more than one measure $\mu$ satisfying (8), but $\int F(\varphi) \mathcal{D}_Q \, \varphi$ does not depend on the choice of $\mu$ satisfying (8). Moreover it is not necessary to identify $\mu$ in order to compute $\int F(\varphi) \mathcal{D}_Q \, \varphi$.

An $A$-type norm can be defined on $\mathcal{F}_Q$ by the equation

$$\|F\|_A := \min_{\mu} \int_{\Phi'} |Z(J)| \, d|\mu|(J) \, . \tag{9}$$

If $W$ is not positive definite, we introduce an auxiliary norm on $\Phi$ in order to define a norm on $\mathcal{F}_Q$. The fact that such an auxiliary norm is not unique does not make the $A$-norm on $\mathcal{F}_Q$ ambiguous.

Set

$$\exp\left(-i\pi Q(\varphi)\right) \mathcal{D}_Q \, \varphi = dw(\varphi) \, ; \tag{10}$$

$w$ is a prodistribution defined by its Fourier transform

$$(\mathcal{F}w)(J) = \exp\left(i\pi W(J)\right) \, . \tag{11}$$

Numerous and nontrivial applications of the use of prodistributions in the calculation of functional integrals can be found in reference 2.

We recall here only the use of linear changes of variables in functional integrals defined with respect to prodistributions. Let $P$ be a linear continuous map between the Banach spaces $\Phi$ and $\Psi$, and $Pw$ be the image of $w$ under $P$, then

$$\int_{\Phi} (F \circ P)(\varphi) \, dw(\varphi) = \int_{\Psi} F(\psi) d(Pw)(\psi) \, . \tag{12}$$

The Fourier transform $\mathcal{F}(Pw)$ satisfies the equation

$$\mathcal{F}(Pw) = \mathcal{F}w \circ \tilde{P} \tag{13}$$

where the transposed $\tilde{P}$ of $P$ is a linear continuous map from the dual $\Psi'$ of $\Psi$ into the dual $\Phi'$ of $\Phi$

$$\left\langle \tilde{P}K, \varphi \right\rangle_{\Phi} = \left\langle K, P\varphi \right\rangle_{\Psi} \, , \qquad \varphi \in \Phi, \qquad K \in \Psi' \, . \tag{14}$$

If it happens that $\Psi$ is a finite dimensional space, then the computation of the left-hand side of (12) reduces to the computation of ordinary integrals. If $\Psi$ is infinite dimensional, the manipulation (12) is often used to separate out interesting finite dimensional integrals from infinite dimensional trivial ones.

## 2.   $\mathscr{D}_{\Theta,Z}$ Integrators Satisfying (4)

Equations (8) and (9) can be generalized:

- The space $\mathscr{F}_{\Theta,Z}$ of complex valued functionals $F$ on $\Phi$, such that (8) is satisfied, is a space of functionals integrable with respect to $\mathscr{D}_{\Theta,Z}$.

- The $A$-type norm defined by (9) makes it possible to state a Fubini theorem for iterated integrations.

If it happens that $\mu$ which satisfies (8) is explicitly known, then

$$\int_{\Phi} F(\varphi)\mathscr{D}_{\Theta,Z}\,\varphi \;=\; \int_{\Phi}\left(\int_{\Phi'}\Theta(\varphi,J)\,d\mu(J)\right)\mathscr{D}_{\Theta,Z}\,\varphi$$

$$=\; \int_{\Phi'} Z(J)\,d\mu(J) \tag{15}$$

the right-hand side being an integral in the traditional sense.

Work on constructing an axiomatic basis for functional integration with respect to integrators satisfying (4) is in progress.

## 3.   Quantum Field Theory Integrators

In Quantum Field Theory, the situation is different. Let

$$\Theta(\varphi,J) = \exp\left(\frac{i}{\hbar}\left(S(\varphi) - \hbar\langle J,\varphi\rangle\right)\right) \tag{16}$$

where $S$ is the bare action, $\varphi$ is a self interacting field, or a collection of interacting fields, and $J$ is an external source; the generating functional $Z(J)$ cannot be chosen a priori because it is an experimental quantity in the following sense. Let

$$\exp iW(J) := Z(J)/Z(0) \tag{17}$$

and let $\Gamma(\bar{\varphi})$ be the Legendre transform of $W$

$$\Gamma(\bar{\varphi}) := W(J) - \hbar\langle J,\bar{\varphi}\rangle, \qquad \bar{\varphi} := \frac{\delta W(J)}{\delta J}, \tag{18}$$

then $\Gamma(\bar{\varphi})$ is the effective action which can be used to compute observable quantities. We cannot use (4) to identify $\mathscr{D}_{\Theta,Z}$ explicitly in Quantum Field Theory but we shall use (4) to impose the following conditions on $\mathscr{D}_{\Theta,Z}$; they are chosen so as to make integration by parts and linear change of variable of integration straightforward manipulations:  namely the infinitesimal version of the invariance of $\mathscr{D}_{\Theta,Z}$ under translation

$$\int_{\Phi}\frac{\delta}{\delta\varphi}\left(\Theta(\varphi,J)\right)\mathscr{D}_{\Theta,Z}\,\varphi = 0, \tag{19}$$

and the fundamental property,

$$\text{if}\quad \psi = M\varphi, \qquad \text{then} \qquad \mathscr{D}_{\Theta,Z}\,\psi = \det M\,\mathscr{D}_{\Theta,Z}\,\varphi. \tag{20}$$

The choice of functionals $F \in \mathscr{F}_{\Theta,Z}$ integrable by $\mathscr{D}_{\Theta,Z}$ must be compatible with (19).

Given an integrator $\mathscr{D}_{\Theta,Z}$ satisfying (19) and (20) and a map $\Theta(\varphi, J)$ of the type (16), one can investigate a number of problems; for instance:

i) Computing $\int_\Phi F(\varphi) \exp\left(\frac{i}{\hbar} S(\varphi)\right) \mathscr{D}_{\Theta,Z}\,\varphi$ for several functionals $F$.

ii) Low energy limit of the effective action $\Gamma(\bar\varphi)$ given a bare action $S(\varphi)$. When is the low energy limit of $\Gamma(\bar\varphi$, observable constants) equal to $S(\bar\varphi$, observable constants)?

iii) High energy limit of the effective action. When is a theory "asymptotically free"?

iv) Stationary phase approximation of $\int_\Phi F(\varphi) \exp\left(\frac{i}{\hbar} S(\varphi)\right) \mathscr{D}_{\Theta,Z}\,\varphi$.

In the years to come, we will write a book on functional integration and test the proposed framework on many problems of physical interest.

## References

[1] Cartier P., DeWitt-Morette C., *Intégration Fonctionnelle; éléments d'axiomatique.* To appear in the Comptes Rendus de l'Académie des Sciences.

[2] DeWitt-Morette C., *Functional Integration; a multipurpose tool.* To appear in the Proceedings of the International Colloquium in Honour of Yvonne Choquet-Bruhat, Eds. Flato M., Kerner R., Lichnerowicz A. Kluwer Academic Publishers, Dordrecht. Available as an IHES reprint (IHES, 35 Route de Chartres, F-91440 Bures-sur-Yvette).

# SOLUTION OF THE COUPLED EINSTEIN CONSTRAINTS
# ON ASYMPTOTICALLY EUCLIDEAN MANIFOLDS

Yvonne Choquet-Bruhat

Université Paris 6

dedicated to Dieter Brill

Abstract. We consider the coupled elliptic system of constraints on the Cauchy data for Einstein's equations obtained by the conformal method when the initial manifold has non constant mean extrinsic curvature $\tau$. We prove that this system admits asymptotically euclidean solutions when $\nabla\tau$ is small in an appropriate norm and the scalar curvature of the conformal metric is either everywhere non negative, or everywhere negative with fall off conditions that we specify.

Résumé. Nous considérons le système couplé des contraintes sur les données de Cauchy obtenu par la méthode conforme quand la variété initiale a une courbure extrinsèque moyenne $\tau$ qui n'est pas constante. Nous montrons que ce système elliptique admet des solutions aymptotiquement euclidiennes quand $\nabla\tau$ est assez petit dans une norme appropriée et que la courbure scalaire de la métrique conforme est soit partout non négative, soit partout négative avec un comportement à l'infini que nous explicitons.

## INTRODUCTION.

An initial data set in General Relativity is a pair $(\overline{g},K)$ , $\overline{g}$ a riemannian metric and K a symmetric covariant 2-tensor on a 3-dimensional manifold V. As a consequence of Einstein equations they must satisfy the constraints

(1) $\qquad \overline{R} \_ K.K + (\text{tr}K)^2 = E$

(2) $\qquad \overline{\nabla}.K - \overline{\nabla}(\text{tr}K) = J$

where E, a scalar non negative function, and J, a vector field, are zero in vacuum.

The standard procedure to solve the constraints is the use of the conformal method inaugurated by Lichnerowicz, Li., and developped by Choquet-Bruhat and York (cf. for instance the review article C.B-Yo). The constraints are found to decouple into a linear elliptic system for a vector field on V and a non linear elliptic equation, called the

Lichnerowicz equation, for the scalar conformal factor when trK = constant, i.e. when the initial submanifold V will be a submanifold of constant mean extrinsic curvature of the einsteinian space time. Exhaustive results have been found for the Lichnerowicz equation, for a compact V (cf C.B, or C.B-Yo completed in Is), or an asymptotically euclidean V (cf. C.S-C.B, Ca, or C.B-Yo).

It seems interesting, however, to obtain solutions of the constraints with non constant trK, since there are space times which do not admit such submanifolds (cf. Ba, Br). In a previous paper, in collaboration with J. Isenberg and V. Moncrief, we have proved the existence of solutions of the constraints on compact manifolds with non constant mean curvature $\tau$ under some hypothesis of smallness on $\nabla\tau$, with the additional hypothesis that $\tau$ does not vanish if $g$, the given metric conformal to $\bar{g}$, is in the negative Yamabe class. In this paper we consider the asymptotically euclidean case and prove some existence theorems for solutions of the constraints on non maximal submanifolds.

## 1. EQUATIONS

We recall the system obtained by the conformal method. When g is a given riemannian metric on a given $C^\infty$ 3-manifold V with scalar curvature R = 8r, while A is a given traceless divergence free symmetric covariant 2 tensor, $q = \frac{\rho}{8} \geqslant 0$ is a given function, j a given vector ($\rho$ and j are determined by the sources, vanishing with them), $b = \frac{\tau^2}{12}$ is given ($\tau$ is the mean extrinsic curvature of V as a submanifold of the physical space time) the equations are

$$(1.1) \qquad \Delta_g \varphi - r\varphi + a_w \varphi^{-7} + q\varphi^{-3} - b\varphi^5 = 0$$

$$(1.2) \qquad \delta_g L_g W = \frac{2}{3} \varphi^6 \nabla\tau + j \ .$$

where $L_g$ is the conformal Killing operator

$$L_g W = \pounds_w g - \frac{2}{3} g \, \delta_g W, \quad \text{i.e.} \quad (L_g W)_{ij} = \nabla_i W_j + \nabla_j W_i - \frac{2}{3} g_{ij} \nabla_l W^l$$

and

$$a_w = \frac{1}{8} (A + L_g W) \cdot (A + L_g W) \geqslant 0, \qquad \text{dot: scalar product in g.}$$

The initial data set $(\bar{g}, K)$ on V, metric and symmetric covariant 2-tensor, for the physical space time is then

$$\bar{g} = \varphi^4 g, \qquad K = \varphi^{-2}(A + L_g W) + \frac{1}{3} g\tau .$$

## 2. ASYMPTOTICALLY EUCLIDEAN MANIFOLDS. WEIGHTED SOBOLEV SPACES.

A $C^\infty$ , n-dimensional, riemannian manifold $(V,e)$ is called <u>euclidean at infinity</u> if there exists a compact subset K of V such that V − K is the disjoint union of a finite number of open sets $U_i$, and $(U_i,e)$ is isometric to the exterior of a ball of $\mathbb{R}_n$ with its canonical euclidean metric.

The <u>weighted Sobolev space</u> $W^p_{s,\delta}$ , $1 < p < \infty$, $s \in \mathbb{N}_+$, $\delta \in \mathbb{R}$ , is the closure of $C_0^\infty$ ($C^\infty$ functions with compact support in V) in the norm

$$\|u\|_{W^p_{s,\delta}} = \left\{ \sum_{0 \leqslant m \leqslant s} \int_V |D^m u|^p \, \sigma^{p(\delta+m)} d\mu \right\}^{1/p},$$

D and $|\ |$ denote the covariant derivative and the norm in the metric e, while $\sigma \equiv (1 + d^2)^{\frac{1}{2}}$, d the distance in the metric e from a point of V to a fixed point; if $(V,e)$ is the euclidean space one can choose d = r, the euclidean distance to the origin.

We recall the multiplication and imbedding properties (cf. CB-Ch)

$$W^p_{s_1,\delta_1} \times W^p_{s_2,\delta_2} \subset W^p_{s,\delta} \text{ if } s \geqslant s_1, s_2 , \quad s < s_1 + s_2 - \frac{n}{p}, \quad \delta < \delta_1 + \delta_2 + \frac{n}{p} .$$

$$W^p_{s,\delta} \subset C^m_\beta \text{ if } m < s - \frac{n}{p}, \quad \beta < \delta + \frac{n}{p}, \quad \|u\|_{C^m_\beta} \equiv \sum_{0 \leqslant \ell \leqslant m} \text{Sup}(|D^\ell u|\sigma^{\beta+m}).$$

We also have

$$\frac{1}{\sigma^\beta} \in W^p_{s,\delta} \qquad \text{if} \quad \beta > \delta + \frac{n}{p}, \quad s \geqslant 0 .$$

## 3. LERAY SCHAUDER THEORY. THEOREMS ON ELLIPTIC SYSTEMS.

To solve the non linear system (1.1), (1.2) we shall use the Leray-Schauder theory: a solution will be a fixed point, in a bounded open subset $\Omega$ of some Banach space $\mathcal{H}$, of a compact mapping $\mathcal{F}: \Omega \to \mathcal{H}$ . The existence of a fixed point is proved by constructing a continuous family of compact maps $\mathcal{F}_t$, $t \in [0,1]$, such that $\mathcal{F}_1 \equiv \mathcal{F}$ and that $\mathcal{F}_0$ has one and only one fixed point in $\Omega$ (then Id $- \mathcal{F}_0$ has degree $\pm 1$ in $\Omega$) and no $\mathcal{F}_t$ has a fixed point on the boundary $\partial\Omega$ of $\Omega$ (then all the maps $\mathcal{F}_t$ have the same degree in $\Omega$).

Before defining the mappings $\mathcal{F}$ and $\mathcal{F}_t$ we recall the following theorems.

**Theorem** 1. (CB-Ch,Ca). Let g be a riemannian metric on $V_3$ with g − e $\in W^p_{\sigma,\rho}$

and let $\alpha \in W^p_{\sigma-2,\rho+2}$, $\sigma > 2 + \dfrac{3}{p}$, $-\dfrac{3}{p} < \rho$, and $\alpha \geqslant 0$. Then the equation

(3.1)    $\Delta_g u - \alpha u = f$,    with $f \in W^p_{\sigma-2,\delta+2}$ given, $-\dfrac{3}{p} < \delta < 1 - \dfrac{1}{p}$ and $\delta \leqslant \rho$

has one and only one solution $u \in W^p_{\sigma,\delta}$ .
The solution of (3.1) satisfies the inequalities

(3.2)        $\|u\|_{W^p_{s,\delta}} \leqslant C \|f\|_{W^p_{s-2,\delta+2}}$ ,  $s = 2, \ldots, \sigma$ .

where the C's are constants depending only on V, e, g, $\alpha$.

**Theorem 2.** Let **g** be a riemannian metric on $(V_3,e)$ such that $g - e \in W^p_{\sigma,\rho}$,
$\sigma > 3 + \dfrac{3}{p}$, $\rho > -\dfrac{3}{p}$ , then **(V,g) admits no conformal Killing vector field**
and the system

(3.3)                $\delta_g L_g W = h$,    with $h \in W^p_{\sigma'-2,\delta+2}$ given,

$2 \leqslant \sigma' \leqslant \sigma$ , $-\dfrac{3}{p} < \delta < 1 - \dfrac{3}{p}$ and $\delta \leqslant \rho$

has one and only one solution $W \in W^p_{\sigma',\delta}$ . **It satisfies the inequalities**

(3.4)        $\|W\|_{W^p_{s,\delta}} \leqslant C \|h\|_{W^p_{s-2,\delta+2}}$ ,  $s = 2,\ldots,\sigma'$ .

Proof. The non existence of a conformal Killing vector field is
proven in Ch-OM. For existence of W and inequalities, cf CB-Ch. and Ca.

Definition. We say in the following that the data g, A, q, j, $\tau$ on (V, e)
are p-$\rho$- **regularly asymptotically euclidean** if $g - e \in W^p_{\sigma,\rho}$, $\sigma > 3 + \dfrac{3}{p}$ ,
$A \in W^p_{2,\rho+1}$, q, $j \in W^p_{1,\rho+2}$ and $\tau \in W^p_{2,\rho+1}$ , with $\rho + \dfrac{3}{p} > 0$. It results that
$R \in W^p_{\sigma-2,\rho+2}$, while A.A , $b \equiv \dfrac{\tau^2}{12} \in W^p_{1,\rho+1}$.

## 4. CASE R $\geqslant$ 0.

The construction of the compact maps $\mathcal{F}_t$, choice of $\Omega$ and proof of non
existence of fixed points on $\partial\Omega$ will be different depending on the sign of
R. We first consider the case R $\geqslant$ 0 on V.

Let there be given a function $\psi \equiv 1 + v$ and a vector field $\omega$ on V. We
consider the linear system of differential equations depending on a real
parameter t, with unknown $\varphi$ and W:

(4.1)        $\Delta_g \varphi - r\varphi = - t(a_\omega \psi^{-7} + q\psi^{-3} - b\psi^5)$,

i.e.         $\Delta_g u - ru = r + tf(v,\omega)$ ,      $u \equiv \varphi - 1$

with         $f(v,\omega) \equiv a_\omega \psi^{-7} + q\psi^{-3} - b\psi^5$ ,      $\psi \equiv 1 + v$

and                    $a_\omega \equiv \dfrac{1}{8} (A + L_g\omega).(A + L_g\omega)$,

(4.2)        $\delta_g L_g W = \dfrac{2}{3} t\psi^6 \nabla\tau + j$.

**Lemma 1.** Let g, A, q, j, $\tau$ be p-$\rho$-regularly asymptotically euclidean on (V,e) with p > 3 .
Let $\mathcal{H}$ be the Banach space of pairs (v , $\omega$) such that $v \in W^p_{1,\delta}$, $\omega \in W^p_{2,\delta}$.
Let $\Omega$ be the bounded open subset of $\mathcal{H}$ defined by the inequalities

(4.3)                    $\|v\|_{W^p_{1,\delta}} < k$,

(4.4)                    $\|\omega\|_{W^p_{2,\delta}} < k'$,

where k and k' are some given positive numbers, and

(4.5)                    $\underset{V}{\text{Inf}}(1 + v) > \ell > 0$

with $\ell$ some number such that $0 < \ell < 1$.
Then, if $-\dfrac{3}{p} < \delta < 1 - \dfrac{3}{p}$, $\delta < \rho$ the resolution of the linear system (4.1),
(4.2) on $V_3$ defines for each $t \in \mathbb{R}$ a compact map $\mathcal{F}_t$:

$$\mathcal{F}_t : \bar{\Omega} \to \mathcal{H} \quad \text{by} \quad (v \equiv \psi-1, \ \omega) \mapsto (u \equiv \varphi-1, \ W),$$

**Proof.** $\Omega$ is open in $\mathcal{H}$ because on (V,e) we have the continuous inclusion $W^p_{1,\delta} \subset C^0_b$, space of continuous and bounded functions, if $1 > \dfrac{3}{p}$, $\delta > -\dfrac{3}{p}$.
We recall that $W^p_{s',\delta'} \subset W^p_{s,\delta}$ if $s' \geqslant s$ and $\delta' \geqslant \delta$, with compact imbedding if both inequalities are strict.
Imbedding and multiplication properties of $W^p_{s,\delta}$ spaces show that under the hypothesis made

$$a_\omega \in W^p_{1,\delta+1} \times W^p_{1,\delta+1} \subset W^p_{1,\delta'+2}, \qquad b \in W^p_{2,\delta+1} \times W^p_{2,\delta+1} \subset W^p_{2,\delta'+2}$$

if $\qquad \delta < \delta' < \delta + (\delta + \dfrac{3}{p})$ .

On the other hand $v \in W^p_{1,\delta}$ with $1 + v \geqslant \ell > 0$ implies that $(1+v)^N$ is continuous and uniformly bounded on $V_3$ for any positive or negative integer N, while $\nabla(1+v)^N \in W^p_{0,\delta+1}$, from which results since $\delta + \dfrac{3}{p} > 0$ that $f \in W^p_{1,\delta'+2}$ if we impose moreover

$$\delta' \leqslant \rho .$$

Analogous  considerations show that $h \in W^p_{1,\delta'+2}$.

The  linear system  (4.1), (4.2)  has therefore one and only one solution (u, W) which is not only in $W^p_{1,\delta} \times W^p_{2,\delta}$ but also in $W^p_{3,\delta} \times W^p_{3,\delta'}$, therefore the mapping $\mathcal{F}_t: (v,\omega) \mapsto (u,v)$ is a compact mapping from $\Omega$ into $\mathcal{H}$.

**Lemma**  **2.** Let g, A, q, j, $\tau$ be p-$\rho$ regularly asymptotically euclidean, with $R \equiv 8r \geqslant 0$ Suppose p > 3. Denote by $\delta$ and $\delta'$ numbers such that

$$-\dfrac{3}{p} < \delta < \delta' \leqslant \rho , \qquad \delta' < 1 - \dfrac{3}{p} .$$

Then

1) The solution $\varphi_0 \equiv 1 + u_0$ , $u_0 \in W^p_{3,\rho} \subset W^p_{3,\delta}$ , of

(4.6) $\qquad \Delta_g \varphi_0 - r\varphi_0 = 0$

is such that there exists a number $\alpha > 0$ and so that on V

$$0 < \alpha \leqslant \varphi_0 \leqslant 1 .$$

2) Let there be given $b_0 \geqslant 0$ ,$b_0 \in W^p_{1,\delta'+2}$ . The equation

(4.7) $\qquad \Delta_g \Phi_0 - r\Phi_0 - b_0 \Phi_0^5 = 0$

admits  a solution  $\Phi_0 \equiv 1 + U_0$, $U_0 \in W^p_{3,\delta'} \subset W^p_{3,\delta}$. This solution is unique and there exists $\ell > 0$ such that

$$0 < 2\ell \leqslant \Phi_0 \leqslant \varphi_0 \qquad \text{on } V_3$$

3) Suppose $a_1 \in W^p_{1,\delta'+2}$ , $a_1 \geqslant 0$ , be given. The equation

(4.8) $\qquad \Delta_g \Phi_1 - r\Phi_1 + a_1 \Phi_1^{-7} + q\Phi_1^{-3} = 0$

admits a solution $\Phi_1$ such that $\Phi_1 \equiv 1 + U_1$, $U_1 \in W^P_{3,\delta+2}$. This solution is unique and satisfies

$$\varphi_0 \leqslant \Phi_1 .$$

4) If $a_1 \geqslant a_w$ and $b_0 \geqslant b$, then any solution $\varphi > 0$, with $\varphi - 1 \in W^P_{3,\delta}$, of

$$\Delta_g \varphi - r\varphi = -t(a_w \varphi^{-7} + q\varphi^{-3} - b\varphi^5)$$

is such that

(4.9)                    $\Phi_0 \leqslant \varphi \leqslant \Phi_1 .$

Proof. It makes repeated use of the maximum principle: if $\varphi \in C^2$ satisfies the linear inequality on V

$$\Delta_g \varphi - h \varphi \geqslant 0 , \qquad \text{with } h \geqslant 0 \text{ and bounded}$$

then $\varphi$ cannot attain a non negative maximum M except if $\varphi \equiv M$.

1) Since $\varphi_0$ is $C^2$ and satisfies (4.6) it is constant if it attains a non negative maximum. Since $\varphi$ tends to 1 at infinity on $V_3$ it attains such a maximum on $V_3$ if it takes values greater than 1, hence $\varphi_0 \leqslant 1$. An analogous reasonning applied to $-\varphi_0$ shows that $\varphi_0$ cannot attain a non positive minimum.

2) We solve (4.7) by using the Leray-Schauder theory. We consider the compact mappings $F_t: W^P_{1,\delta} \to W^P_{3,\delta'} \subset W^P_{1,\delta}$ by $v \mapsto u$ obtained by resolution of the linear equation

$$\Delta_g \Phi - r\Phi = t \, b_0 \Psi^5 , \qquad \Psi \equiv 1 + v , \quad \Phi \equiv 1 + u$$

in the bounded open set $\Omega$

$$\|v\|_{W^P_{1,\delta}} < k , \quad k > 0 \text{ some given number,}$$

$$\underset{x \in V}{\text{Sup }} \Phi < 1 + \epsilon , \quad \epsilon > 0 \text{ some given number}$$

The mapping $F_0$ has one fixed point in $\Omega$, $u_0 \equiv \varphi_0 - 1$, if k is chosen such that $k > \|u_0\|_{W^p_{1,\delta}}$ .

All mappings $F_t$ are such that if $v \in \bar{\Omega}$ then $\|u\|_{W^p_{1,\delta}} \leq \|u\|_{W^p_{2,\delta}} \leq K$ where K depends only on $\epsilon$ and on the $W^p_{0,\delta}$ norms of r and $b_0$ .

The maximum principle shows on the other hand that a fixed point $\Phi$ of $F_t$ in $\bar{\Omega}$ cannot take non positive values, i.e. $\Phi > 0$ on $V_3$, because $\Phi$ tends to 1 at infinity and satisfies the linear equation

$$\Delta_g \Phi - (r + tb_0 \Phi^4)\Phi = 0.$$

The inequality $\Phi \leq \varphi_0$ is then a consequence of the equation

$$\Delta_g (\Phi - \varphi_0) - r(\Phi - \varphi_0) = t\, b_0 \Phi^5$$

and the fact that $\Phi - \varphi_0$ tends to zero at infinity.

These results show that we can choose k and take $\epsilon > 0$ arbitrary such that no $\mathcal{F}_t$ has a fixed point on $\partial\Omega$, intersection of $\bar{\Omega}$ with one or the other of the two sets

$$\|u\|_{W^p_{1,\delta}} = k \qquad \text{and} \qquad \text{Sup}_V \Phi = 1+\epsilon .$$

We denote by $\Phi_0$ the solution of the original equation (4.7) and by $2\ell$ its strictly positive infimum on $V_3$. The uniqueness of this positive solution is again a consequence of the maximum principle.

3) The equation for $\Phi_1$ is the usual Lichnerowicz equation on maximal V, the existence and uniqueness of $\Phi_1 \geq \varphi_0$ is well known (cf. CB.Yo)

4) A solution $\varphi > 0$ of the equation

$$\Delta_g \varphi - r\, \varphi = -t(a_w \varphi^{-7} + q\varphi^{-3} - b\varphi^5)$$

satisfies the linear equation

$$\Delta_g (\Phi_0 - \varphi) - \{r + bt(\Phi_0^4 + \Phi_0^3 \varphi + \ldots + \varphi^4)\}(\Phi_0 - \varphi) = \Phi_0^5 (b_0 - tb) + t(a_w \varphi^{-7} + q\varphi^{-3}) \geq 0 .$$

from which we deduce since $\varphi > 0$ and $\Phi_0 - \varphi$ tends to zero at infinity that

$$\Phi_0 - \varphi \leq 0 \qquad \text{on } V_3 .$$

A similar proof shows that

$$\varphi - \Phi_1 \leqslant 0 \qquad \text{on } V_3 .$$

We now prove an existence theorem for the coupled system.

**Theorem. Let g, A, q, j, $\tau$ satisfy the hypothesis of lemma 2. There exists a number $\eta > 0$ such that if $\|\nabla\tau\|_{W^p_{0,\delta+2}} < \eta$ for some $\delta$ with $-\frac{3}{p} < \delta < 1 - \frac{3}{p}$ the constraints admit a regularly asymptotically euclidean solution on V.**

Proof.

Let $\Omega$ be the bounded open set of $W_{1,\delta} \times W_{2,\delta}$ defined by

$$(4.10) \qquad \|\varphi - 1\|_{W^p_{1,\delta}} < k , \qquad \|W\|_{W^p_{2,\delta}} < k' ,$$

$$(4.11) \qquad \underset{V}{\text{Inf}} \ (\varphi - \Phi_0) > - \ell , \qquad \underset{V}{\text{Inf}} \ (\Phi_1 - \varphi) > - \ell .$$

where $\Phi_0$ is the solution of an equation (4.7) with $b_0 \geqslant b$ , hence $\Phi_0$ is known from the data, while $\Phi_1 = 1 + u_1$, $u_1 \in W^p_{3,\delta}$ is solution of the following equation, where M is some positive number to be chosen later (note that by a result quoted in §2 we have $\frac{M}{\sigma^{\beta+2}} \in W^p_{1,\delta+2}$)

$$\Delta_g \Phi_1 - r\Phi_1 + \frac{M}{\sigma^{\beta+2}} \Phi_1^{-7} + q\Phi_1^{-3} = 0,$$

with $\beta$ some number such that

$$\delta + \frac{3}{p} < \beta < \delta' + \frac{3}{p} .$$

The supremum of $\Phi_1$ on $V_3$ is some increasing function of M, $\mu(M)$. We denote $m(M) = \mu(M) + \ell$ . We have for $\varphi \in \bar{\Omega}$

$$\varphi \leqslant m(M) \qquad \text{on } V_3.$$

A fixed point will not lie in the boundary $\partial\Omega$ if it satisfies the strict inequalities (4.10), (4.11).

A fixed point (u, W) of $\mathcal{F}_t$ satisfies the inequalities

$$\|u\|_{W^p_{1,\delta}} \leqslant \|u\|_{W^p_{2,\delta}} \leqslant C \{\|r\|_{W^p_{0,\delta}} + t \|f(u,W)\|_{W^p_{0,\delta}}\}$$

$$\|W\|_{W^p_{2,\delta}} \leqslant C\{\|j\|_{W^p_{0,\delta+2}} + t \|\varphi^6 \nabla \tau\|_{W^p_{0,\delta+2}}$$

We deduce from the expression of $f$ that if $(u, W) \in \bar{\Omega}$

$$\|f(u,W)\|_{W^p_{0,\delta+2}} \leqslant \ell^{-7}\|a_w\|_{W^p_{0,\delta+2}} + \ell^{-3}\|q\|_{W^p_{0,\delta+2}} + m(M)^5\|b\|_{W^p_{0,\delta}} .$$

The norms of $q$ and $b$ are given. The norm of $a_w$ is estimated by, due to multiplication properties of weighted Sobolev spaces

$$\|a_w\|_{W^p_{0,\delta+2}} \leqslant C \; (\|A + L_g W\|_{W^p_{1,\delta+1}})^2 \leqslant C' \; (\|A\|^2_{W^p_{1,\delta+1}} + \|W\|^2_{W^p_{2,\delta}})$$

with C and C' constants depending only on $(V_3, g)$.

Finally,  in $\bar{\Omega}$, with K and K' constants depending on the data g, A, q, j, $\tau$ and on $m(M)$

$$\|f(u, W)\|_{W^p_{0,\delta+2}} \leqslant K + K'(\|W\|_{W^p_{2,\delta}})^2 .$$

On the other hand, if $u \equiv 1 - \varphi \in \bar{\Omega}$, and W satisfies (4.2) with $\varphi = \psi$ then

(4.12) $$\|W\|_{W^p_{2,\delta}} \leqslant C(m(M)^6\|\nabla\tau\|_{W^p_{0,\delta+2}} + \|j\|_{W^p_{0,\delta+2}}) .$$

These inequalities show that when M is chosen the strict inequalities (4.10) and (4.11) will be verified by all fixed points of $\mathcal{F}_t$ in $\bar{\Omega}$ by choosing k and k' large enough.

We know from lemma 2 that a fixed point of $\mathcal{F}_t$ in $\bar{\Omega}$ satisfies

$$\varphi \geqslant \Phi_0$$

therefore it satisfies the first of the strict inequality (4.11).
To show that it satisfies the second strict inequality (4.11) if $\nabla\tau$ is small enough we shall show that if

$$\gamma \equiv \|\nabla\tau\|_{W^p_{0,\delta+2}}$$

is  small enough  then we  can choose  M such  that $\varphi \leqslant \Phi_1$. Indeed we have proved in lemma 2 that this inequality holds if

(4.13)
$$a_W \leqslant \frac{M}{\sigma^{\beta+2}} \quad , \quad \text{i.e.} \quad \|a_W\|_{C^0_{\beta+2}} \leqslant M .$$

By inclusion properties of weighted Sobolev spaces we have

$$\|a_W\|_{C^0_{\beta+2}} \leqslant C\|a_W\|_{W^p_{1,\delta'+2}} \quad , \quad \text{since } \beta < \delta' + \frac{3}{p} .$$

We know that if $\delta' < \delta + (\delta + \frac{3}{p})$ we have

$$\|a_W\|_{W^p_{1,\delta'+2}} \leqslant C(\|W\|^2_{W^p_{2,\delta}} + \|A\|^2_{W^p_{1,\delta+1}})$$

Therefore if $(1 - \varphi, W) \in \bar{\Omega}$ we see, using the inequality (4.12), that (4.13) is satisfied if

$$m(M)^{12}\|\nabla\tau\|^2_{W^p_{0,\delta+2}} + \|j\|^2_{W^p_{0,\delta+2}} + \|A\|^2_{W^p_{1,\delta+1}} \leqslant C M.$$

i.e.
$$\gamma^2 \leqslant m(M)^{-12}\{C M - (\|j\|^2_{W^p_{0,\delta+2}} + \|A\|^2_{W^p_{1,\delta+1}}\}$$

We choose M such that the number between parenthesis is stictly positive, and then $\gamma > 0$ small enough to satisfy the above inequality.

We have proved that no mapping $\mathcal{F}_t: \bar{\Omega} \to \mathcal{H}$ , $t \in [0,1]$ has a fixed point on $\partial\Omega$.

To complete the proof it is sufficient to show that $\mathcal{F}_0$ has one fixed point in $\Omega$. Since $\mathcal{F}_0$ is the constant map $(v,\omega) \mapsto (u_0,W_0)$ it wil be the case if k and k' are chosen such that $(u_0,W_0)$ satisfy the inequalities (4.10), because we already know that $\varphi_0$ satisfy (4.11).


5. CASE R < 0.

When $R \equiv 8r < 0$ the mappings $\mathcal{F}_t: (v,\omega) \mapsto (u,W)$ will be defined by resolution of the linear system

(5.1)  $\Delta_g u = tf(v,\omega)$ , $f(v,\omega) \equiv r\psi - a_\omega\psi^{-7} - q\psi^{-3} + b\psi^5$, $\psi \equiv 1 + v$

(5.2)  $\delta_g L_g W = \frac{2}{3} \psi^6\nabla\tau + j$

with $(v,\omega) \in \Omega$, an open bounded subset of the space $\mathcal{H}$ of lemma 2 in §4.

longer monotonic in $\varphi$ another version of the maximum principle has to
be used to prove that no $\mathcal{F}_t$ has a fixed point on $\partial\Omega$, namely:
If $u \in C^2$ , it cannot attain a maximum at a point x where $(\Delta_g u)(x) > 0$
[resp., a minimum at a point where $(\Delta_g u)(x) < 0.$]

**Theorem.** Let $g,A,q,j,\tau$ be p-$\rho$ regularly asymptotically euclidean, $p > 3$.
**Suppose $R \equiv 8r < 0$ and (recall $b = \dfrac{\tau^2}{12}$)**

$$(5.3) \qquad \text{Inf } (b \ \sigma^{\beta+2}) = B > 0$$
$$\quad V$$

$$(5.4) \qquad \text{Inf } \dfrac{|r|}{b} = \lambda > 0 ,$$
$$\quad V$$

**Then there exists $\eta > 0$ such that the system (1.1), (1.2) has a solution
$(\varphi > 0, W)$ with $1 - \varphi$, $W \in W^p_{3,\delta}$, $-\dfrac{3}{p} < \delta < \text{Inf}(\rho, 1 - \dfrac{3}{p})$ if $\|\nabla\tau\|_{W^p_{0,\rho+2}} < \eta$
and $\beta$ is such that $\delta + \dfrac{3}{p} < \beta < \delta' + \dfrac{3}{p}$ with $\delta < \delta' < \text{Inf}(\rho, 1 - \dfrac{3}{p})$ .**

Proof. The mappings $\mathcal{F}_t$ are now the family of compact maps defined by
resolution of the linear system (5.1), (5.2) in the closure of the bounded
open set $\Omega \subset \mathcal{H}$ defined by the inequalities (4.10) and

$$(5.4) \qquad \text{Inf } \varphi > \ell , \qquad \text{Sup } \varphi < m , \qquad \varphi = 1 + u$$
$$\quad V \qquad\qquad\quad V$$

where $\ell$ and m are numbers to be chosen later and are such that

$$(5.5) \qquad 0 < \ell < 1 < m .$$

The mapping $\mathcal{F}_0$: $(v,\omega) \mapsto (0,W_0)$ has one fixed point in $\Omega$ if k' is large
enough for the second inequality (4.10) to be satisfied by $W_0$.

All mappings $\mathcal{F}_t$: $\bar{\Omega} \to \mathcal{H}$ will have a range satisfying the strict inequalities
(4.10) if $k > K$, $k' > K'$ where K and K' are numbers depending on the data
and on $\ell$, m .

To show that a fixed point of $\mathcal{F}_t$ satisfies the strict inequalities (5.4) we
start in the usual way, used also in CB-Is-Mo in the case of a compact $V_3$,
namely we consider the polynomial

$$P(z) \equiv bz^3 - |r|z^2 - qz - a_w$$

A fixed point $\varphi \equiv 1 + u$ of $\mathcal{F}_t$, $t \in (0,1]$, in $\bar{\Omega}$ satisfies

$$\Delta_g \varphi = t \, \varphi^{-7} P(z) \, , \qquad \ell^4 \leqslant z \equiv \varphi^4 \leqslant m^4 \, .$$

To show that $\varphi$ satisfies the strict inequalities (5.4) it is sufficient to show that at a maximum where $\varphi = m$ [resp. at a minimum where $\varphi = \ell$] we have

$$P(m^4) > 0 \, , \qquad [\text{resp. } P(\ell^4) < 0 \,] \, .$$

Since $q \geqslant 0$, $a_w \geqslant 0$ we shall have $P(\ell^4) < 0$ a fortiori if

$$\ell > 0 \quad \text{and} \quad \ell < \lambda = \underset{V}{\text{Inf}} \, \frac{|r|}{b} \, .$$

On the other hand we shall have $P(m^4) > 0$ if $m^4 > \mu(A,B)$ with $\mu(A,B)$ the positive root of the equation

$$B\mu^3 - \rho\mu^2 - Q\mu - M = 0$$

with

$$\rho \geqslant \|r\|_{C^0_{\gamma+2}} \, , \qquad Q \geqslant \|q\|_{C^0_{\gamma+2}} \, ,$$

(5.6)

$$M \geqslant \|a_w\|_{C^0_{\beta+2}} \, .$$

Numbers $B > 0$, $\rho > 0$ and $Q \geqslant 0$ depending only on the data exist by the hypothesis made on these data and on $\beta$. A number M exists for each fixed point of a mapping $\mathcal{F}_t$ because then $W \in W^p_{1,\delta}$..

The following choice of $\ell$ and $m$ satisfy the requirements

(5.7)        $0 < \ell < \text{Inf} \, (\lambda,1), \qquad m > \text{Sup}(\mu(M,B)^{1/4},1),$

We have only to check, as in §4, that, for a fixed point of $\mathcal{F}_t$ in $\overline{\Omega}$, $a_w$ satisfies (5.6). We are led to an inequality of the same form as (4.13) with now m(M) given by the inequality (5.7). The size of $\gamma$ can be estimated as in CB-Is-Mo.


REFERENCES

[Ba]   R. Bartnik   "Remarks on constant mean curvature surfaces in cosmological space times", Comm. Math. Phys. **117** p.615-624, 1988.

[Br]  D. Brill  "On space-times without maximal surfaces" in Proceedings of the  third Marcel  Grossmann meeting, Beijing 1982 H Ning ed. North Holland 1983.

[Ca]  M.  Cantor  "The  existence  of  asymptotically flat initial data for vacuum space- times" Comm. Math. Phys. 57 p.83-96, 1977.

[C.S-C.B]  A. Chaljub-Simon  and Y. Choquet-Bruhat "solution of the problem of  constraints on  open and closed manifolds" Jour. Gen. Rel. and Grav. 5 p.47- 64, 1974.

[C.B] Y. Choquet-Bruhat "Global solution of the problem of constraints on a closed manifold"
Acta di Istituto di Alta Matematica, Rome, 12 p.317-325, 1973

[C.B-Ch] "Elliptic systems in $H_{s,\delta}$ spaces" Acta Matematica 146 p129-150, 1981.

[C.B-Yo]  Y. Choquet-Bruhat  and J.  York "The  Cauchy problem" in "General Relativity", A. Held ed. Plenum 1979.

[C.B-Is-Mo]  Y. Choquet-Bruhat,  J. Isenberg  and V.  Moncrief "Solution of constraints for Einstein equations" C.R. Ac. Scie. Paris, to appear 1992.

[Ch-OM] D. Christodoulou and N. O'Murchada "The boost problem in General Relativity" Comm. in Math. Phys. 80 p271-300, 1981.

[Is]  "Parametrization of  the  space  of solutions to Einstein equations" Phys. Rev. letters 59 p2389-2392, 1987.

[Li]  A. Lichnerowicz  "Sur l'intégration des équations d'Einstein" J.Math. Pures et Ap. 23 p.37-63, 1944.

# Compact Cauchy Horizons and Cauchy Surfaces

*P.T. Chruściel* [*]      *J. Isenberg* [†]

### Abstract

We show that if a globally hyperbolic spacetime $(M, g)$ extends to a non globally hyperbolic spacetime $(M', g')$, and if the Cauchy horizon $\mathcal{H}$ for $M$ in $M'$ is compact, then the Cauchy surfaces for $(M, g)$ must be diffeomorphic to $\mathcal{H}$. As a corollary to this result, we show that if a $(2+1)$-dimensional spacetime has compact Cauchy surfaces with topology other than $T^2$, then it cannot be extended to a spacetime with a compact Cauchy horizon.

## 1.  Introduction

Dieter Brill and one of us (JI) used to talk a lot about Mach's Principle. We both were of the Wheeler school, so our Machian discussions often focussed on issues involving the initial value formulation of Einstein's theory. One such issue was the following question: If a spacetime $(M, g)$ is known to be globally hyperbolic, how can one tell (from intrinsic information) if a given embedded spacelike hypersurface $\Sigma$ is a Cauchy surface for $(M, g)$? The answer to this question is important if one wants to know what minimal information about the universe "now" is needed to determine the spacetime metric (and hence its inertial frames) for all time in $(M, g)$.

It turns out [1] that if a spacelike hypersurface $\Sigma$ embedded in a globally hyperbolic spacetime is compact (without boundary), then it must be a Cauchy surface. So, whatever the topology of $\Sigma$, all Cauchy surfaces in $(M, g)$ must be diffeomorphically equivalent to $\Sigma$; and indeed it follows [2] that $M$ must be diffeomorphic to $\Sigma \times \Re$.

[*]Alexander von Humboldt fellow, on leave of absence from the Institute of Mathematics of the Polish Academy of Sciences. Max-Planck-Institut für Astrophysik, Karl Schwarzschild Str. 1, D8046 Garching bei Munich, Germany. This author was partially supported by the National Science Foundation under grant # PHY-902301 at the University of Oregon and by a KBN grant # 2 1047 9101. e-mail: piotr@ibm-1.mpa.ipp-garching.mpg.de.

[†]Department of Mathematics and Institute for Theoretical Science University of Oregon, Eugene, OR 97403, USA. This author was partially supported by the National Science Foundation under grant # PHY-902301 at the University of Oregon.

Let us say we consider a spacetime $(M', g')$ which is not globally hyperbolic; however, in $(M', g')$ there is a globally hyperbolic region $(M, g)$ with the boundary of $M$ in $M'$ forming a Cauchy horizon $\mathcal{H}$. Is there any necessary topological relationship between the Cauchy horizon $\mathcal{H}$ and the Cauchy surfaces in $(M, g)$? In this paper, we show (Theorem 3.1) that if $\mathcal{H}$ is compact (without boundary) then the Cauchy surfaces in $(M, g)$ must be diffeomorphic to $\mathcal{H}$, and hence $M = \mathcal{H} \times \Re$.

It follows from this result that a spacetime which is generated from initial data specified on a noncompact Cauchy surface cannot be extended across a compact Cauchy horizon. If we consider $(2 + 1)$-dimensional spacetimes, Theorem 3.1 even limits the class of spacetimes with compact Cauchy surfaces which may be extended across a compact Cauchy horizon. We recall [3] the fact that a Cauchy horizon is necessarily ruled by a congruence of null geodesics. It follows by a simple argument that $\mathcal{H}$ admits a nowhere vanishing vector field. But the only compact (orientable) two-dimensional manifold which admits a nonvanishing vector field is the two-torus, $T^2$. Hence, if a $(2+1)$ dimensional globally hyperbolic spacetime has Cauchy surfaces which are not two-tori, then it cannot be extended across a compact Cauchy horizon (Theorem 4.2).

Theorem 4.2 is relevant to a question which arose among a number of us (Brill, Eardley, Horowitz, Moncrief, and the two authors) at Aspen in 1990. Recall the Misner $(1+1)$-dimensional, flat cylinder, spacetimes [4]. Often used to model the behavior of Taub-NUT spacetimes, these Misner spacetimes extend across a compact $(S^1)$ Cauchy horizon. Besides these standard $(1+1)$-dimensional Misner spacetimes, there are $(2+1)$, $(3+1)$, and indeed $(n+1)$ (any $n$) dimensional versions of them. Some of the generalized Misner spacetimes are obtained simply by taking the direct product of a $(1+1)$–Misner spacetime with a flat $(n-1)$–torus. Their Cauchy surfaces are $n$-dimensional tori, and they extend across compact Cauchy horizons (also $T^n$'s). But there are other, more perplexing $(n+1)$–dimensional generalized Misner spacetimes which are obtained by taking the interior of the future light cone through a point in $(n+1)$–dimensional Minkowski spacetime, and then quotienting by the action of a certain discrete subgroup of the Lorentz group so as to spatially identify the spacetimes. For appropriate subgroups of the Lorentz group the resulting space–time will be spatially compact. The Lorentz subgroup action which is quotiented out preserves the constant mean curvature ("mass hyperboloid") hypersurfaces in the interior of the future causal cone; so these negatively-curved hypersurfaces serve as Cauchy surfaces for the resulting spacetimes. Can these generalized Misner spacetimes be extended across a Cauchy horizon?

Noting that the spacetimes are flat, and geodesically incomplete, one might expect that a Taub-NUT-like (and Misner-like) extension can be carried out. However, at least in the $(2+1)$-dimensional case, it follows from our results that if an extension can be made, then the resulting Cauchy horizon must be noncompact. For if the

Cauchy horizon were compact, it and therefore the Cauchy surfaces would have to be two–tori. But the compact "mass hyperboloid" Cauchy surfaces have constant negative curvature, so they must be two-manifolds of genus greater than or equal to two. It remains to be seen if one can extend these Misner spacetimes across a *non* compact Cauchy horizon of some sort.

This paper is primarily devoted to proving Theorem 3.1. In preparation for doing this, we discuss definitions and conventions, and recall some properties of globally hyperbolic spacetimes in Section 2. The statement and proof of Theorem 3.1 appears in Section 3. Then in Section 4. we discuss its corollaries (including Theorem 4.2) and make some concluding remarks.

## 2.   Preliminaries

We work here with spacetimes which are $C^k$ Lorentz manifolds $(M, g)$ for $k \geq 3$. Generally we assume that $M$ is a $C^k$, paracompact, Hausdorff, connected, $(n + 1)$-dimensional manifold $(n \geq 1)$, and that $g$ is a $C^{k-1}$ Lorentz-signature metric on $M$.

At any point $p \in M$, the metric $g$ divides the set of tangent vectors $V \in T_pM$ into subsets which are timelike (if $g(V, V) < 0$), spacelike (if $g(V, V) > 0$), or null (if $g(V, V) = 0$). The timelike vectors in $T_pM$ split into two disconnected families, which one can arbitrarily label "future timelike" and "past timelike." These designations have nonlocal meaning iff the spacetime is *time-orientable*. By definition, a spacetime is time-orientable iff for any continuous map $\Gamma : S^1 \to M$ of the circle $S^1$ into the spacetime manifold M, and for any choice of (i) a point $p_0 = \Gamma(\theta_0)$ in the image of the map and (ii) a timelike vector $V_0 \in T_{p_0}M$, there exists a continuous map $\tilde{\Gamma} : S^1 \to TM$ such that $\tilde{\Gamma}(\theta) \in T_{\Gamma(\theta)}M$, and $\tilde{\Gamma}(\theta_0) = V_0$, with $\tilde{\Gamma}(\theta)$ timelike (non-vanishing) for all $\theta$. So if a spacetime is time-orientable, there is a continuous global distinction between future timelike and past timelike vectors. One easily verifies that if $(M, g)$ is time-orientable, there exists a $C^{k-1}$ vector field $X$ on $M$ which is timelike everywhere (and either future timelike or past timelike everywhere). Note that, by definition, such a vector field is nowhere vanishing.

Time orientability is a mild restriction on a spacetime $(M, g)$; indeed, if $(M, g)$ is not time orientable, it always has a double cover which is. The condition of global hyperbolicity, on the other hand, is quite strong. If $(M, g)$ is globally hyperbolic, then all of the following must be true [2, 3]:

a)   $(M, g)$ must be time orientable.

b)   There exists a $C^k$ embedded Cauchy surface in $(M, g)$; *i.e.*, there exists an $n$-dimensional hypersurface $\Sigma$ embedded in $M$ such that every inextendible causal path in $M$ intersects $\Sigma$ once and only once.

**c)**  There exists a $C^k$ function $T : M \to \Re$ with nowhere vanishing gradient such that the level sets of $T$ are all Cauchy surfaces $\Sigma_T$ in $(M,g)$. This "time function" can be chosen so that its image includes all of the reals, and then it follows that for any future-directed inextendible causal path $\gamma : \Re \to M$, one has $\lim_{s\to\infty} T \circ \gamma(s) = +\infty$ and $\lim_{s\to-\infty} T \circ \gamma(s) = -\infty$.

**d)**  Any pair of $C^m$ Cauchy surfaces in $(M,g)$ for $0 \le m \le k$, are $C^m$ diffeomorphic.

It is an immediate consequence of these properties of globally hyperbolic spacetimes that every Cauchy surface is a closed subset of $(M,g)$ and $M$ is $C^k$ diffeomorphic to $\Sigma \times I$, where $I$ is an interval (possibly infinite) in $\Re$. Further, one readily verifies that if one does not assume a priori that $M$ is connected, then — assuming that $(M,g)$ is globally hyperbolic — $M$ is connected iff its Cauchy surfaces $\Sigma$ are connected.

In this paper, we are interested in globally hyperbolic spacetimes $(M,g)$ which can be extended past their domains of dependence to some non globally hyperbolic spacetime $(M',g')$. In particular, we are interested in studying the boundary of $\partial M$ of $M$ in $M'$, which is called the Cauchy horizon of $M$.

Let us recall the topological nature of the boundary $\partial A$ of any set $A$ in $B$. If we presume that $A$ is open in $B$ (as we shall for $M$ in $M'$), then by definition $\partial A$ consists of all those points $p \in B \setminus A$ for which there exists a sequence $p_i$ which lives in $A$ and converges to $p$. The boundary $\partial A$ inherits a natural topology from $B$: a subset $\mathcal{O}$ of $\partial A$ is open in $\partial A$ iff there exists an open subset $\mathcal{U}$ of $B$ such that $\mathcal{O} = \partial A \cap \mathcal{U}$. It follows that if $C \subset \partial A$ is a connected component of $\partial A$, then there exists an open set of $\mathcal{U} \subset B$ such that $\mathcal{U} \cap \partial A = C$. This fact plays a role in the proof of Theorem 3.1, below.

Before we state and prove our main result, we wish to establish a technical result concerning flows on spacetime. Let $\phi_s(p)$ denote the flow of a vector field $X$, as defined by the differential equation $\frac{d}{ds}\phi_s(p) = X \circ \phi_s(p)$. We shall need the following.

**Lemma 2.1**  *Let $Q$ be a compact subset of a $C^k$ spacetime $(M,g)$ (with $k \ge 2$) and let $U$ be an open neighborhood of $Q$. Let $X$ be a $C^\ell$ vector field on $M$, with $\ell \ge 1$. There exists $s_0 > 0$ such that for all $s \in [-s_0, +s_0]$, one has $\phi_s(Q) \subset U$.*

**Proof :**  Since $M$ is Hausdorff and paracompact, there exists a $C^{k-1}$ *Riemannian* metric $h$ on $M$ such that (1) $(M,h)$ is a complete Riemannian manifold, and (2) the metric topology induced on $M$ by the distance function $d_h$ associated to $h$ coincides with the original topology on $M$ [5].

Now, for each point $p \in Q$, there exists a positive number $r_p$ such that the metric balls $B(p, 3r_p) := \{q \in M \mid d_h(p, q) < 3r_p\}$ are open, have compact closure, and are contained in U. Since $Q$ is compact, the covering $\{B(p, r_p)\}_{p \in Q}$ of $Q$ has a finite subcover $\{B(p_i, r_{p_i})\}_{i \in [1,2...,I]}$. If we now choose $d_o := \min_{i \in I}\{r_{p_i}\}$ and define the set

$$K := \{m \in M \mid d_h(m, Q) \leq d_o\}$$

then we find that $K \subset U$. To see this, we calculate

$$
\begin{aligned}
d_h(m, p_i) &\leq d_h(m, p) + d_h(p, p_i) \qquad \text{for any point } p \in B(p_i, r_{p_i}) \\
&\leq d_o + r_{p_i} \\
&\leq 2r_{p_i} ,
\end{aligned}
$$

and recall that $B(p, 3r_{pi}) \subset U$ for all $i \in I$. Note that since $K \subset \cup_{i=1}^{I} \bar{B}(p_i, 3r_{pi})$, K is compact.

Consider the function

$$
\begin{aligned}
f : K &\rightarrow \Re \\
p &\mapsto \parallel X(p) \parallel_h
\end{aligned}
$$

where $\parallel X(p) \parallel_h$ is the norm of the vector $X(p)$ with respect to the inner product $h$ at $p$. Since $X$ is $C^\ell$ and $h$ is $C^{k-1}$, $f$ is a continuous function on $K$. Thus, since $K$ is compact, we have

$$C_0 := \sup_{p \in K} f(p) < \infty$$

Now, for each $p \in Q$, consider the flow path $\phi_s(p)$. If we choose $t$ so that $\phi_s(p)$ remains in $K$ for all $s \in [0, t]$, then the length $L(t)$ of the path $\phi_s(p)$ for $s \in [0, t]$ satisfies the inequality

$$L(t) = \int_0^t \parallel X(\phi_s(p)) \parallel_h ds \leq C_0 t .$$

Hence, if we choose $|t| \leq d_o/C_o$ we shall have $d_h(\phi_t(p), Q) \leq d_o$, so $\phi_t(p) \in U$ for all $p \in Q$. □

## 3.  Main Result

We now state and prove our main result.

**Theorem 3.1** *Let $(M', g')$ be a $C^k$ (with $k \geq 3$), $(n+1)$–dimensional, time orientable spacetime. Let $(M, g)$ with $g = g'|_M$, be a globally hyperbolic spacetime open in $M'$. If there exists a connected component $\mathcal{H}$ of $\partial M$ which is a $C^\ell$ (with $1 \leq \ell \leq k - 2$) compact, submanifold of $M'$, then every $C^\ell$ Cauchy surface $\Sigma$ in $(M, g)$ is $C^\ell$ diffeomorphic to $\mathcal{H}$.*

**Remark:** It should be noted that the result above is somewhat more general than the case in which $\mathcal{H}$ is a Cauchy horizon: Indeed, the result holds for $M$'s which do not necessarily coincide with the interior of $D(\Sigma; M')$, where $D(\Sigma; M')$ denotes the domain of dependence of the hypersurface $\Sigma$ in $M'$. In fact, it even holds when the hypersurface $\Sigma$ above is not achronal as a subset of $M'$.

**Proof :** The basic idea of the proof is to use a timelike vector field in $M'$ to drag $\mathcal{H}$ diffeomorphically into the globally hyperbolic region $M$, and then show that the dragged $\mathcal{H}$ is diffeomorphic to some Cauchy surface in $M$. Since the spacetime $(M', g')$ is, by assumption, time orientable, there always exists a globally defined (nonvanishing) time-like vector field on $M'$, which we shall call $X$. Let $\phi_s$ be the flow generated by $X$ (*cf.* Section 2.).

Before studying the action of $\phi_s$, on $\mathcal{H}$, we need to note two important properties of $\mathcal{H}$. These properties follow from the fact that $\mathcal{H}$ lies in the boundary of the globally hyperbolic spacetime $(M, g)$, and the fact that the topology of $\mathcal{H} \subset \partial M$ guarantees that there exists a neighborhood of $\mathcal{H}$ in $M'$ containing no points of $\partial M$ other than those in $\mathcal{H}$. Using arguments much like those in [2], we establish the following:

**Lemma 3.2** *Under the hypotheses of Theorem 3.1 we have:*

(a)    *The dimension of $\mathcal{H}$ is one less than that of its ambient manifold $M'$.*

(b)    *For any point $p \in \mathcal{H}$ and any timelike vector $V \in T_p M'$, $V$ is transverse to $\mathcal{H}$ (i.e., the projection of $V$ into $T_p \mathcal{H} \subset T_p M'$ does not equal $V$ itself).*

Since $X$ is timelike and therefore everywhere transverse to $\mathcal{H}$, we expect either the flow $\phi_s$ — or $\phi_{-s}$ — to move $\mathcal{H}$ into $M$. Let us assume, without loss of generality, that we have chosen $X$ so that at least for some specified point $p \in \mathcal{H}$, $\phi_s(p) \in M$ for $s_0 < s < 0$, for some $s_0$. We now verify the following:

**Lemma 3.3** *There exists $\delta > 0$ such that for $-\delta \leq s < 0$*

(a)    $\phi_s(\mathcal{H}) \subset M$

(b)    $\phi_s(\mathcal{H})$ *is a compact $C^\ell$ submanifold of $M$ (where $0 \leq \ell \leq k - 2$, and we presume that $\mathcal{H}$ is a $C^\ell$ submanifold of $M'$.)*

**Proof :** Since $X$ is everywhere timelike and transverse to $\mathcal{H}$, and since $\mathcal{H}$ is a $C^\ell$ submanifold of $M'$, it follows from the properties of flows that there exists some $s_0 > 0$ such that if $|s| < s_0$, then $\phi_s(\mathcal{H})$ is a $C^\ell$ submanifold of $M'$ as well.

Now let us define the sets

$$K_{s_1}^{s_2} := \cup_{s_1 \le s \le s_2} \phi_s(\mathcal{H})$$

for any pairs $s_1 < s_2$ with $|s_1| < s_0$ and $|s_2| < s_0$. One readily verifies (from the properties of flows) that $K_{s_1}^{s_2}$ is $C^\ell$ diffeomorphic to $[s_1, s_2] \times \mathcal{H}$. Since $\mathcal{H}$ is compact and connected, it follows that $K_{s_1}^{s_2}$ is compact and connected in $M'$.

It is useful to define on $K_{s_1}^{s_2}$ the function

$$\tau : K_{s_1}^{s_2} \rightarrow \Re$$

$$p \mapsto s \quad \text{such that } p \in \phi_s(\mathcal{H}).$$

This is a well-defined function, and one verifies that it is $C^\ell$ when $\ell \le k - 2$.

Since $\mathcal{H}$ is a connected component of the boundary of the open set $M$ in $M'$, we find (as noted in Section 2.) that there exists an open set W in $M'$ such that $\mathcal{H} \subset W$, yet $W \cap (\partial M \setminus \mathcal{H}) = \phi$. Using Lemma 2.1, we then find that for some $\delta > 0$, the set $K_{-\delta}^\delta$ is contained in W. It follows that $K_{-\delta}^\delta \cap \partial M = \mathcal{H}$. This implies in particular that for $|s| \le \delta$, if $\phi_s(\mathcal{H}) \cap \partial M \ne \emptyset$, then $s = 0$.

Now let $p$ be a point in $\mathcal{H}$. Since $\mathcal{H} \subset \partial M$, it follows from the definition of topological boundary that there exists a sequence $p_i \in M$ which converges to $p$. By the definition of convergence, there exists some $\hat{\imath}$ such that if $i \ge \hat{\imath}$, then $p_i \in K_{-\delta}^\delta$, and $p_i \in K_{-\delta}^0$. Further, we can always find a subsequence $p_{i_n}$ of $p_i$ such that (i) $p_i$ is the first member of the subsequence — i.e., $p_{i_0} = p_i$; (ii) $p_{i_n} \in K_{-\delta}^0$; and (iii) $\tau_n := \tau(p_{i_n})$ is monotonically increasing. Possibly decreasing the value of $\delta$, we may also guarantee that $p_{i_0} \in \phi_{-\delta}(\mathcal{H})$.

We now consider the sequence of sets

$$\mathcal{O}_n := K_{\tau_n}^{\tau_{n+1}} \cap M$$

Since $p_{i_n}$ and $p_{i_{n+1}}$ are both contained in $\mathcal{O}_n$, the sets $\mathcal{O}_n$ are non-empty. Since $M$ is open, $\mathcal{O}_n$ is open in $K_{\tau_n}^{\tau_{n+1}}$. We wish to argue that $\mathcal{O}_n$ is closed in $K_{\tau_n}^{\tau_{n+1}}$ as well. Suppose this were not true; then there must be a sequence of points $q_m \in \mathcal{O}_n$ which converges to $q_\infty$, with $q_\infty \in K_{\tau_n}^{\tau_{n+1}}$ but $q_\infty \notin \mathcal{O}_n$. So $q_\infty$ must be an element of $M' \setminus M$. But if $q_\infty$ is the limit of a sequence in $\mathcal{O} \subset M$ and $q_\infty \in M' \setminus M$, it follows that $q_\infty \in \partial M$. Since $q_\infty$ is supposed to be an element of $K_{\tau_n}^{\tau_{n+1}}$ as well, this tells us that $K_{\tau_n}^{\tau_{n+1}} \cap \partial M$ is not empty. But $K_{\tau_n}^{\tau_{n+1}} \subset K_{-\delta}^\delta$, so $K_{\tau_n}^{\tau_{n+1}} \cap \partial M \subset \mathcal{H}$. And, we noted above that for $0 < |s| < \delta$, we have $\phi_s(\mathcal{H}) \cap \partial M = \emptyset$. So we have a contradiction; it follows that $\mathcal{O}_n$ is closed in $K_{\tau_n}^{\tau_{n+1}}$.

Since $K_{\tau_n}^{\tau_{n+1}}$ is connected, and since $\mathcal{O}_n$ is nonempty, open in $K_{\tau_n}^{\tau_{n+1}}$, and closed in $K_{\tau_n}^{\tau_{n+1}}$, it follows that $\mathcal{O}_n = K_{\tau_n}^{\tau_{n+1}}$. Thus $K_{\tau_n}^{\tau_{n+1}}$ is a closed subset of M. It follows that for all values of $s$ for which $\tau_n < s < \tau_{n+1}$, one has $\phi_s(\mathcal{H}) \subset M$. Since $\tau_1 = -\delta$ and since $\tau_{n+1}$ approaches zero, in fact $\phi_s(\mathcal{H}) \subset M$ for all $s$ satisfying $-\delta \le s < 0$.

To show that, for these values of $s$, $\phi_s(\mathcal{H})$ is compact in $M$, we use the facts (established above) that the sets $K^{\tau_{n+1}}_{\tau_n}$ are compact in $M$, and that for $\tau_n \leq s \leq \tau_{n+1}$, $\phi_s(\mathcal{H})$ is a closed subset of $K^{\tau_{n+1}}_{\tau_n}$. This completes the proof of Lemma 3.3.          □

Let us return to the proof of Theorem 3.1. We have now determined that for each value of s satisfying $-\delta < s \leq 0$ ($\delta > 0$ defined in Lemma 3.3), the set $\phi_s(\mathcal{H})$ is a compact $C^\ell$ submanifold of the globally hyperbolic spacetime $M$. If, in addition, we knew that $\phi_s(\mathcal{H})$ were spacelike, then it would follow [1] that $\phi_s(\mathcal{H})$ must be a compact Cauchy surface for $M$; we would be essentially done with the proof of the theorem. However we do not know a priori that $\phi_s(\mathcal{H})$ is spacelike, so there is more work to do.

Since $(M, g)$ is globally hyperbolic, there exists a globally defined time function $T : M \to \Re$, the level sets of which are $C^\ell$ Cauchy surfaces in $(M, g)$ (cf. Section 2.). Changing time orientation if necessary we may assume that $X$ is future oriented. Let us moreover assume that $T$ has been chosen so that the range of $T$ is the whole of $\Re$, and let us define

$$T_+ := \sup\{t(p) \mid p \in \phi_{-\delta}(\mathcal{H})\}.$$

Since $T$ is a continuous function on $M$ and since $\phi_s(\mathcal{H})$ is compact in $M$, it follows that $T_+$ is finite. Thus the set

$$\Sigma_+ := \{p \in M \mid T(p) = T_+\}$$

is a $C^\ell$ Cauchy surface.

To show that $\Sigma_+$ is compact in $M$, we shall first show that $\Sigma_+ \cap K^0_{-\delta}$ is compact in $M$, and then later prove that $\Sigma_+ \cap K^0_{-\delta} = \Sigma_+$. Since we know that $\Sigma_+$, as a Cauchy surface, is a closed set in $M$, it follows that if we can show that

$$\Sigma_+ \cap K^0_{-\delta} \subset K^{-\eta}_{-\delta} \tag{1}$$

for some $0 < \eta < \delta$, then since $K^{-\eta}_{-\delta}$ is compact in M, it follows that $\Sigma_+ \cap K^0_{-\delta}$ must be compact as well.

Suppose that the inclusion (1) is false. Then there must be a sequence $s_i$ converging to 0 and a choice of $p_i \in \phi_{s_i}(\mathcal{H})$ such that $p_i \in \Sigma_+$. These $p_i$ are contained in $K^0_{-\delta}$, which is compact in $M'$ (not in $M$); so there must be a point $p \in \mathcal{H}$ and a subsequence $s_{i_n}$ of $s_i$ such that $p_{i_n}$ converges to $p$. Now consider a timelike inextendible curve $\Gamma$ in $M$ such that

$$\Gamma \cap K^0_{-\delta} = \{\phi_s(p) \mid -\delta \leq s < 0\}.$$

Since $\Sigma_+$ is a Cauchy surface, there exists a point $q \in \Sigma_+$ such that $q = \Sigma_+ \cap \Gamma$. It follows that $p$ (to which $\Gamma$ converges in $M'$) lies in the future set $I^+(q)$ of $q \in \Sigma_+$. But the sequence $p_{i_n} \in \Sigma_+$ converges to $p$, so for sufficiently large $n$, $p_{i_n}$ must be in

$I^+(q)$ as well. This contradicts the achronality of $\Sigma_+$, so the inclusion (1) must hold. Hence, we conclude that $\Sigma_+ \cap K^0_{-\delta} = \Sigma_+ \cap K^{-\eta}_{-\delta}$ is compact in M.

We wish to now define a $C^\ell$ diffeomorphism from $\phi_{-\delta}(\mathcal{H})$ — which is $C^\ell$ diffeomorphic to $\mathcal{H}$ — and $\Sigma_+$. To do this, let $\pi$ be any point in $\phi_{-\delta}(\mathcal{H})$ and define the function

$$\sigma_\pi : [0, \delta) \;\rightarrow\; \Re$$
$$s \;\mapsto\; T(\phi_s(\pi)).$$

We verify that $\sigma_\pi$ is $C^\ell$ (and hence continuous), that $\sigma_\pi(0) = T(\pi) \leq T_+$, and that $\sigma_\pi$ diverges to $+\infty$ as $s$ approaches $\delta$ (since the time function T diverges to $+\infty$ as one approaches the horizon). Hence, by the Intermediate Value Theorem, there is some $s_\pi$ for which $\sigma(s_\pi) = T_+$; and therefore $\phi_{s_\pi}(\pi) \in \Sigma_+$. This defines the $C^\ell$ map

$$\psi : \phi_{-\delta}(\mathcal{H}) \;\rightarrow\; \Sigma_+ \cap K^{-\eta}_{-\delta}$$
$$\pi \;\mapsto\; \phi_{s_\pi}(\pi).$$

The map $\psi$ is injective, since the timelike vector field $X$ is transverse to $\phi_{-\delta}(\mathcal{H})$ by definition. It is surjective as a consequence of the definition of $K^{-\eta}_{-\delta}$, and because $X$ is transverse to the spacelike $\Sigma_+$. Hence $\psi$ is a $C^\ell$ diffeomorphism.

It remains to show that $\Sigma_+ \cap K^{-\eta}_{-\delta} = \Sigma_+$. Since $\Sigma_+ \cap K^{-\eta}_{-\delta}$ is the image of $\phi_{-\delta}(\mathcal{H})$ under the diffeomorphism $\psi$, $\Sigma_+ \cap K^{-\eta}_{-\delta}$ is an open subset of $\Sigma_+$. Since $K^{-\eta}_{-\delta}$ is closed in $M$, it follows that $\Sigma_+ \cap K^{-\eta}_{-\delta}$ is a closed subset of $\Sigma_+$. Hence, $\Sigma_+ \cap K^{-\eta}_{-\delta}$ is a connected component of $\Sigma_+$. But $\Sigma_+$ (being a Cauchy surface) is connected iff $M$ is connected, and $M$ is connected by hypothesis, so $\Sigma_+ \cap K^{-\eta}_{-\delta} = \Sigma_+$.

Thus $\Sigma_+ = \Sigma_+ \cap K^{-\eta}_{-\delta}$ is compact in $M$, and $\psi$ is a $C^\ell$ diffeomorphism from $\phi_{-\delta}(\mathcal{H})$ to $\Sigma_+$. $\qquad\square$

## 4.   Consequences of the Main Result

Theorem 3.1 has an immediate consequence for spacetimes constructed from initial data. Recall that if one specifies a Riemannian metric $\gamma$ and a symmetric tensor $K$ on an $n$–dimensional manifold $\Sigma$ with $\gamma$ and $K$ satisfying the Einstein vacuum constraints, then (i) there exists a globally hyperbolic $(n+1)$-dimensional spacetime $(M = \Sigma \times I, g)$ which satisfies the vacuum Einstein equations, and (ii) there exists an embedding $i : \Sigma \rightarrow M$ such that $i(\Sigma)$ has intrinsic metric $\gamma$ and extrinsic curvature $K$ [6]. The hypersurface $i(\Sigma)$ is a Cauchy surface for the spacetime $(M, g)$, which is called a development of the Cauchy data $(\Sigma; \gamma, K)$. For a given set of Cauchy data, there are many non isometric developments; however, one finds [7] that there always exists a unique maximal globally hyperbolic spacetime development in which all others may be isometrically embedded.

There is a Corollary of Theorem 3.1 concerning non globally hyperbolic extensions of maximal globally hyperbolic developments. Recall that $(M', g'; \psi)$ is called an extension of $(M, g)$ if there exists an isometric embedding $\psi$ of $M$ into $M'$. We have

**Corollary 4.1** *Let $(\Sigma; \gamma, K)$ be a $C^\infty$ constraint–satisfying initial data set, and let $(M, g)$ be its maximal globally hyperbolic development. If there exists a $C^k$ spacetime extension $(M', g'; \psi)$ of $(M, g)$ such that $\mathcal{H} := \partial(\psi(M))$ is a $C^\ell$ $(1 \le \ell \le k - 2)$ compact submanifold of $M'$, then $\mathcal{H}$ is $C^\ell$ diffeomorphic to $\Sigma$. In particular, if $\Sigma$ is not compact, then $\mathcal{H}$ cannot be compact.*

For Cauchy data specified on a 3-dimensional compact manifold $\Sigma$, there is no strict relationship known to exist between the topology of $\Sigma$ and the possibility that there is an extension of the maximal globally hyperbolic development with consequent formation of a compact Cauchy horizon. On the other hand, as noted in the introduction, the situation is very different for Cauchy data specified on a 2-dimensional manifold. We find the following

**Theorem 4.2** *Let $(M, g)$ be a $(2+1)$–dimensional globally hyperbolic spacetime with Cauchy surface $\Sigma$. If $(M, g)$ can be extended across a compact Cauchy horizon to a spacetime $(M', g')$ which — together with $(M, g)$ — satisfies the hypotheses of Theorem 3.1, then $\Sigma$ must be a two–torus.*

As discussed in the Introduction, Theorem 4.2 is an immediate consequence of Theorem 3.1, together with the fact that a smooth Cauchy horizon must admit a nowhere vanishing vector field [2], and the fact that the only compact orientable 2-dimensional manifold which admits a nowhere vanishing vector field is the two-torus. Note that, to a certain extent, Theorem 4.2 generalizes to $(2m + 1)$–dimensional spacetimes for $m > 1$. Specifically, since Cauchy horizons of any dimension are null hypersurfaces generated by null geodesics, a compact Cauchy horizon must have Euler characteristic zero. Hence if a $(2m + 1)$-dimensional globally hyperbolic spacetime has Cauchy surfaces with nonvanishing Euler characteristic, then it cannot be extended across a compact Cauchy horizon.

In thinking about spacetimes with Cauchy horizons, it is important to recall that while many spacetimes are known to have them [4, 8], the Strong Cosmic Censorship Conjecture [9] claims that they do not generically occur in physically significant spacetimes. There has been some progress in recent years in our understanding of issues related to this conjecture [10, 11, 12]. However, the important question it raises regarding Cauchy horizons (compact or otherwise) remains essentially unresolved and in need of much further study.

# References

[1] Budic R., Isenberg J., Lindblom L., Yasskin P., *On the Determination of Cauchy Surfaces from Intrinsic Properties.* Comm. Math. Phys. 61 (1978) 87-95.

[2] Geroch R., *The Domain of Dependence.* Jour. Math. Phys. 11 (1970) 437-439.

[3] Hawking S., Ellis G., *The Large Structure of Spacetime.* Cambridge U. Press, 1973.

[4] Misner C., *Taub-NUT space as Counterexample to Almost Everything.* Lectures in Math., Vol. 8, pp. 160-169.

[5] Hicks N., *Notes on Differential Geometry.* Van Nostrand, 1965.

[6] Choquet-Bruhat Y., York J., *The Cauchy Problem.* In *General Relativity and Gravitation* (ed. A. Held), *Plenum,* 1980.

[7] Choquet-Bruhat Y., Geroch R., *Global Aspects of the Cauchy Problem.* Comm. Math. Phys. 14 (1969) 329-335.

[8] Moncrief V., *Infinite Dimensional Family of Vacuum Cosmological Models with Taub-NUT Type Extensions.* Phys. Rev. D23 (1981) 312–315.

[9] Penrose R., *Spacetime Singularities.* Proceedings of the First Marcel Grossman Meeting on General Relativity (ed. R. Ruffini), North Holland, 1977.

[10] Moncrief V., Isenberg J., *Symmetries of Cosmological Cauchy Horizons.* Comm. Math. Phys. 89 (1983) 387-413.

[11] Chruściel P.T., Isenberg J., Moncrief V., *Strong Cosmic Censorship in Polarized Gowdy Spacetimes.* Class. Qtm. Grav. 7 (1990) 1671-1679.

[12] Chruściel P.T., *On Uniqueness in the Large of Solutions of Einstein's Equations ("Strong Cosmic Censorship").* Proceedings of the CMA, Vol. 27, Australian National University, Canberra 1991.

# The Classical Electron

*J.M. Cohen\* and E. Mustafa \**

We resolve the longstanding paradox that in classical electrodynamics the energy and linear momentum of the Abraham-Lorentz (classical) electron do not transform as 4-vector components under Lorentz transformations. In our treatment these quantities transform properly and remain finite in the point particle limit.

## 1. Introduction

In Classical Electrodynamics the energy and linear momentum of the extended classical electron do not transform as components of a 4-vector (Leighton, 1959); also infinities arise in the point charge limit.

Abraham (1905) and Lorentz (1909) proposed the classical model of the electron. Lorentz suggested a model in which the electron consisted of a thin, uniformly charged shell. Poincaré (1909) added a stress in order to stabilize the electron. However, he used an expression for the cohesive stress-energy tensor which led to difficulties (Fermi, 1922; Pais, 1948).

Dirac (1938a, 1938b), Matthison (1931, 1940, 1942) and others have proposed point particle models for the classical electron. The lack of covariance associated with the Abraham-Lorentz model and other extended models of the electron does not arise in such models or in classical models based on non local generalisations of Maxwell electromagnetism; see for example, Bopp (1942), McManus (1948) and Feynmann (1948)[1]. Erber (1961) reviews much of the literature on models of the classical electron.

According to classical theory (Leighton, 1959) the expressions for the energy and linear momentum of the classical electron (modeled as a shell charge $q$ and radius $r_0$, moving with velocity $v$ in the z-direction with $\gamma \equiv (1 - v^2)^{-\frac{1}{2}}$) are

$$E_{e\&m} = \frac{\gamma q^2}{2r_0}\left(1 + \frac{v^2}{3}\right)$$

---

\*Physics Department, University of Pennsylvania, Philadelphia, PA, USA. Supported in part by NSF.

[1]In addition to the question of covariance other issues must be considered and resolved satisfactorily (in any classical model of the electron) before a model can be regarded as self-consistent.

$$P^z_{e\&m} = \frac{2\gamma q^2 v}{3r_0} \tag{1}$$

which do not form components of a 4-vector.

To avoid this difficulty, various authors have altered the definition of the energy-momentum vector (Kwal, 1949; Rohrlich, 1960; Jackson, 1975). The root of this difficulty was not clear and one might wonder whether it is due to the infinite extent of the electromagnetic fields. However, this is not the case as we show in Appendix A.

## 2.   Charged Shell

Using the standard definitions of energy and linear momentum, we give a shell model for the classical electron which removes the difficulties encountered by earlier extended models. We base our analysis (Cohen, 1968; Cohen & Mustafa, 1986) on the conserved quantity

$$I = \int_{t\ const} \xi_\mu T^{\nu 0} d\sigma_0 \tag{2}$$

where $\xi_\mu$ is a Killing vector, $T^{\mu\nu}$ is the stress-energy and $d\sigma_0$ is the volume element in 3-space. This gives the energy

$$E = \int_{t\ const} T^{00} d\sigma_0$$

and linear momentum

$$P^s = \int_{t\ const} T^{30} d\sigma_0$$

which follow from the timelike and spacelike Killing vectors, $\partial_t$ and $\partial_s$.

The cohesive stress makes a contribution to the energy-momentum vector; it can be found from a free body diagram or from the weak-field limit of the corresponding general relativistic expression. The energy and linear momentum corresponding to the cohesive stress, with $q \gg m$ ($m$ defined below), are given by (Cohen & Mustafa, 1986):

$$E_{cohes} = -\frac{\gamma q^2 v^2}{6r_0}$$

$$P^z_{cohes} = -\frac{\gamma q^2 v}{6r_0} \tag{3}$$

which also do not transform as components of a 4-vector. However, the sum of Eqs. 1 and 3 yields

$$E_{e\&m} + E_{cohes} = \frac{\gamma q^2}{2r_0}$$

$$P^z_{e\&m} + P^z_{cohes} = -\frac{\gamma q^2 v}{2r_0} \tag{4}$$

which do form components of a 4-vector. If the shell contains matter of density $\kappa\delta(r - r_0)$, the energy and linear momentum of the matter are given by

$$E_{matter} = \gamma\kappa$$

$$P^z_{matter} = \gamma\kappa v \tag{5}$$

the sum of Eqs. 1, 3 and 5 gives

$$E = \gamma m$$

and

$$P^z = \gamma m v \tag{6}$$

which are components of a 4-vector. (In Eq. 6, the total mass is $m = \kappa + q^2/(2r_0)$). Note that the total mass $m$ is independent of the shell radius, and thus the point charge limit can be taken without infinities. [The difficulties with the classical electron also can be circumvented via quantum electrodynamics (Schwinger 1958)].

# A  Appendix

## A1.  The balloon model

We consider here a gas filled balloon of radius $r_0$, moving with velocity $v$ in the z-direction. The pressure P and gas density $\rho$ are taken to be uniform inside the balloon. If the mass distribution in the surface of the balloon is given by the density $m\delta(r - r_0)$, the expressions for the energy and linear momentum of the interior gas and surface matter are

$$E_{matter} = \gamma\left(m + m' + \frac{4\pi r_0^3}{3}pv^2\right)$$

$$P^z_{matter} = \gamma v\left(m + m' + \frac{4\pi r_0^3}{3}P\right) \tag{7}$$

where $m' = \frac{4\pi r_0^3}{3}\rho$ is the rest mass of the interior gas. (Note that $E_{matter}$ and $P^z_{matter}$ do not form components of a 4-vector.)

The outward acting pressure is balanced by a surface tension / cohesive stress whose contributions to the energy and linear momentum of the balloon are (Cohen & Mustafa, 1986):

$$E_{cohes} = \frac{2}{3}\gamma v^2 S$$

$$P^z_{cohes} = \frac{2}{3}\gamma v S \tag{8}$$

with $S = -2\pi r_0^3 P$ for an uncharged balloon. (Here also, $E_{cohes}$ and $P^z_{cohes}$ do not form components of a 4-vector.)

The expressions for the total energy and linear momentum of the system (given by the sum of Eqs. 7 and 8) are

$$E - \gamma(m + m') = \gamma M$$

$$P^z = \gamma(m + m')v = \gamma M v \tag{9}$$

where $m + m'$ represents the total rest mass $M$ of the system (the terms in P and S cancel). It is interesting to note that the total energy $E$ and linear momentum $P^z$ transform as components of a 4-vector, whereas neither $E_{cohes}$ and $P^z_{cohes}$ nor $E_{matter}$ and $P^z_{matter}$ separately form a 4-vector.

The balloon model is an example of a completely mechanical system for which the non-cohesive contributions to the energy and linear momentum do not constitute a 4-vector. The balloon model demonstrates that the problem is not unique to the classical electron, and therefore does not result from the infinite spatial extent of the classical electron's electromagnetic fields.

# B   Appendix

## B1.   The charged balloon

The classical electron and the balloon model are special cases of a more general system consisting of a charged balloon supported by an interior gas pressure and/or a cohesive surface stress. Adding the contributions from Eqs. 1, 7 and 8, we find that the total energy and linear momentum of this system are given by

$$E = \gamma \left( M + \frac{v^2}{3} \left( 4\pi r_0^3 P + 2S + \frac{q^2}{2r_0} \right) \right)$$

$$P^z = \gamma \left( M + \frac{1}{3} \left( 4\pi r_0^3 P + 2S + \frac{q^2}{2r_0} \right) \right) v \tag{10}$$

where

$$M = \kappa + \frac{4\pi r_0^3 \rho}{3} + \frac{q^2}{2r_0}.$$

If the body is in equilibrium we find

$$4\pi r_0^3 P + 2S + \frac{q^2}{2r_0} = 0. \tag{11}$$

We observe that Eq. 11 ensures that the total energy and linear momentum of the system (Eq. 10) transform as a 4-vector. The balloon model (Appendix A) is recovered by setting $q = 0$ in Eqs. 10 and 11. In addition, the classical electron model of a charged mass shell is recovered if the balloon's interior is empty, i.e. $P = 0$ and $\rho = 0$.

We also note from Eq. 11 that in the absence of a cohesive surface stress, i.e. $S = 0$, the balloon can be supported by a negative internal pressure

$$P = -\frac{q^2}{8\pi r_0^4} \tag{12}$$

as Poincaré attempted to show.

## Acknowledgement

For stimulating discussions, we thank Professors. R. H. Dicke, W. Drechsler, J. Ehlers, P. Havas and J. A. Wheeler.

## References

Abraham, M., Theorie der Electrizitāt, Vol. 2, Teubner, Leipzig, (1905).

Bopp, F., Ann. Phys., 42 573 (1942).

Cohen, J. M., J. Math. Phys. 9, 905 (1968).

Cohen, J. M. and Mustafa, E., Int. J. Theor. Phys. 25, 717 (1986).

Dirac, P. A. M., Proc. Roy. Soc., 167A, 148 (1938a).

Dirac, P. A. M., Ann. Inst. Poincaré, 9, 13 (1938b).

Erber, T., Fortschr. Phys., 9, 343 (1961).

Fermi, E., Physik. Zeitschr., 23, 340 (1922).

Feynmann, R. P., Phys. Rev., 74, 939 (1948).

Jackson, J. D., Classical Electrodynamics, Wiley & Sons, New York (1975), p. 793.

Kwal, B., J. Phys. Radium, 10, 103 (1949).

Leighton, R. B., Principles of Modern Physics, McGraw-Hill, New York (1959), p. 52.

Lorentz, H. A., The Theory of Electrons, Teubner, Leipzig (1909).

Matthison, M., Zeitschr. Phys., 67, 826 (1931).

Matthison, M., Proc. Camb. Phil. Soc., 36, 331 (1940).

Matthison, M., Proc. Camb. Phil. Soc., 38, 40 (1940).

McManus, H., Proc. Roy. Soc., 195A, 323 (1948).

Pais, A., Developments in the Theory of The Electron, Institute for Advanced Study, and Princeton University, Princeton, New Jersey (1948), pp. 5-7.

Poincaré, H. Rend. Circ. Mat. Palermo, 21, 129 (1906).

Rohrlich. F., Am. J. Phys., 28, 639 (1960).

Schwinger, J., Quantum Electrodynamics, Dover Publications, Inc., N. Y. (1958).

# Gauge (In)variance, Mass and Parity in D=3 Revisited

*S. Deser* *

## Abstract

We analyze the degree of equivalence between abelian topologically massive, gauge-invariant, vector or tensor parity doublets and their explicitly massive, non-gauge, counterparts. We establish equivalence of field equations by exploiting a generalized Stueckelberg invariance of the gauge systems. Although the respective excitation spectra and induced source-source interactions are essentially identical, there are also differences, most dramatic being those between the Einstein limits of the interactions in the tensor case: the doublets avoid the discontinuity (well-known from D=4) exhibited by Pauli–Fierz theory.

It is a pleasure to dedicate this work to Dieter Brill on the occasion of his 60th birthday. I have learned much from him over the years, not least during our old collaborations on general relativity. I hope he will be entertained by these considerations of related theories in another dimension.

## 1.  Introduction

Perhaps the most paradoxical feature of topologically massive (TM) theories [1, 2] is that their gauge invariance coexists with the finite mass and single helicity, parity violating, character of their excitations. This phenomenon, common to vector (TME) and tensor (TMG) models, is special to 2+1 dimensions because in higher (odd) dimensions the operative Chern–Simons (CS) terms are of at least cubic order in these fields and so do not affect their kinematics; only higher rank tensors could acquire a topological mass there. Equally surprisingly, TM doublets, with mass parameters of opposite sign, are not only invariant under combined parity and field interchanges[1] but they are equivalent in essential respects to non-gauge vector or tensor models

---

*The Martin Fisher School of Physics, Brandeis University, Waltham, MA 02254, USA. This work was supported by the National Science Foundation under grant #PHY88–04561.

[1]Doublets of pure (non-propagating) vector CS models have also been studied [3].

with ordinary mass terms: Both their excitation contents and the interactions they induce among their sources are essentially identical, as originally indicated in [2].

To be sure, other mass generation mechanisms exist in gauge theories: in D=2, the Schwinger model — spinor electrodynamics — yields a massive photon, but not at tree level. In D=4, in addition to the Higgs mechanisms, there is one involving a mixed CS-like term bilinear in antisymmetric tensor and vector fields [4]. However, none function, as this one does, purely with a single gauge field. It is also well-known that apparently different free-field theories can have the same excitation spectrum. Indeed, there even exist self-dual formulations of TM theories [5, 6, 7, 8] involving a single field, in which gauge invariance is hidden; however, their minimal couplings to a given source are then inequivalent.

My purpose here is to review and make more explicit how there can be coexistence between those equivalences and the differences in gauge character, as well as to point out that traces of these differences nevertheless remain. On the equivalence side, we will see that the TM field equations can be written in the non-gauge form because their gauge freedom can be hidden through a generalized Stueckelberg field redefinition. We will also discuss three ways in which differences manifest themselves. First, because gauge invariance implies the existence of Gauss constraints, long-range potentials carrying information about the sources survive in the massive doublets, but have no counterpart in the non-gauge models. Second, we will compare the massive excitation spectra with those of the limiting massless theories and explain why normally massive theories always differ from the latter, while TME (but not TMG) is one example in which the number of excitations (if not their helicities) agree with massless theory. Third, the Einstein limits of the tensor models differ dramatically: As is well-known [9] in D=4, massive tensors have components which do not decouple from the sources in this limit and lead to source-source interactions different from those of Einstein theory; TMG, on the other hand, will be shown to avoid this discontinuity. The present analysis is restricted to abelian models; their nonlinear generalizations are not obviously amenable to these considerations.[2] Extension to their supersymmetric generalizations [1, 10], on the other hand, is straightforward.

## 2. Vector Theories

Consider first a topologically massive vector doublet, which will be compared with massive vector (Proca) theory. The respective free Lagrangians are, in $(-++)$ signature,

$$L_{\text{VD}} = L_{\text{TME}}\left(A^1_\mu, m\right) + L_{\text{TME}}(A^2_\mu, -m) - \tfrac{1}{\sqrt{2}}\, j^\mu(A^1_\mu + A^2_\mu)\,,$$

$$L_{\text{TME}}\left(A_\mu, m\right) \equiv -\tfrac{1}{4}\, F^2_{\mu\nu}(A) + \tfrac{1}{4}\, m\, \epsilon^{\alpha\mu\nu} A_\alpha\, F_{\mu\nu}(A) \tag{1}$$

---

[2]Their nonlinear parts fail to superpose, and the respective sources cannot simultaneously be covariantly conserved with respect to each gauge field and to their sum.

and

$$L_P(A_\mu) = -\tfrac{1}{4} F_{\mu\nu}^2 - \tfrac{1}{2} m^2 A_\mu^2 - j^\mu A_\mu \,. \tag{2}$$

By gauge invariance, the current must be conserved in TME and the comparison can only be made if, as we assume henceforth, $\partial_\mu j^\mu = 0$. Let us review what is known [2]: Invariance of $L_D$ under combined parity and field interchange conjugations is manifest, since the parity transform is equivalent to a change of sign of $m$, which is just cancelled by the interchange $A_\mu^1 \leftrightarrow A_\mu^2$. Next, the equivalence of the effective current-current interactions generated by (1) and by (2) follows from the forms of the respective propagators, dropping all (irrelevant) terms proportional to $p_\mu p_\nu$: That of each TME is the sum of an ordinary (even) Proca term, $\eta_{\mu\nu}[p^2 + m^2]^{-1}$, and a mass- and parity-odd one $\sim (m\,\epsilon_{\mu\nu\alpha}p^\alpha)p^{-2}(p^2 + m^2)^{-1}$, which cancels out in the $\pm m$ doublet. Hence the interaction is of the usual finite-range form $\sim j^\mu(p^2 + m^2)^{-1} j_\mu$ in both theories.

It is instructive to establish how equivalence is displayed at the level of the field equations, and especially how the manifest TME gauge invariance can be "hidden" in the process. The field equations are

$$\partial_\mu F^{\mu\nu}(A_1) + m\,{}^*F^\nu(A_1) = \tfrac{1}{\sqrt{2}}\, j^\nu \tag{3a}$$

$$\partial_\mu F^{\mu\nu}(A_2) - m\,{}^*F^\nu(A_2) = \tfrac{1}{\sqrt{2}}\, j^\nu \,. \tag{3b}$$

Here ${}^*F^\nu \equiv \tfrac{1}{2}\,\epsilon^{\nu\alpha\beta}F_{\alpha\beta}$ is the dual field strength, with $F_{\mu\nu} = -\epsilon_{\mu\nu\alpha}\,{}^*F^\alpha$. In terms of the sums and differences, $A_\mu^\pm = \tfrac{1}{\sqrt{2}}(A_\mu^1 \pm A_\mu^2)$, we have

$$\partial_\mu F^{\mu\nu}(A_+) + m\,{}^*F^\nu(A_-) = j^\nu \tag{4a}$$

$$\partial_\mu F^{\mu\nu}(A_-) + m\,{}^*F^\nu(A_+) \equiv \epsilon^{\nu\mu\alpha}\partial_\mu[{}^*F_\alpha(A_-) + mA_\alpha^+] = 0 \,. \tag{4b}$$

The general solution of (4b) is

$${}^*F_\alpha(A^-) + m\,A_\alpha^+ = \partial_\alpha\Lambda + F_\alpha(V) \tag{5}$$

where $V_\mu$ is a solution of the homogeneous Maxwell equation, $\epsilon^{\mu\nu\alpha}\epsilon^{\alpha\lambda\sigma}\partial_{\nu\lambda}^2 V_\sigma = 0$. Inserting (5) into (4a) yields the Proca equation in (not quite) Stueckelberg form,

$$\partial_\mu F^{\mu\nu}(A_+) - m^2[A_+^\nu - m^{-1}\partial^\nu\Lambda - m^{-1}\,{}^*F^\nu(V)] = j^\nu \,. \tag{6}$$

The gauge-freedom carried by $\Lambda$ can be field-redefined away by $A_+^\nu \to A_+^\nu - m^{-1}\partial^\nu\Lambda$ as usual. But so can the additional $V$ term, by $A_+^\nu \to A_+^\nu - m^{-1}\,{}^*F^\nu(V)$ since $\partial_\mu F^{\mu\nu}({}^*F(V))$ is just $\Box\,{}^*F^\nu(V)$, which vanishes for a field strength ${}^*F^\nu(V)$ obeying Maxwell's equations. So the $A_+$ field's invariance is absorbed by an extended Stueckelberg transformation, leaving the Proca form. Once the $A_+$-field is determined (up to this gauge) by (6), we can use (4b) to determine $A_-$ (up to a gauge), with $m^* F^\nu(A_+)$ as the effective current determining $F^{\mu\nu}(A_-)$ through an ordinary massless Maxwell equation.

The equivalence of the two systems' degrees of freedom goes as follows. Each TME embodies a single massive particle with helicity $m/|m|$. Proca theory obviously has $(D{-}1) = 2$ massive degrees of freedom, and they too have helicities $\pm 1$, though this fact requires the same analysis of the full Lorentz algebra, (rather than merely that of the rotation generator $M^{ij}$), as was required for TME [2]; as we shall see, the naive $M^{ij}$ is purely orbital for both systems. For completeness, we identify the excitations in terms of the canonical formulation of both models, using first-order form of (1),(2) where $(A_\alpha,\ F^{\mu\nu})$ are initially independent variables. The free Lagrangian

$$L = -\tfrac{1}{2} F^{\mu\nu}(\partial_\mu A_\nu - \partial_\nu A_\mu) + \tfrac{1}{4} F^{\mu\nu} F^{\alpha\beta} \eta_{\mu\alpha}\eta_{\nu\beta} + \tfrac{\alpha}{2} m\, \epsilon^{\mu\alpha\nu} A_\mu \partial_\alpha A_\nu - \tfrac{\beta}{2} m^2 A_\mu^2 \quad (7)$$

represents TME or Proca theory, as $\alpha = 1,\ \beta = 0$ or $\alpha = 0,\ \beta = 1$ respectively.[3] The stress tensor associated to (7) is the usual (massive) vector one,

$$T_{\mu\nu} = F_\mu{}^\alpha F_{\nu\alpha} - \tfrac{1}{4} \eta_{\mu\nu} F_{\alpha\beta}^2 + \beta m^2 (A_\mu A_\nu - \tfrac{1}{2}\eta_{\mu\nu} A_\alpha^2)\ ; \quad (8)$$

the CS term (being metric-independent) does not contribute. Let us briefly recapitulate the canonical reduction of (7) in terms of the space-time decomposition of its variables. Our conventions and notations are $\epsilon^{012} = \epsilon^{12} = +1 = -\epsilon_{012}$,

$$F^{0i} = F_{i0} = E^i\ , \qquad B \equiv \tfrac{1}{2}\epsilon^{ij} F_{ij}\ , \quad (9)$$

while the transverse-longitudinal decomposition of a spatial vector is expressed by

$$V^i = V_T^i + V_L^i = \epsilon^{ij}\xi^{-1}\partial_j v_T + \xi^{-1}\partial_i v_L\ , \quad \xi = (-\nabla^2)^{1/2}\ . \quad (10)$$

In both theories, one may eliminate the trivial constraint

$$F_{ij} = \partial_i A_j - \partial_j A_i = \epsilon^{ij} B = \epsilon^{ij}\xi a_T\ . \quad (11)$$

However, in TME, only the gauge-independent pair $(e_T,\ a_T)$ remains, whereas the longitudinal components also survive in Proca theory. The difference lies in the fact that $A_0$ is respectively a Lagrange multiplier or an auxiliary field in the two systems; that is, its variation yields the constraint

$$-\nabla\cdot E + \tfrac{\alpha}{2} mB + \beta m^2 A_0 = j^0\ . \quad (12)$$

Dropping the source for the moment, we then find for the free actions, which began as

$$I = -\int d^3x [E \cdot (\dot{A} - \nabla A_0) + \tfrac{1}{2} E^2 + \tfrac{1}{2} B^2(A) + \alpha m\{\epsilon^{ij} A_i \dot{A}_j - \tfrac{1}{2} A_0 B(A)\}$$

$$+ \tfrac{\beta}{2} m^2 (A^2 - A_0^2)]\ , \quad (13)$$

---

[3] Keeping both terms results in a parity-violating theory with two different helicities, as does taking the CS term to be of the form $\epsilon^{\alpha\mu\nu} A_\alpha F^{\lambda\sigma}\eta_{\mu\lambda}\eta_{\nu\sigma}$, which is no longer gauge-invariant, nor metric-independent.

the final forms

$$I_{\text{TME}} = \int d^3x [(-e_T)\dot{a}_T - \tfrac{1}{2}e_T^2 - \tfrac{1}{2}a_T(m^2 - \nabla^2)a_T] , \tag{14}$$

$$I_P = I_{\text{TME}} + \int d^3x [\bar{a}_L\dot{\bar{e}}_L - \tfrac{1}{2}\bar{a}_L^2 - \tfrac{1}{2}\bar{e}_L(m^2 - \nabla^2)\bar{e}_L] . \tag{15}$$

The doublet action is just the sum of two independent terms (14). To reach (14), we eliminated the longitudinal electric field using the Gauss constraint (12). To obtain (15), we used (12) for $\alpha = 0$, $\beta = 1$, to eliminate $A_0$, and rescaled $\bar{a}_L \equiv m^{-1}a_L$, $\bar{e}_L \equiv me_L$. The free field content then, is that both theories have two massive degrees of freedom. To determine their helicities requires, in both cases, study of the full Lorentz algebra, for the rotation generators are ostensibly purely orbital because of the peculiarities of 2-space. That is, from its definition,

$$M = \int d^2r\, \epsilon_{ij}\, x^i\, T_0^j , \tag{16}$$

it follows from (8) and the canonical reductions that

$$M = \int d^2r \sum_{i=1}^{2} (-e_T^i)\partial_\theta a_T^i \tag{17a}$$

for the TME doublet and

$$M = \int d^2r [(-e_T)\partial_\theta a_T + \bar{a}_L\partial_\theta\bar{e}_L] \tag{17b}$$

for the massive vector, as is to be expected since the canonical variables are spatial scalars.

We now reinstate the $j^\mu A_\mu$ term to recover the effective current-current coupling in the present context. The canonical TME doublet action (14) contains a term $\sim \tfrac{1}{\sqrt{2}}j_T(a_T^1 + a_T^2)$ together with the longitudinal interactions, which for our conserved currents would look like

$$\rho\frac{\Box}{(m^2 - \Box)}\nabla^{-2}\rho \sim \rho(-\nabla^{-2})\rho + \rho\frac{(m^2/\nabla^2)}{m^2 - \Box}\rho .$$

This is the usual sum of a Coulomb term and a (retarded) Yukawa interaction, whereas the $j_T a_T$ coupling above results in the purely retarded form $j_T(m^2 - \Box)^{-1}j_T$. The Proca action would have, in addition to the $j^T a_T$ and $j^L a_L$ terms, the remnant of $\rho A_0$, which together of course yield the same $\rho - \rho$ interactions as above for TME doublets. Hence the total current-current coupling is of course the same $j^\mu(m^2 - \Box)^{-1}j_\mu$ in both cases, and limits smoothly to the Maxwell one as $m \to 0$.

## 3.   Tensor Theories

The gravitational case is similar to the vector in most respects. A doublet of opposite mass TMG has the same excitation content as a single massive, Pauli–Fierz, tensor theory and leads to essentially equivalent induced source-source interactions [2].

The doublet's action is [2]

$$I_{TD} = I_{\text{TMG}}(h^1_{\mu\nu}, \mu) + I_{\text{TMG}}(h^2_{\mu\nu}, -\mu) + \tfrac{1}{\sqrt{2}} \kappa T^{\mu\nu}(h^1_{\mu\nu} + h^2_{\mu\nu})$$

$$I_{\text{TMG}}(h_{\mu\nu}, \mu) = -\kappa^{-2} \int d^3x R^Q(h) + 1/2\mu \int d^3x \epsilon^{\mu\alpha\beta} G^\nu_\alpha \partial_\mu(\kappa h_{\beta\nu}) , \qquad (18)$$

where $Q$ indicates that only terms quadratic in $\kappa h_{\mu\nu} \equiv (g_{\mu\nu} - \eta_{\mu\nu})$ are to be kept in the expansion of the Einstein actions and $G^\nu_\alpha$ is linearized. Again, parity conservation is restored by including the interchange $h^1_{\mu\nu} \leftrightarrow h^2_{\mu\nu}$. The sign of the Einstein term is "ghost-like," i.e., opposite of the conventional one, in order that the TMG excitations be non-ghost. The Einstein constant $\kappa^2$ has dimensions of length, while $\mu$ is dimensionless since the CS term is of third derivative order; the mass $m$ is $\mu\kappa^{-2}$, but the "massless" Einstein limit is clearly $\mu \to \infty$. The Pauli–Fierz action is as in D=4,

$$I_{PF} = + \int d^3x R^Q(k) - \frac{m^2}{2} \int d^3x (k^2_{\mu\nu} - k^2) + \kappa \int d^3x k_{\mu\nu}T^{\mu\nu} , \quad k \equiv k^\alpha_\alpha . \qquad (19)$$

The couplings in both cases are the usual minimal ones, and $T^{\mu\nu}$ is necessarily conserved for consistency with the (linearized) gauge invariance of TMG; henceforth we assume[4] $\partial_\mu T^{\mu\nu} = 0$. The propagator of a single TMG field (neglecting terms proportional to $p^\mu p^\nu$, which vanish for conserved sources) differs from that of the massive one in two respects: it contains, in addition, a term with numerator $\sim \mu(\epsilon^{\mu\alpha\beta}p^\gamma+$ symm) and also a term independent of $\mu$, corresponding to a free (ghost) Einstein field propagator. In the doublet then, the odd terms cancel, leaving two interaction terms, the usual Pauli–Fierz finite range one,

$$\kappa^2 \int d^3x [T^{\mu\nu}(-\Box + m^2)^{-1}T_{\mu\nu} - \tfrac{1}{2} T^\mu_\mu(-\Box + m^2)^{-1}T^\nu_\nu] \qquad (20a)$$

and the extra ghost Einstein part,

$$-\kappa^2 \int d^3x [T^{\mu\nu}(-\Box)^{-1}T_{\mu\nu} - T^\mu_\mu(-\Box)^{-1}T^\nu_\nu] . \qquad (20b)$$

[The coefficients of the $T^\mu_\mu T^\nu_\nu$ terms are $(D-1)^{-1}$ and $(D-2)^{-1}$ respectively for general D.] The Einstein "interaction" is, however an artifact, consisting entirely of contact

---

[4]Of course, just as in normal general relativity, a dynamical $T_{\mu\nu}$ will no longer be conserved as a result of its coupling and the full nonlinear theory will be required.

terms when conservation of $T^{\mu\nu}$ is taken into account [11]. The finite-range interaction (20a) is then effectively the entire residue in both systems, and equivalence of interactions is established modulo the (trivial) Einstein part (20b) in the doublet.

Now let us analyze the field equations of the TMG doublet, as we did for TME. They are

$$-G_{\mu\nu}(h_1) + m^{-1}C_{\mu\nu}(h_1) = \tfrac{1}{\sqrt{2}}\,\kappa\,T_{\mu\nu} \tag{21a}$$

$$-G_{\mu\nu}(h_2) - m^{-1}C_{\mu\nu}(h_2) = \tfrac{1}{\sqrt{2}}\,\kappa\,T_{\mu\nu}\,, \tag{21b}$$

where $G_{\mu\nu}$ is the linearized Einstein tensor and $C_{\mu\nu}$ is the third-derivative order linearized Cotton–Weyl tensor $C^{\mu\nu} \equiv \epsilon^{\mu\alpha\beta}\partial_\alpha(R^\nu_\beta - \tfrac{1}{4}\delta^\nu_\beta R)$; it is symmetric, traceless and (identically) conserved. We again take the sum and difference of (21a,b) in terms of $h^\pm_{\mu\nu} = \tfrac{1}{\sqrt{2}}(h^1_{\mu\nu} \pm h^2_{\mu\nu})$,

$$-G_{\mu\nu}(h^+) + m^{-1}C_{\mu\nu}(h^-) = \kappa\,T_{\mu\nu} \tag{22a}$$

$$-G_{\mu\nu}(h^-) + m^{-1}C_{\mu\nu}(h^+) = 0\,. \tag{22b}$$

One may solve (22b) for either $h^-_{\mu\nu}$ or $h^+_{\mu\nu}$ and insert into (22a) to get an equation in terms of the other; this will yield third or fourth derivative inhomogeneous equations. I will sketch the results, but to save space will omit all the "Stueckelberg" gauge parts as well as required symmetrizations that arise in the process; I will also drop scalar curvatures since it is the Ricci tensor that counts (actually the trace of (22b) shows that $R(h^-)$ vanishes). All "equalities" below that are subject to these caveats will carry $\approx$ signs. Using the fact that $G^{\mu\nu}(h) = -\tfrac{1}{2}\,\epsilon^{\mu\alpha\beta}\epsilon^{\nu\lambda\sigma}\partial^2_{\alpha\lambda}h_{\beta\sigma}$, we learn from (22b) that, in the obvious notation $R^+_{\nu\beta} \equiv R_{\nu\beta}(h^+)$,

$$\epsilon^{\mu\alpha\beta}\partial_\alpha(m^{-1}R^+_{\nu\beta} + \tfrac{1}{2}\,\epsilon^{\nu\lambda\sigma}\partial_\lambda h^-_{\beta\sigma}) \approx 0\,. \tag{23}$$

Apart from homogeneous (gauge) terms, then, we may eliminate $R^+_{\nu\beta}$ in terms of first derivatives of $h^-_{\beta\sigma}$ in (22a) to find the third order equation

$$\epsilon^{\mu\lambda\sigma}\partial_\lambda(2R^-_{\sigma\nu} + m^2 h^-_{\sigma\nu}) \approx 2m\kappa\,T_{\mu\nu}\,. \tag{24a}$$

The left side is the gauge-invariant curl of a Klein–Gordon form; in harmonic gauge where $2R^-_{\sigma\nu} = -\Box h^-_{\sigma\nu}$, we have

$$\epsilon^{\mu\lambda\sigma}\partial_\lambda(-\Box + m^2)h^-_{\sigma\nu} \approx 2m\kappa\,T_{\mu\nu}\,. \tag{24b}$$

Alternatively, we may take the curl of (22b) to learn that $mC^-_{\mu\nu} = \Box R^+_{\mu\nu}$, and hence we may write (22a) as

$$(m^2 - \Box)R^+_{\mu\nu} \approx -\kappa m^2\,T_{\mu\nu}\,. \tag{25a}$$

This is a (fourth order) Klein–Gordon equation which states that, in harmonic gauge,

$$\Box(m^2 - \Box)h^+_{\mu\nu} \approx 2\kappa m^2\,T_{\mu\nu}\,. \tag{25b}$$

[Of course (24b) and (25b) just reflect the propagator structure of TMG and its doublets that was given earlier.]

Let us now compare the above TMG equations with the Pauli–Fierz ones,

$$G_{\mu\nu}(k) + \tfrac{m^2}{2}(k_{\mu\nu} - \eta_{\mu\nu}\, k) = \kappa\, T_{\mu\nu} \,. \tag{26}$$

The analysis goes as in D=4, namely one first notes that (for conserved $T_{\mu\nu}$) the divergence of (26) implies $\partial^\mu(\eta_{\mu\nu}\, k - k_{\mu\nu}) = 0$; the double divergence is just the scalar curvature $R(k)$, which then also vanishes. Using these constraints, we may write $G_{\mu\nu}(k)$ as follows:

$$\begin{aligned} G_{\mu\nu}(k) \; = \; R_{\mu\nu}(k) \; &= \; -\tfrac{1}{2}(\Box k_{\mu\nu} - \partial^2_{\alpha\mu}k^\alpha_\nu - \partial^2_{\alpha\nu}k^\alpha_\mu + \partial^2_{\mu\nu}k) \\ &= \; -\tfrac{1}{2}(\Box k_{\mu\nu} - \partial^2_{\mu\nu}k) \,. \end{aligned} \tag{27}$$

Now perform the "Stueckelberg" field redefinition, $k_{\mu\nu} \to k_{\mu\nu} - m^{-2}\partial^2_{\mu\nu}k$ in (26), under which $G_{\mu\nu}$ is of course invariant. This change in form of the mass term just cancels the $\partial^2_{\mu\nu}k$ in (27) and yields the standard Klein–Gordon equation

$$(m^2 - \Box)(k_{\mu\nu} - \eta_{\mu\nu}k) = 2\kappa T_{\mu\nu} \,. \tag{28}$$

Comparing with the corresponding (schematic!) equations for the TMG doublet, (24b) or (25b), we see there the same basic Klein–Gordon propagation, modified by the higher order character of TMG. Had we kept the $R^+$ terms, the gauge parts, performed Stueckelberg shifts, etc., we would have again deduced from these field equations the equivalence of source-source interactions modulo the extra Einstein "coupling" in TMG.

We will not reproduce the canonical analysis of TMG, which may be found in [2], nor that of the massive theory, well-known from D=4. The pattern is clear from the vector story: in both cases there are two massive excitations, whose helicity $\pm 2$ character cannot be determined from the rotation generators alone.

## 4.   Differences

The correspondence between the TM gauge doublets and their normally massive counterparts is not complete. We discuss three types of differences here. The first deals with existence of long-range Coulomb-like (but locally pure gauge) potentials in the gauge doublets — despite the finite range character of the excitations and interactions — this is the most manifest aspect of their gauge invariance, with no massive counterpart. Secondly, we analyse the zero mass limits of the free excitation spectra, and thirdly, that of the source-source interactions. The latter aspect displays particularly striking differences in the tensor case: TMG leads to a smooth (and hence trivial) Einstein limit, whereas the Pauli–Fierz discontinuity of D=4 [9] not only persists in D=3 but now gives a non-trivial, and therefore especially "different," interaction from that of Einstein gravity.

The Proca and Pauli–Fierz equations consist entirely of a set of Klein–Gordon equations of the form $(m^2 - \Box)\phi_i = \rho_i$, with no differential constraints and hence no long-range components. The TM theories, however, do contain Gauss constraints, but with additional terms. In the vector case, (12) shows (for $\beta = 0$) that it is now $mB$, rather than the short-range $\nabla \cdot E$ which carries the asymptotic information [2], because $mB$ is a curl whose spatial integral is proportional to the total charge $Q = \int d^2r\, j^0$. We therefore expect that in our doublet, it is the difference, $B(A_-)$, which is relevant. This is indeed the case: while the $A_\mu^+$-field is short-range, obeying as it does the Proca equation (6), it follows from (5) that $A_\mu^-$ is determined then by $A^+$. Specifically, $*F^0(A_-) = B(A_-)$ is proportional to the sum of $\nabla \cdot E(A_+)$ and the charge density $j^0$. Consequently, since $E(A_+)$ is short-range, it is $A_-$ at spatial infinity that is proportional to $Q = \int d^2r\, j^0$. This can also be read off from the time component of (5): the right hand side's spatial integral vanishes, while the integral of $A_0^+$ is proportional to $Q$, by the Proca equation.

A very similar picture emerges in the tensor case: the lower derivative part, $G_0^0$, of the TMG field equations (21) carries the asymptotic information about the energy of the source [2], just as it does in Einstein gravity. But this time it is the term even in $m$, and hence $G_0^0(h^+)$ that is relevant. Now $G_0^0(h^+)$ is (like $B(A)$) a total spatial divergence, and the source energy is the spatial integral of $T_0^0$, so we need only integrate the (00) components of (22a) over space, noting that the $C^-$ term falls off too rapidly to contribute at infinity.

Next we compare the massive and massless excitation spectra, using the well-known fact that for each gauge symmetry broken by a lower derivative term, one new degree of freedom arises (only one, because the former Lagrange multiplier, unlike the gauge variable, is merely promoted to be an auxiliary field). Thus massive vector theory acquires one degree of freedom beyond the (D−2) of Maxwell theory because $m^2 A_\mu^2$ breaks its gauge invariance. The massive tensor model has (D−1) additional excitations beyond the $\frac{1}{2}$ D(D−3) of Einstein theory because $m^2(h_{\mu\nu}^2 - h^2)$ breaks all but the longitudinal one of the D gauges of linearized gravity; so there is always a discontinuity (except at D=2, where the Einstein kinetic term is absent altogether). On the other hand, in (singlet) TME, the CS term shares the Maxwell invariance and there is no discontinuity in the number, but only in the helicity, of the single excitation (massless particles always have vanishing helicity [12]). [This method of counting also applies in vector theory when a "mass" term is added to a pure CS term [5], breaking the latter's gauge invariance to give rise to a single massive degree of freedom; it is in fact equivalent to TME [6], as is also the case if the mass term arises through a Higgs mechanism [7].] In TMG, while the Einstein term also preserves gauge invariance of the CS term, it breaks the latter's conformal invariance, thereby generating the massive graviton discontinuously from its two separately nondynamical parts.

More dramatic are the differences between source-source interactions in our models and in their $m = 0$ or $m = \infty$ counterparts. Whereas the degree of freedom count is a more formal difference – for example, it has long been known from D=4 that the extra degree of freedom in Proca theory decouples from the current as $m \to 0$, so that the limit is (all but gravitationally) indistinguishable from electrodynamics, there is a discontinuity in the induced source couplings already in D=4 between Einstein and Pauli–Fierz models [5], precisely because not all the extra modes decouple from the stress tensor in the massless limit. Here TMG provides a clear difference from the massive model: As we have seen by considering the propagator, the infinite mass limit[5] of the TMG doublet (where its free action reduces to the Einstein part) leads to just the same $T_{\mu\nu} - T_{\mu\nu}$ coupling (or rather lack of coupling!) as in Einstein theory. On the other hand, the massless ("Einstein") limit of Pauli–Fierz coupling differs from the Einstein form by the famous $(D-1)^{-1}$ versus $(D-2)^{-1}$ coefficient of the $-T_\mu^\mu \Box^{-1} T_\nu^\nu$ term of (20). This is especially dramatic in D=3 where it means in particular the difference between, respectively, existence or absence of a Newtonian limit!

## 5.  Summary

We have reviewed the resemblances and differences between the two ways of giving mass to gauge fields in D=3, through gauge-invariant Chern–Simons terms or through explicit mass terms. The equivalences between these ways, despite their different gauge properties, were followed also at the level of field equations where the "fading" of gauge invariance into ordinary Klein–Gordon-like expressions was understood as a manifestation of Stueckelberg mechanisms. Most dramatic among the differences was the fact that TMG provides the only example of a smooth limit for the interactions from massive to Einstein form, in contrast to the discontinuity in the explicitly massive tensor theory.

## References

[1] Siegel W., *Nucl. Phys.* **B156** (1979) 135; Jackiw R. and Templeton S., *Phys. Rev.* **D23** (1981) 2291; Schonfeld J., *Nucl. Phys.* **B185** (1981) 157.

[2] Deser S., Jackiw R., and Templeton S., *Ann. Phys.* **140** (1982) 372.

[3] Hagen C.R., *Phys. Rev. Lett.* **68** (1992) 3821.

[4] Allen T.J., Bowick M.J., and Lahiri A., *Mod. Phys. Lett.* **A6** (1991) 559.

---

[5] For completeness, we note that the $m \to 0$ limit of TMG is to be contrasted with pure CS, rather than Einstein, gravity. Here our doublet is now "discontinuous," since it leads to a net pure trace, $\sim T_\mu^\mu \Box^{-1} T_\nu^\nu$ coupling, whereas pure CS gravity of course cannot even couple to sources whose traces fail to vanish; continuity is of course restored if the source is traceless. The Pauli–Fierz theory (like Proca theory) of course leaves no interactions at all in the infinite mass limit.

[5] Townsend P., Pilch K., and van Nieuwenhuizen P., *Phys. Lett.* **136B** (1984) 38.

[6] Deser S. and Jackiw R., *Phys. Lett.* **139B** (1984) 371.

[7] Deser S. and Yang Z., *Mod. Phys. Lett.* **A4** (1989) 2123.

[8] Deser S. and McCarthy J., *Phys. Lett.* **245B** (1990) 441; Aragone C. and Khoudeir A., *Phys. Lett.* **B173** (1986) 141.

[9] Van Dam H. and Veltman M., *Nucl. Phys.* **B22** (1970) 397; Boulware D. and Deser S., *Phys. Rev.* **D6** (1972) 3368.

[10] Deser S. and Kay J.H., *Phys. Lett.* **120B** (1983) 97; Deser S. in *Quantum Theory of Gravity*, Christensen S., ed. *(Adam Hilger, Bristol 1984)*.

[11] Deser S., Jackiw R., and 't Hooft G., *Ann. Phys.* **152** (1984) 220.

[12] Binegar B., *J. Math. Phys.* **23** (1982) 1511; Deser S. and Jackiw R., *Phys. Lett.* **B263** (1991) 431.

# Triality, exceptional Lie groups and Dirac operators

Frank Flaherty
Department of Mathematics
Oregon State University
Corvallis, OR, 97331-4605, USA

## Abstract

In dimension eight there are three basic representations for the spin group. These representations lead to a concept of triality and hence lead to the construction of two exceptional commutative algebras: The Chevalley algebra $\mathcal{A}$ of dimension twenty-four and the Albert algebra $\mathcal{J}$ of dimension twenty-seven. All of the exceptional Lie groups can be described using triality, the octonians, and these two algebras.

On a complex four-manifold, with triality a parallel field, the Dirac and associated twistor operators can be constructed on either bundle of exceptional algebras. The geometry of triality leads to refinement of duality common to four dimensions.

In nine dimensions there is a weaker notion of triality which is related to several additional multiplicative structures on $\mathcal{J}$.

## Introduction

Cartan's classification of simple Lie groups yields all Lie groups with some exceptions. In his thesis Cartan described these exceptional groups but, save the one of smallest rank $F_2$, he was unable to describe the geometry of the groups. In 1950 Chevalley and Schafer successfully identified $F_4$ and $E_6$ as structure groups for the exceptional Jordan algebra and the Freudenthal cross product on $\mathcal{J}$, respectively. Later Freudenthal successfully identified the geometry associated with the remaining exceptional groups, $E_7$ and $E_8$. All of the exceptional groups have low rank and are, in one way or another, related to the octonians (Cayley numbers), the principle of triality, and the exceptional algebras.

With the advent of superstring theory and the work of Witten and others, it is necessary to understand the exceptional Lie groups and the corresponding differential operators. To this end, the Dirac operator is described explicitly while the twistor operator is merely indicated without detail.

## Dirac operators

To define the Dirac operator on a manifold one needs the global notion of a spin manifold. Without going into details this is a $(C^\infty)$ orientable Riemannian manifold $M$ for which the second Stiefel-Whitney class vanishes.

The bundle of Clifford Algebras over $M$ is then constructed using $T_p(M)$ and the Riemannian quadratic form $g_p(x, x)$. The resulting vector bundle of Clifford algebras will be denoted by $Cl(M)$. The resulting spinor bundle will be denoted by $S$, with $S_o$, $S_1$ the even and odd spinors (if dim $M = 2m$).

Denoting the Riemannian connection on $M$ by $D$, it can be shown that on a spin manifold $D$ extends as a derivation to the bundle $S$, regarded as $Cl(M)$ module. In other words,

$$D(xu) = (Dx)u + xDu$$

$$u \in \Gamma(Cl(M)), \quad x \in \Gamma(S)$$

The notation $\Gamma$ means take the sections of the bundle that follows.

There is then a first order differential operator $\mathcal{D} : \Gamma(S) \to \Gamma(S)$ which locally is defined by $\mathcal{D}s = \sum e_i D_{e_i} s$ for $(e_i)$ a local orthonormal frame field, $x \in \Gamma(U, S)$. The sections are local sections of $S$ over the chart $U$. Note that the symbol of $\mathcal{D} = \sum e_i . \zeta_i$ at the cotangent vector, $\zeta = (\zeta_i)$ is $\zeta .$ , left Clifford multiplication by $\zeta$, and hence, it follows that

$$symb(\mathcal{D}^2) = -\zeta . \zeta =\parallel \zeta \parallel$$

The self-adjoint first order operator $\mathcal{D}$ is called the Dirac operator.

**Theorem.** *Let $\mathcal{D}$ be the Dirac operator on a compact spin manifold $M$ then*

$$ker\mathcal{D} = ker\mathcal{D}^2$$

*and $ker\mathcal{D}$ is finite dimensional.*

*Proof.* Finite dimensionality follows from ellipticity.

$$\mathcal{D}^2 u = 0 \Rightarrow \parallel \mathcal{D}u \parallel = 0 \Rightarrow \mathcal{D}u = 0.$$

In the next section, the Dirac operator is extended to the bundle of algebras $\mathcal{J}$.

## The exceptional group $F_4$

The exceptional Lie group $F_4$ is defined to be the group of automorphisms of the exceptional Jordan algebra $\mathcal{J}$. The Chevalley algebra, $\mathcal{A}$ sits as a subalgebra of $\mathcal{J}$. There is an exact sequence displaying ordinary 3-space as a coextension of $\mathcal{J}$ via $\mathcal{A}$. The Lie algebra structure of $F_4$ can then be displayed as a vector space sum of $spin(8)$, and the three basic

epresentations in dimension eight. The bracket is defined in terms of the triality operator
n $\mathcal{A}$.

More precisely, the Lie algebra of $F_4$ can be written in the form:

$$f_4 = \mathfrak{g} \oplus V \oplus S_o \oplus S_1$$

a which $\mathfrak{g} = \mathrm{spin}(8)$. Using triality, the bracket operation may be defined as:

$$[V, V] = [S_o, S_o] = [S_1, S_1] = \mathfrak{g}$$

hile $[\mathfrak{g}, V] = V$ and cyclically. Finally, $[V, S_o] = S_1$, and cyclically. The Lie algebra of $E_8$
an be given in a similar manner. I am indebted to Bert Kostant for this description of $F_4$
nd $E_8$. The group $F_4$ serves as the group of the bundle of exceptional Albert algebras.

### Dirac operators on the exceptional algebras

The Dirac operator can be introduced on the bundle of Albert (Chevalley) algebras
n a complex four-fold using the corresponding multiplications in place of the Clifford
nultiplication. For the Chevalley algebra, this construction is due to Troy Warwick. In
ddition, the twistor operator can be defined. The question of integrabilty of complex
tuctures on twistor space can then phrased in terms of vanishing of curvature. Finally,
Bochner and Lichnerowicz type identities are derived, giving rise to a vanishiing theorem
or harmonic sections of the bundle of Albert algebras. A sketch of these ideas follows.

Consider a Kaehler manifold $M$ of complex dimension four on which the triality map is
, parallel filed. Over $M$ there is the bundle of exceptional algebras $\mathcal{J}$. A section of $\mathcal{J}$ is
f the form

$$X = \begin{pmatrix} \xi_1 & x_3 & \overline{x_2} \\ \overline{x_3} & \xi_2 & x_1 \\ x_2 & \overline{x_1} & \xi_3 \end{pmatrix}$$

n which the the $\xi$s are real and the $x$s are octonians. The basic operation on these matrices
s the Jordan multiplication defined by:

$$X \circ Y = \frac{1}{2}(XY + YX)$$

n which the matrix multiplication on the right hand side uses octonian multiplication of
he entries. It can be verified that $\mathcal{J}$ is a Jordan algebra, that is, in addition to being
ommutative, the multiplication rule $\circ$ satisfies the identity

$$X \circ (Y \circ X^2) = X^2 \circ (Y \circ X)$$

or all $X, Y \in \mathcal{J}$. Note that in physicists notation the exceptional Jordan algebra $\mathcal{J}$ is
usually denoted by $M_3^8$ after Jordan, von Neumann, and Wigner.

The Dirac operator can then defined by

$$\mathcal{D} = \sum e_i \circ D_i.$$

The $e$s are embedded in the right corner of the matrix in the off-diagonal position.

**Theorem.** *If $X$ is a harmonic section of $\mathcal{J}$ for which $\xi_1$ vanishes and if the Cayley curvature of $\mathcal{J}$ is non-negative then $X$ is parallel. Further, if the Cayley curvature of $\mathcal{J}$ is positive somewhere on $M$ then the any harmonic section of $\mathcal{J}$ in which $\xi = 0$ must vanish.*

*Proof.* A Bochner-Lichnerowicz formula for $\mathcal{D}$ yields

$$\langle \mathcal{D}^2(X), X \rangle = \text{Laplacian terms in X} \oplus \text{Cayley curvature terms in X}$$

The argument then follows in a trivial manner.

Note that the curvature terms must be made more precise. It is not clear what role any vanishing theorem plays in physics although they contribute to Penrose's twistor theory in a basic way. The notion of a Killing section of $\mathcal{J}$ is obvious and these sections generate the solutions to the twistor equation. Although the twistor operator equation has not been explicitly defined here, it follows from representation theory that there is a projection as in dimension four and the definition is clear.

## Triality

The geometry of triality is somewhat analogous to the more familiar concept of duality. Denote the triality map by $\tau$. The order of $\tau$ is then three and $\tau$ permutes the component spaces of $\mathcal{A}$ cyclically. Moreover, $\mathcal{A}$ may be decomposed according to the eigenvalues of $\tau$ thus giving rise to notions of invariance and conjugation. Finally, if one projectivizes the components of $\mathcal{A}$, which are the basic representations of the spin group, then these projective 7-spaces each contain quadrics in which the notions of point, $\alpha$ plane, and $\beta$ plane are trial to each other. Trial means that these ideas can be permuted. The generators of the permutations are $\tau$ and $\sigma$, in which $\sigma$ is defined by the decomposition map of $\mathcal{A}$ by the third roots of unity.

As an example of the direction of this research the twistor operator $\mathcal{T}$ is defined using the orthogonal decopmosition of the tensor product of the even and odd spinors. In fact, the Dirac operator needed for this construction i s

$$\mathcal{D} = \sum \tau(e_i) \circ \tau^2 D_i.$$

Finally, there is an intriguing possibility that the 26 dimensions of string theory can be treated by considering a 24 dimensional triality manifold, similar to complex manifolds or twistor spaces. In summary, the strange dimensions of exceptional algbras may be intimately related to physical theory.

# THE REDUCTION OF THE STATE VECTOR AND LIMITATIONS ON MEASUREMENT IN THE QUANTUM MECHANICS OF CLOSED SYSTEMS

## J.B. Hartle

Department of Physics, University of California Santa Barbara, CA 93106-9530

*"... persuaded of these principles, what havoc must we make?"* – Hume

## ABSTRACT

Measurement is a fundamental notion in the usual approximate quantum mechanics of measured subsystems. Probabilities are predicted for the outcomes of measurements. State vectors evolve unitarily in between measurements and by reduction of the state vector at measurements. Probabilities are computed by summing the squares of amplitudes over alternatives which could have been measured but weren't. Measurements are limited by uncertainty principles and by other restrictions arising from the principles of quantum mechanics. This essay examines the extent to which those features of the quantum mechanics of measured subsystems that are explicitly tied to measurement situations are incorporated or modified in the more general quantum mechanics of closed systems in which measurement is not a fundamental notion. There, probabilities are predicted for decohering sets of alternative time histories of the closed system, whether or not they represent a measurement situation. Reduction of the state vector is a necessary part of the description of such histories. Uncertainty principles limit the possible alternatives at one time from which histories may be constructed. Models of measurement situations are exhibited within the quantum mechanics of the closed system containing both measured subsystem and measuring apparatus. Limitations are derived on the existence of records for the outcomes of measurements when the initial density matrix of the closed system is highly impure.

## 0 PREFACE

In 1959, then an undergraduate at Princeton in search of a senior thesis topic, I was introduced by John Wheeler to his young colleague, Dieter Brill. This was fortunate from my point of view, for Dieter proved to have the patience, time and talent not only to introduce me to the beauties of Einstein's general relativity but also give me instruction and guidance in the practice of research. Our subject – the method of the self-consistent field in general relativity and its application to the gravitational geon – was also fortunate. Through it we helped lay the foundations for the short wavelength approximation for gravitational radiation (Brill and Hartle, 1964). In particular, building on ideas of Wheeler (1964), we introduced what Richard Isaacson (Isaacson 1968ab) was later kind enough to call the "Brill-Hartle" average for the effective stress-energy tensor of short wavelength radiation, and which was to to prove such a powerful tool when made precise in his hands in his general theory of this approximation. It would be difficult to imagine a more marvelous introduction to research. I have not written a paper with Dieter Brill since, but each day I use the lessons learned from him so long ago. It is a pleasure to thank him with this small essay on the occasion of his 60th birthday.

## 1 INTRODUCTION

"Measurement" is central to the usual formulations of quantum mechanics. Probabilities are predicted for the outcomes of measurements carried out on some subsystems of the universe by others. In a Hamiltonian formulation of quantum mechanics, states of a subsystem evolve unitarily in between measurements and by reduction of the state vector at them. In a sum-over-histories formulation, amplitudes are squared and summed over alternatives "which could have been measured but weren't" to calculate the probabilities of incomplete measurements. In these and other ways the notion of measurement plays a fundamental role in the usual formulations of quantum theory.

The quantum mechanics of a subsystem alone, of course, does not offer a quantum mechanical description of the workings of the measuring apparatus which acts upon it, but it does limit what can be measured. We cannot, for instance, carry out simultaneous ideal measurements of the position and momentum of a particle to arbitrary accuracies. Ideal measurements are defined to leave the subsystem in eigenstates of the measured quantities and there are no states of the subsystem for which position and momentum are specified to accuracies better than those allowed by the Heisenberg uncertainty principle. Analyses of the workings of measuring apparatus and subsystem as part of a single quantum system reveal further quantum mechanical limitations on ideal measurements, as in the work of Wigner (1952) and Araki and Yanase (1960).

Cosmology is one motivation for generalizing the quantum mechanics of measured subsystems to a quantum mechanics of closed systems in which measurement plays no fundamental role. Simply providing a more coherent and precise formulation of quantum mechanics, free from many of the usual interpretive difficulties, is motivation enough for many. Today, because of the efforts of many over the last thirty-five years, we have a quantum mechanics of closed systems.* In this formulation, it is the internal consistency of probability sum rules that determines the sets of alternatives of the closed system for which probabilities are predicted rather than any external notion of "measurement" (Griffiths, 1984; Omnès, 1988abc; Gell-Mann and Hartle, 1990). It is the absence of quantum mechanical interference between the individual members of a set of alternatives, or decoherence, that is a sufficient condition for the consistency of probability sum rules. It is the initial condition of the closed system that, together with its Hamiltonian, determines which sets of alternatives decohere and which do not. Alternatives describing a measurement situation decohere, but an alternative does not have to be part of a measurement

---

* A pedagogical introduction to the quantum mechanics of closed systems can be found in the author's other contribution to these volumes, (Hartle, 1993a), where references to some of the literature may be found.

situation in order to decohere. Thus, for example, with an initial condition and Hamiltonian are such that they decohere, probabilities are predicted for alternative sizes of density fluctuations in the early universe or alternative positions of the moon whether or not they are ever measured.

The familiar quantum mechanics of measured subsystems is an approximation to this more general quantum mechanics of closed systems. It is an approximation that is appropriate when certain approximate features of measurement situations can be idealized as exact. These include the decoherence of alternative configurations of the apparatus in which the result of the measurement is registered, the correlation of these with the measured alternatives, the short duration of certain measurement interactions compared to characteristic dynamical time scales of the measured subsystems, the persistence of the records of measurements, etc. etc.* The question naturally arises as to the extent to which those features of the quantum mechanics of measured subsystems that were tied to measurement situations are incorporated, modified, or dispensed with in the more general quantum mechanics of closed systems. Are two laws of evolution still needed? Is there reduction of the state vector, and if so, when? What becomes of a rule like "square amplitudes and sum over probabilities that one could have measured but didn't"? What becomes of the limitations on measurements in a more general theory where measurement can be described but does not play a fundamental role. This essay is devoted to some thoughts on these questions.

## 2 THE REDUCTION OF THE STATE VECTOR

In the approximate quantum mechanics of measured subsystems the Schrödinger picture state of the subsystem is described by a time-dependent vector, $|\psi(t)\rangle$, in the subsystem's Hilbert space. In between measurements the state vector evolves unitarily:

$$i\hbar \frac{\partial |\psi(t)\rangle}{\partial t} = h|\psi(t)\rangle . \tag{1}$$

If a measurement is carried out at time $t_k$, the probabilities for its outcomes are

$$p(\alpha_k) = \left\| s^k_{\alpha_k} |\psi(t_k)\rangle \right\|^2 . \tag{2}$$

Here, the $\{s^k_{\alpha_k}\}$ are an exhaustive set of orthogonal, Schrödinger picture, projection operators describing the possible outcomes. The index $k$ denotes the *set* of outcomes at time $t_k$, for example, a set of ranges of momentum, or a set of ranges of position, etc. The index $\alpha_k$ denotes the particular alternative within the set – a *particular* range of momentum, a *particular* range of position, etc. If the measurement was an "ideal" one, that "disturbed the system as little as possible", the state vector is reduced at $t_k$ by the projection that describes the outcome of the measurement:

$$|\psi(t_k)\rangle \rightarrow \frac{s^k_{\alpha_k} |\psi(t_k)\rangle}{\| s^k_{\alpha_k} |\psi(t_k)\rangle \|} . \tag{3}$$

---

* For more discussion of ideal measurement models in the context of the quantum mechanics of closed systems see Section IV and Hartle (1991a)

This is the "second law of evolution", which together with the first (1), can be used to calculate the probabilities of sequences of ideal measurements.

The two laws of evolution can be given a more unified expression. For example, in the Heisenberg picture, the joint probability of a sequence of measured outcomes is given by the single expression[*]:

$$p(\alpha_n, \ldots, \alpha_1) = \|s^n_{\alpha_n}(t_n) \cdots s^1_{\alpha_1}(t_1)|\psi\rangle\|^2 \tag{4}$$

where $|\psi\rangle$ is the Heisenberg state vector and

$$s^k_{\alpha_k}(t_k) = e^{iht_k/\hbar} s^k_{\alpha_k} e^{-iht_k/\hbar} \tag{5}$$

are the Heisenberg picture projection operators with $h$ the Hamiltonian of the subsystem. Nevertheless, even in such compact expressions one can distinguish unitary evolution from the action of projections at an "ideal" measurement.

One gains the impression from parts of the literature that some think the law of state vector reduction to be secondary in importance to the law of unitary evolution. Perhaps by understanding the quantum mechanics of large, "macroscopic" systems that include the measuring apparatus the second law of evolution can be derived from the first. Perhaps the law of the reduction of the state vector is unimportant for the calculation of realistic probabilities of physical interest. No ideas could be further from the truth in this author's opinion. Certainly the second law of evolution is less precisely formulated that the law of unitary evolution because the notion of an "ideal" measurement is vague and many realistic measurements are not very ideal. However, as shown conclusively by Wigner (1963), the second law of evolution is not reducible to the first and it is essential for the calculation of probabilities of realistic, everyday interest as we shall now describe.

Scattering experiments can perhaps be said to involve but a single measurement of the final state once the system has been prepared in an initial state. Many everyday probabilities, however are for *time sequences* of measurements. For instance, in asserting that the moon moves on a certain classical orbit one is asserting that successions of suitably crude measurements of the moon's position and momentum will be correlated in time by Newton's deterministic law. Thus, measured classical behavior involves probabilities for time sequences like (4). Successive state vector reductions are essential for their prediction as well as many other questions of interest in quantum mechanics.

Since the state vector of a subsystem evolves unitarily except when th at subsystem is measured by an external device, some have argued that one could dispense with the "second law of evolution" in the quantum mechanics of a closed system. All predictions would be derived from a state vector, $|\Psi(t)\rangle$, of the closed system that

---

[*] The utility of the Heisenberg picture in giving a compact expression for the two laws of evolution has been noted by many authors, Groenewold (1952) and Wigner (1963), among the earliest. Similar unified expressions can be given in the sum-over-histories formulation of quantum mechanics (Caves 1986, 1987 and Stachel 1986)

evolves in time only according to the Schrödinger equation (Everett, 1957; DeWitt, 1970). However, a state vector is a function of one time and can, therefore, be used to predict only the probabilities of alternatives that are at one time according to the generalization of (2)

$$p(\alpha_k) = \left\| P_{\alpha_k}^k |\Psi(t_k)\rangle \right\|^2 . \tag{6}$$

Here, the $\{P_{\alpha_k}^k\}$ are an exhaustive set of orthogonal, Schrödinger picture, projection operators representing alternatives of the closed system at a moment of time. For instance, in a description of the system in terms of hydrodynamic variables they might represent alternative ranges of the energy density averaged over suitable volumes. In a description of a measurement situation, the $P's$ might represent alternative registrations of that variable by an apparatus.

The restriction to a unitary law of evolution and the action of projections at a single time as in (6) would rule out the calculation of probabilities for time histories of the closed system. Some have suggested that probabilities at the single marvelous moment of time "now" are enough for all realistic physical prediction and retrodiction.[*] In this view, for example, probabilities referring to past history are more realistically understood as the probabilities for correlations among present records. However, just to establish whether a physical system is a good record, one needs to examine the probability for the correlations between the present value of that record and the past event it was supposed to have recorded. That is a probability for correlation between alternatives at two times — the probability of a history. For this and other reasons, probabilities of histories are just as essential in the quantum mechanics of closed systems as they were in the quantum mechanics of measured subsystems.

There is a natural generalization of expressions like (4) to give a framework for predicting the joint probabilities of time sequences of alternatives in the quantum mechanics of closed systems (Griffiths, 1984; Omnès, 1988abc; Gell-Mann and Hartle, 1990). The joint probability of a history of alternatives is

$$p(\alpha_n, \ldots, \alpha_1) = \left\| P_{\alpha_n}^n(t_n) \cdots P_{\alpha_1}^1(t_1) |\Psi\rangle \right\|^2 \tag{7}$$

where the Heisenberg $P$'s evolve according to

$$P_{\alpha_k}^k(t_k) = e^{iHt_k/\hbar} P_{\alpha_k}^k(0) \, e^{-iHt_k/\hbar} . \tag{8}$$

and the times in (7) are ordered with the earliest closed to $|\Psi\rangle$. Here, projection operators, state vectors, the Hamiltonian $H$, etc all refer to the Hilbert space of a closed system, containing both apparatus and measured subsystem if any. This is most generally the universe, in an approximation in which gross quantum fluctuations in the geometry of spacetime can be neglected.[†] The Heisenberg state vector $|\Psi\rangle$ represents the initial condition of the closed system, assumed here to be a pure state for simplicity.

---

[*] For a recent expression of this point of view, see Page and Wootters (1983).

[†] For a generalized quantum mechanics of closed systems that includes quantum spacetime see Hartle (1993b) and the references therein.

The occurrence of the projections in (7) can be described by saying that the state vector is "reduced" at each instant of time where an alternative is considered. However, the important point for the present discussion of state vector reduction is that the projections in (7) are not, perforce, associated with a measurement by some external system. This is a quantum mechanics of a closed system! The $P's$ can represent any alternative at a moment of time. Measurement situations within a closed system of apparatus and measured subsystem can be described by appropriate $P's$ (see Section IV) but the $P's$ do not necessarily have to describe measurement situations. They might describe alternative positions of the moon whether or not it is being observed or alternative values of density fluctuations in the early universe where ordinary measurement situations of any kind are unlikely to have existed. Thus, the state vector can be said to be reduced in (7) by the action of the projections and one might even say that there are "two laws of evolution" present, but those reductions and evolutions have nothing to do, in general, with measurement situations. In the author's view, it is clearer not to use the language of "reduction" and "two laws of evolution", but simply to regard (7) as the law for the joint probability of a sequence of alternatives of a closed system. Projections occur therein because they are the way alternatives are represented in the quantum mechanics of closed systems.

There is a good reason why the probabilities (4) of a sequence of alternatives of a subsystem refer only to the results of *measured* alternatives. It would be inconsistent generally to calculate probabilities of histories that have not been measured because the sum rules of probability theory would not be satisfied as a consequence of quantum mechanical interference. In the two-slit experiment, for instance, the probability to arrive at a point on the screen is not the sum of the probabilities to go through the alternative slits and arrive at that point unless the alternative passages have been measured and the interference between them destroyed. Thus, probabilities are not predicted for all possible sets of histories of a subsystem but only those which have been "measured".

Probabilities are not predicted for every set of alternative histories of a closed system either. But it is not an external notion of "measured" that discriminates those sets for which probabilities are predicted from those which are not. Rather, it is the internal consistency of the probability sum rules that distinguishes them (Griffiths, 1984). Probabilities are consistent for a set of histories, when, in a partition of the set of histories into an exhaustive set of exclusive classes, the probabilities of the individual classes are the sums of the probabilities of the histories they contain for all allowed partitions. A sufficient condition for the consistency of probabilities is the absence of interference between the individual histories in the set as measured by the overlap

$$\langle \Psi | P_{\alpha'_n}^n(t_n) \cdots P_{\alpha'_1}^1(t_1) \cdot P_{\alpha_1}^1(t_1) \cdots P_{\alpha_n}^n(t_n) | \Psi \rangle \propto \delta_{\alpha'_1 \alpha_1} \cdots \delta_{\alpha'_n \alpha_n}. \tag{9}$$

Sets of histories that satisfy (9) are said to *decohere*.* Decoherence implies the consistency of the probability sum rules. In the quantum mechanics of closed systems, probabilities are predicted for just those sets of alternative histories that decohere

---

* There are several possible decoherence conditions. This is *medium decoherence* in the terminology of Gell-Mann and Hartle (1990b).

according to (9) as a consequence of the system's Hamiltonian and initial quantum state $|\Psi\rangle$.

## 3 UNCERTAINTY PRINCIPLES

The state of a single particle cannot be simultaneously an eigenstate of position and momentum. It follows from their commutation relations that position and momentum cannot be specified to accuracies greater than those allowed by the Heisenberg uncertainty principle

$$\Delta x \Delta p \geq \tfrac{1}{2}\hbar \ . \tag{10}$$

Following the standard discussion, we infer from the mathematical inequality (10) that it is not possible to simultaneously perform ideal measurements of position and momentum to accuracies better than that allowed by the uncertainty principle (10). There can be no such ideal measurement because there is no projection operator, $s$, that could represent its outcome in (3).

The limitations on ideal measurements implied by the uncertainty principle (3) are usually argued to extend to non-ideal measurements as well. Examination of quantum mechanical models of specific measurement situations have for the most part verified the consistency of this extension although some have maintained otherwise.[*] No such elaborate analysis is needed to demonstrate the impossibility of ideal measurements of position and momentum to accuracies better than those allowed by the uncertainty principle. That limitation follows from the quantum mechanics of the subsystem alone.

The mathematical derivation of the uncertainty relation (10) is , of course, no less valid in the Hilbert space of a closed system than it is for that of a subsystem. In the quantum mechanics of a closed system, however, the absence of projection operators that specify $x$ and $p$ to accuracies better than (10) is not to be interpreted as a limitation on external measurements of this closed system. By hypothesis there are none! Rather, in the quantum mechanics of closed systems, uncertainty relations like (10) are limitations on how a closed system can be *described*. There are no histories in which position and momentum can be simultaneously specified to accuracies better than allowed by Heisenberg's principle.

Although there are no projection operators that specify position and momentum *simultaneously* to accuracies better than the limitations of the uncertainty principle, we can consider histories in which position is specified sharply at one time and momentum at another time. Let $\{\Delta_\alpha\}$ be an exhaustive set of exclusive position intervals, $\{\tilde{\Delta}_\beta\}$ be an exhaustive set of exclusive momentum intervals, and $\{P_\alpha(t')\}$ and $\{\tilde{P}_\beta(t)\}$ be the corresponding Heisenberg picture projection operators at times $t'$ and $t$ respectively. An individual history in which the momentum lies in the

---

[*] For example, Margenau (1958) and Prugovečki (1967). We are not discussing here, nor do we discuss later, "unsharp" observables or "effects". For those see *e.g.*, Busch (1987).

interval $\tilde{\Delta}_\beta$ at time $t$ and the position in interval $\Delta_\alpha$ at a *later* time $t'$ would correspond to a branch of the initial state vector of the form:

$$P_\alpha(t')\tilde{P}_\beta(t)|\Psi\rangle \ . \tag{11}$$

As $\alpha$ and $\beta$ range over all values, an exhaustive set of alternative histories of the closed system is generated. Probabilities are assigned to these histories when the set decoheres, that is, when the branches (11) are sufficiently orthogonal according to (9).

Nothing prevents us from considering the case when $t'$ coincides with $t$. If the alternative histories decohere, one would predict the joint probability $p(\alpha,\beta)$ that the momentum is in the interval $\tilde{\Delta}_\beta$ at one time and *immediately* afterwards the position is in the interval $\Delta_\alpha$. That would give a different meaning to the probability of a simultaneous specification of position and momentum.

Even if the intervals $\{\Delta_\alpha\}$ and $\{\tilde{\Delta}_\beta\}$ are infinitesimal, corresponding to a sharp specification of position and momentum there are some states $|\Psi\rangle$ for which these alternatives decohere. Eigenstates of momentum provide one example. However, for no state $|\Psi\rangle$ will the marginal probability distributions of position and momentum have variances that violate the uncertainty principle. That is because decoherence implies the probability sum rules so that

$$p(\alpha) \equiv \sum_\beta p(\alpha,\beta) = \|P_\alpha(t)|\Psi\rangle\|^2 \tag{12a}$$

and

$$p(\beta) \equiv \sum_\alpha p(\alpha,\beta) = \|\tilde{P}_\beta(t)|\Psi\rangle\|^2 \tag{12b}$$

where, in each case, the last equality follows from decoherence. However, the left-hand sides of (12) are just the usual probabilities for position and momentum computed from a single state. Their variances must satisfy the uncertainty principle.

One would come closer to the classical meaning of simultaneously specifying the position and momentum if histories of coincident position and momentum projections decohered independently of their order. That is, if the set of histories

$$\tilde{P}_\beta(t)P_\alpha(t)|\Psi\rangle \tag{13}$$

were to decohere in addition to the set defined by (11) with $t' = t$. In that case it is straightforward to show that the joint probabilities $p(\alpha,\beta)$ are independent of the order of the projections as a consequence of decoherence. Whether states can be exhibited in which both (11) and (13) decohere is a more difficult question.

## 4 LIMITATIONS ON IDEAL MEASUREMENTS

While measurement is not fundamental to a formulation of the quantum mechanics of a closed system, measurement situations can be described within it. That is because we can always consider a closed system consisting of measuring apparatus and measured subsystem or most generally and accurately the entire universe. Roughly speaking, a *measurement situation* is one in which a variable of the measured subsystem, perhaps not normally decohering, becomes correlated with high probability with a variable of the apparatus that decoheres because of *its* interactions with the rest of the universe. The variable of the apparatus is called a *record* of the measurement outcome. The decoherence of the alternative values of this record leads to the decoherence of the measured alternatives because of their correlation. Measurement situations can be described quantitatively in the quantum mechanics of closed systems by using the overlap (9) to determine when measured alternatives decohere and using the resulting probabilities to assess the degree of correlation between record and measured variable. By such means, any measurement situation, ideal or otherwise, may be accurately handled in the quantum mechanics of the closed system containing both measuring apparatus and measured subsystem.

Conventional discussions of measurement in quantum mechanics often foc us on *ideal measurement models* in which certain approximate features of realistic measurement situations are idealized as exact.* In particular an ideal measurement is one that leaves a subsystem that is initially in an eigenstate of a measured quantity in that same eigenstate after the measurement. The subsystem is thus "disturbed as little as possible" by its interaction with the apparatus. Of course, not very many realistic measurements are ideal in this sense. Typically, after a measurement, subsystem and apparatus are not even in a product state for which it makes sense to talk about the "state of the subsystem". Probably the reason for the focus on ideal measurement models is that they are models of the sorts of measurements for which the "reduction of the state vector" could accurately model the evolution of the measured subsystem interacting with the measuring apparatus. In particular, a reduction of the state of the apparatus will leave the subsystem in the correlated eigenstate of the measured variable.

Quantum mechanics severely restricts the possible ideal measurement situations. Wigner (1952) and Araki and Yanase (1960) showed that, even given arbitrary latitude in the choice of Hamiltonian describing the combined system of apparatus and measured subsystem, only quantities that commute with additive, conserved quantities can be ideally measured. This is a very restrictive conclusion. It rules out, for example, precise, ideal measurements of the position and momentum of a particle with a realistic Hamiltonian. (They do not commute with the additive, conserved angular momentum.) Araki and Yanase showed that, in a certain sense, ideal measurements were *approximately* possible for such quantities, but in a strict sense quantum mechanics prohibits them.

Impurity of the initial state of a closed system limits ideal measurements in another way. To derive this limitation it is necessary to discuss more precisely ideal measurement models in the quantum mechanics of closed systems.† We will use the more of the formulation of the quantum mechanics of closed systems than has

---

* Some classic references are von Neumann (1932), London and Bauer (1939), and Wigner (1963) or see almost any text on quantum mechanics.

† For more detail than can be offered here see Hartle (1991a), Section II.10.

been developed here. The reader can find the necessary background in the author's other contribution to these volumes, (Hartle 1993a).

We consider a closed system in which we can identify alternatives of a subsystem that are to be "measured". Let $\{S_{\alpha_k}^k(t_k)\}$, $\alpha_k = 1, 2, 3, \cdots$ be the Heisenberg projection operators corresponding to these alternatives at a set of times $\{t_k\}$, $k = 1, 2, 3 \cdots$. In a more detailed ideal measurement model we might assume that the Hilbert space of the closed system is a tensor product of a Hilbert space $\mathcal{H}^s$ defining the subsystem and a Hilbert space $\mathcal{H}^r$ defining the rest of the universe outside the subsystem. In the Schrödinger picture, projection operators representing the alternatives of the subsystem would have the form $S_{\alpha_k}^k = s_{\alpha_k}^k \otimes I^r$ where the $s's$ act on $\mathcal{H}^s$ alone. However, such specificity is not needed for the result that we shall derive.

Let us consider how a sequence of ideal measurements of alternatives o f the subsystem $S_{\alpha_k}^k(t_k)$ at times $t_1 < \cdots < t_n$ is described. A history of specific alternatives $(\alpha_1, \cdots, \alpha_n) \equiv \alpha$ is represented by the corresponding chain of projections:

$$C_\alpha = S_{\alpha_n}^n(t_n) \cdots S_{\alpha_1}^1(t_1) \ . \tag{14}$$

One defining feature of an ideal measurement situation is that there should exist at a time $T > t_n$ a *record* of the outcomes of the measurement that is exactly correlated with the measured alternatives of the subsystem. That is , there should be a set of orthogonal, *commuting*, projection operators $\{R_\beta(T)\}$ with $R_\beta(T) \equiv R_{\beta_n \cdots \beta_1}(T)$ which are always exactly correlated with the measured alternatives $C_\alpha$ in histories that contain them both, as a consequence of the system's initial condition. The degree of correlation is defined by the decoherence functional $D(\beta'\alpha'; \beta\alpha)$ which measures the interference between a history consisting of a sequence of measured alterhatives $\alpha = (\alpha_1, \cdots, \alpha_n)$ followed by a record $\beta = (\beta_1, \cdots, \beta_n)$ at time $T$ and a similar history with alternatives $\alpha'$ and $\beta'$. If the records are exactly correlated with the measured alternatives.

$$D\left(\beta', \alpha'; \beta, \alpha\right) = Tr[R_{\beta'}(T) C_{\alpha'} \rho C^\dagger_\alpha R_\beta(T)] \propto \delta_{\beta'\alpha'} \delta_{\beta\alpha} \tag{15}$$

where $\rho$ is the Heisenberg picture initial density matrix of the closed system of apparatus and measured subsystem and $\delta_{\beta\alpha}$ means $\delta_{\beta_1\alpha_1}\delta_{\beta_2\alpha_2} \cdots \delta_{\beta_n\alpha_n}$, etc.

The existence of exactly correlated records as described by (15) ensures the decoherence of the histories of the subsystem and permits the prediction of their probabilities. That is because the records are orthogonal and exhaustive:

$$R_\beta(T)R_{\beta'}(T) = \delta_{\beta\beta'} R_\beta(T), \qquad \sum_\beta R_\beta(T) = I \ . \tag{16}$$

These properties together with the cyclic property of the trace are enough to show that

$$Tr[C_{\alpha'}\rho C^\dagger_\alpha] \propto \delta_{\alpha'\alpha} \tag{17}$$

follows from (15). This is the generalization of the decoherence condition (9) for an initial density matrix. The measurement correlation thus effects the decoherence of the measured alternatives.

Of course, much more is usually demanded of an ideal measurement situation than just decoherence of the measured alternatives. There is the idea that ideal measurements "disturb the measured subsystem as little as possible" and, in particular, that values of measured quantities are not disturbed. These are described in more detail in the context of the quantum mechanics of closed systems in Hartle (1991a). For our discussion, however, we need only the feature that ideal measurements assume *exactly* correlated records of measurement outcomes, for we shall now show that if the density matrix is highly impure such records cannot exist for non-trivial sets of measured histories.[*]

We begin by introducing bases of complete sets of states in which the density matrix $\rho$ and the commuting set of projection operators $\{R_\beta(T)\}$ are diagonal, viz. :

$$\rho = \sum_r |r\rangle \pi_r \langle r| , \tag{18}$$

$$R_\beta(T) = \sum_n |\beta, n\rangle \langle \beta, n| . \tag{19}$$

where $\pi_r$ are the diagonal elements of $\rho$. When the "diagonal" elements of the condition (15) (those with $\alpha = \alpha', \beta = \beta'$) are written out in terms of these bases they take the form

$$\sum_{r,n} \pi_r |\langle \beta, n|C_\alpha|r\rangle|^2 \propto \delta_{\alpha\beta} . \tag{20}$$

The left-hand side is a sum of positive numbers so that this implies

$$\langle r|C_\alpha|\beta, n\rangle = 0 , \quad \text{when } \alpha \neq \beta , \tag{21}$$

for all $r$ for which $\pi_r \neq 0$.

If the density matrix $\rho$ is highly impure, so that $\pi_r \neq 0$ for a *complete* set of states $\{|r\rangle\}$, the relation (21) implies the operator condition

$$C_\alpha|\beta, n\rangle = 0 , \quad \alpha \neq \beta . \tag{22}$$

Therefore, $C_\alpha$ is non-zero only on the subspace defined by $R_\alpha(T)$ where $R_\alpha(T)$ is effectively unity. Thus we have

$$R_\beta(T)C_\alpha = \delta_{\beta\alpha}C_\alpha . \tag{23}$$

Summing this relation over $C_\alpha$ and utilizing the fact that $\sum_\alpha C_\alpha = I$, we find

$$C_\alpha = R_\alpha(T) . \tag{24}$$

---

[*]   The argument we shall give is a straightforward extension of that used by M. Gell-Mann and the author to analyze the possibility of "strong decoherence" in Gell-Mann and Hartle (1993a). Thanks are due to M. Gell-Mann for permission to publish here what is essentially a joint result.

which says that the string of projections is itself a projection. This can happen only if the string consists of a single projection or if all the projections in the string commute with each other. To see the latter fact write (24) in detail as

$$S_{\alpha_n}^n(t_n) \cdots S_{\alpha_1}^1(t_1) = R_{\alpha_n \cdots \alpha_1}(T) \ . \tag{25}$$

Summation implies

$$S_{\alpha_k}^k(t_k) = \sum_{\alpha_j \neq \alpha_k} R_{\alpha_n \cdots \alpha_1}(T) \tag{26}$$

but since the $R$'s commute with each other the $S$'s must also. Even in the case that $C_\alpha$ consists of a single projection, (24) shows that record and projection are indentical. If $C_\alpha$ consists of projections that refer to a subsystem defined by a Hilbert space as described above then the records cannot be elsewhere in the universe. Thus, if the initial density matrix is highly impure, in the sense that it has non-zero probabilities for a complete set of states, there cannot be exactly correlated records of measurement outcomes. In particular there cannot be ideal measurements.

Of course, in realistic measurement situations we do not expect to fin d records that are *exactly* correlated with measured variables of a subsystem. Neither do we necessarily expect *exact* decoherence of measured alternatives or many of the other idealizations of the ideal measurement situation as very experimentalist knows! It, therefore, becomes an interesting question to investigate quantitatively the connection between the $\{\pi_r\}$ of the density matrix and the degree to which approximate records defined by a relaxed (15) exist.

## 5 INTERFERING ALTERNATIVES

The starting point for Feynman's sum-over-histories formulation of quantum mechanics is the prescription of the amplitude for an elementary (completely fine-grained) history of a measured subsystem as

$$\exp[iS(\text{history})/\hbar] \tag{27}$$

where $S$ is the action functional summarizing the subsystem's dynamics. As an example, we may think of a non-relativistic particle moving in one dimension. In this case the elementary histories are the possible paths of the particle, $x(t)$, and the action is the usual

$$S[x(\tau)] = \int dt \left[\tfrac{1}{2}m\dot{x}^2 - V(x)\right] \ . \tag{28}$$

We will use this example for all illustrative purposes in what follows.

A given experimental situation determines some parts of the subsystem's path but leaves undetermined many other parts. For instance, consider a measurement that determines whether or not a particle is in a position interval $\Delta$ at time $t$. In that case the measurement leaves undetermined the positions at times other than $t$ and the relative position within $\Delta$ at time $t$. Given an initial state at time $t_0$ represented by a wave function $\psi(x_0)$, we may compute the probabilities for the outcomes that are determined by the measurement as follows: We first divide the *undetermined* alternatives into "interfering" and "non-interfering" (or "exclusive") alternatives according to the experimental situation. We sum amplitudes for histories weighted by the initial wave function over the interfering alternatives, square that, and sum the square over the non-interfering alternatives. The result is the probability for the measured determination. For example, in the case of the measurement mentioned above that localized a particle to an interval $\Delta$ at time $t$, the probability of this outcome is:

$$p(\Delta) = \int_\Delta dx_f \left| \int_{x_f} \delta x \; e^{iS[x(\tau)]/\hbar} \psi(x_0) \right|^2 . \tag{29}$$

The path integral is over all paths in the time interval $[t_0, t]$ that end in $x_f$, and includes an integral over the initial position $x_0$. These are the "interfering alternatives". The square of the amplitude is summed over the final position within $\Delta$. These positions are the "non-interfering" alternatives.

What determines whether an undetermined alternative is interfering or not? Certainly it is not whether it is measured in the experimental situation. In the above example, positions at times other than $t$ were not measured and they were "interfering". But the experiment also did not measure the relative position within $\Delta$ and this was "non-interfering". According to Feynman and Hibbs (1965):

"It is not hard, with a little experience, to tell what kind of alternatives is involved. For example, suppose that information about alternatives is available (*or could be made available without altering the result*) [author's italics], but this information is not used. Nevertheless, in this case a sum of probabilities (in the ordinary sense) must be carried out over *exclusive* alternatives. These exclusive alternatives are those which *could* have been separately identified by the information."

Thus, in the above example, the value of $x$ at a time other than $t$ is an interfering alternative because we could not have acquired information about it without disturbing the later probability that $x$ is in $\Delta$ at $t$. By contrast, the precise value of $x$ within $\Delta$ is a non-interfering or exclusive alternative because we *could* have measured it precisely and left the probability for the particle to lie in $\Delta$ undisturbed. Indeed, one way to determine whether the particle is in $\Delta$ is simply to measure the position at $t$ precisely.

The author has always found this distinction between types of alternatives confusing. He did not doubt Feynman's ability "to tell what kind of alternative is involved", but he was less sure of his own. This was especially the case since the distinction seemed to involve analyzing, not only the particular experiment in question, but also many others that *might* have been carried out. No precise rules for analyzing a given experimental situation seemed to be available. This situation is considerably clarified in the quantum mechanics of closed systems.

In the quantum mechanics of closed systems, we cannot have a fundamental distinction between "interfering" and "non-interfering" alternatives based on different types of measurement situations, because alternatives are not necessarily associated with measurement situations. Whether alternatives interfere with one another, or do not, depends on the boundary conditions and Hamiltonian that define the closed system. A quantitative measure for the degree of interference is provided by the dechoherence functional. To illustrate this idea, let us consider the single particle model we have been discussing on the time interval $[t_0, t]$. The fine-grained histories are the particle paths on this interval. Sets of alternatives correspond to partitions of these paths into an exhaustive set of exclusive classes $\{c_\alpha\}$, $\alpha = 1, 2, \cdots$. The classes are coarse-grained alternatives for the closed system. For example, one could partition the paths by which of an exhaustive set of position intervals they pass through at one time, which of a different set of position intervals they pass through at another time, etc. There are many more general possibilities (see, e.g. Hartle, 1991). The decoherence functional is a complex valued functional on pairs of coarse-grained alternatives defined in a sum-over-histories formulation of the quantum mechanics of a closed system by:

$$D(\alpha', \alpha) = N \int_{c_{\alpha'}} \delta x' \int_{c_\alpha} \delta x \, \rho_f(x_f, x'_f) \exp\left\{ i \left( S[x'(\tau)] - S[x(\tau)] \right) / \hbar \right\} \rho_i(x'_0, x_0).$$
(30)

The first sum is over paths $x'(t)$ in the class $c_{\alpha'}$ and includes a sum over their initial endpoints $x'_0$ and final endpoints $x'_f$. The sum over paths $x(t)$ is similar. The normalization factor is $N = 1/Tr(\rho_f \rho_i)$ where the $\rho$'s are the operators whose matrix elements appear in (30). We have written the decoherence functional for a general, time-neutral, formulation of quantum mechanics* in which both an initial and a final condition enter symmetrically, represented by density matrices $\rho_i(x'_0, x_0)$ and $\rho_f(x'_f, x_f)$ respectively. The final condition which seems to best represent our universe and ensures causality is a final condition of indifference with respect to final state in which the final density operator is $\rho_f \propto I$.

The "off-diagonal" elements of the decoherence functional ($\alpha' \neq \alpha$) are a measure of the degree of interference between pairs of alternatives. When the interference is negligible between all pairs in an exhaustive set, the "diagonal" elements ($\alpha' = \alpha$) are the probabilities of the alternatives and obey the correct probability sum rules as a consequence of the absence of interference. The orthogonality of the branches in (9) is an operator transcription of this condition in the special case that $\rho_f \propto I$.

The important point for a discussion of "interfering" and "non-interfering" alternatives is that all alternatives are potentially interfering in the quantum mechanics of closed systems. For this reason amplitudes are summed over them in the construction of the decoherence functional (30). Whether alternatives are interfering or not depends on the measure of interference provided by (30), but in its construction all sets of alternatives are treated the same. When interference between each pair is negligible the probabilities for coarser-grained alternatives may be constructed either directly from (30) by summing amplitudes, or by summing the probabilities for the finer-grained alternatives in the coarser-grained ones. The equivalence between the two is the content of decoherence.

---

* See, e.g. Aharonov, Bergmann and Lebovitz (1964) in the quantum mechanics of measured subsystems, and Griffiths (1984) and Gell-Mann and Hartle (1993) in the quantum mechanics of closed systems.

Thus, there is no distinction between kinds of alternatives generally in the formalism, but distinctions may emerge between different kinds of alternatives because of particular properties of $\rho_i$ and $\rho_f$. In particular, if $\rho_f \propto I$ any alternatives at the last time will decohere. Thus, indifference with respect to final states is, in a time-neutral formulation of quantum mechanics, the origin of the usual rule that final alternatives are "non-interfering" rather than an analysis of whether "one could have measured them but didn't".

## 6 CONCLUSION

In the quantum mechanics of closed systems, projections act on states in the formula for the probabilities of histories, but those reductions are not necessarily associated with a measurement situation within the system and certainly not with one from without. Uncertainty principles limit what kinds of alternatives a set of projections can describe, but these limitations need not be of our ability to carry out a measurement. Interfering alternatives can be distinguished from non-interfering ones, not by analyzing what might have been measured, but by using the decoherence functional as a quantitative measure of interference. Probabilities can be consistently assigned only to non-interfering sets of alternative histories but decoherence as a consequence of a particular initial condition and Hamiltonian rather than measurement decides which sets these are.

The fundamental role played by measurement in formulating a quantum mechanics of subsystems is replaced by decoherence in the quantum mechanics of a closed system. In the opinion of the author, the result is not only greater generality so that the theory can be applied to cosmology, but also greater clarity. An important reason for this is the disassociation of the notion of alternative from an ideal measurement. As we saw from the work of Wigner, Araki and Yanase, and the argument of Section IV, ideal measurements are almost impossible to realize exactly within quantum mechanics, and are therefore of limited value as approximations to realistic measurement situations. But in the usual quantum mechanics of measured subsystems, the second law of evolution is stated for ideal measurements, not realistic ones. To discuss the evolution under realistic alternatives it appears necessary to consider more and more of the universe beyond the subsystem of interest until one obtains a subsystem large enough such that measurements of *it* may be approximated as "ideal". By contrast, the alternatives used in the quantum mechanics of closed systems are general enough to describe realistic measurement situations. The theory can provide quantitative estimates of their closeness to "ideal" and therefore to how closely the quantum mechanics of measured subsystems approximates the more general quantum mechanics of closed systems.

Thus, little havoc needs be made to achieve a quantum mechanics of closed systems. All that is needed is a more general formulation in which decoherence rather than measurement is fundamental, but in which most features of the approximate quantum mechanics of subsystems that were tied to measurement reëmerge in a more general and conceptually clearer light.

## ACKNOWLEDGMENT

The work described in this essay is an outgrowth of the author's progr am with M. Gell-Mann to explore and clarify quantum mechanics. It is a pleasure to thank him for many discussions that are reflected, in part, in the work presented here. Preparation of this essay was supported, in part, by the National Science Foundation under grant NSF PHY90-08502.

# 7 REFERENCES

Aharonov, Y., Bergmann, P., and Lebovitz, J. (1964) *Phys. Rev.* **B134**, 1410.

Araki, H. and Yanase, M. (1960) *Phys. Rev.* **120**, 622.

Brill, D. and Hartle, J.B. (1964) *Phys. Rev.* B **135**, 271.

Busch, P. (1987) *Found. Phys.* **17**, 905.

Caves, C. (1986) *Phys. Rev. D* **33**, 1643.

———— (1987) *Phys. Rev. D* **35**, 1815.

DeWitt, B. (1970) *Physics Today* **23**, no. 9.

Everett, H. (1957) *Rev. Mod. Phys.* **29**, 454.

Feynman, R.P. and Hibbs, A. (1965) *Quantum Mechanics and Path Integrals*, McGraw-Hill, New York.

Gell-Mann, M. and Hartle, J.B. (1990a) in *Complexity, Entropy, and the Physics of Information, SFI Studies in the Sciences of Complexity*, Vol. VIII, ed. by W. Zurek, Addison Wesley, Reading or in *Proceedings of the 3rd International Symposium on the Foundations of Quantum Mechanics in the Light of New Technology* ed. by S. Kobayashi, H. Ezawa, Y. Murayama, and S. Nomura, Physical Society of Japan, Tokyo.

Gell-Mann, M. and Hartle, J.B. (1990b) in the *Proceedings of the 25th International Conference on High Energy Physics, Singapore, August, 2-8, 1990*, ed. by K.K. Phua and Y. Yamaguchi (South East Asia Theoretical Physics Association and Physical Society of Japan) distributed by World Scientific, Singapore.

Gell-Mann, M. and Hartle, J.B. (1993a) *Classical Equations for Quantum Systems* (to be published in *Phys. Rev. D*).

Gell-Mann, M. and Hartle, J.B. (1993b) in *Proceedings of the 1st International A. D. Sakharov Conference on Physics, USSR, May 27-31, 1991* and in *Proceedings of the NATO Workshop on the Physical Origins of Time Assymmetry, Mazagon, Spain, September 30-October4, 1991* ed. by J. Halliwell, J. Perez-Mercader, and W. Zurek, Cambridge University Press, Cambridge.

Griffiths, R. (1984) *J. Stat. Phys.* **36**, 219.

Groenewold, H.J. (1952) *Proc. Akad. van Wetenschappen*, Amsterdam, Ser. B, **55**, 219.

Hartle, J.B. (1991a) *The Quantum Mechanics of Cosmology*, in *Quantum Cosmology and Baby Universes: Proceedings of the 1989 Jerusalem Winter School for Theoretical Physics*, ed. by S. Coleman, J.B. Hartle, T. Piran, and S. Weinberg, World Scientific, Singapore, pp. 65-157.

Hartle, J.B. (1991b) *Phys. Rev.* D **44**, 3173.

Hartle, J.B. (1993a) *The Quantum Mechanics of Closed systems* in the *Festschrift for C.W. Misner*, ed. by B.-L. Hu, M.P. Ryan, and C.V. Vishveshwara, Cambridge University Press, Cambridge.

Hartle, J.B. (1993b) *Spacetime Quantum Mechanics and the Quantum Mechanics of Spacetime* in *Proceedings of the 1992 Les Houches Summer School Gravitation and Quantizations*, ed. by B. Julia, North Holland, Amsterdam.

Isaacson, R. (1968a) *Phys. Rev.* **166**, 1263.

_____ (1968b) *Phys. Rev.* **166**, 1272.

London, F. and Bauer, E. (1939) *La théorie de l'observation en mécanique quantique*, Hermann, Paris.

Margenau, H. (1958) *Phil. of Sci.* **25**, 23.

Omnès, R. (1988a) *J. Stat. Phys.* **53**, 893.

_____ (1988b) *J. Stat. Phys.* **53**, 933.

_____ (1988c) *J. Stat. Phys.* **53**, 957.

Page, D. and Wootters, W. (1983) *Phys. Rev.* D **27**, 2885.

Prugovčki, E. (1967) *Can. J. Phys.* **45**, 2173.

Stachel, J. (1986) in *From Quarks to Quasars*, ed. by R.G. Colodny, University of Pittsburg Press, Pittsburgh, p. 331ff.

von Neumann, J. (1955) *Mathematische Grundlagen der Quantenmechanik*, J. Springer, Berlin (1932). [English trans. *Mathematical Foundations of Quantum Mechanics*, Princeton University Press, Princeton].

Wheeler, J.A. (1964) in *Relativity Groups and Topology*, ed. by B. DeWitt and C. DeWitt, Gordon and Breach, New York.

Wigner, E.P. (1952) *Zeit. f. Phys.* **131**, 101.

_____ (1963) *Am. J. Phys.* **31**, 6.

# Quantum Linearization Instabilities of de Sitter Spacetime

*Atsushi Higuchi* *

### Abstract

It has been pointed out that the physical states in linearized quantum gravity are required to be invariant under the continuous isometries of the background spacetime if the Cauchy surfaces are compact. This requirement would appear to allow only the vacuum state as the physical state in linearized quantum gravity in de Sitter spacetime. The first step toward resolving this apparent paradox is to construct a new Hilbert space of de Sitter-invariant states. In this article an approach to this task is presented. First de Sitter-invariant states with infinite norm are constructed by smearing the states in the original Fock space of linearized gravity over the de Sitter group. Then a finite inner product of these states is defined by dividing the original inner product by the infinite volume of the de Sitter group. The Hilbert space of de Sitter-invariant states thus obtained is hoped to serve as a starting point toward a meaningful perturbative quantum gravity (at the tree level) in de Sitter spacetime.

## 1. Introduction

In discussing solutions to the linearized Einstein equations, it is important to make sure that they extend to exact solutions. It was found by Professors Brill and Deser [2, 3] that there are spurious solutions in linearized gravity in static flat spacetime with the topology of the 3-torus ($T^3$). Now it is known that the vacuum Einstein equations are linearization stable—i.e., there are no spurious solutions to the linearized Einstein equations—at a spacetime *with compact Cauchy surfaces* only if there are no Killing fields. (It is known that this condition is also sufficient for linearization stability if the cosmological constant is zero. See [14].) If there are Killing fields, a solution to the linearized Einstein equations extends to an exact solution only if it satisfies the so-called linearization stability conditions (LSCs) which are quadratic in perturbation.

---

*Enrico Fermi Institute, University of Chicago, 5640 S. Ellis Avenue, Chicago, IL 60637, USA. This author was supported in part by the National Science Foundation under Grant No. PHY89-18388.

146

(Again, these LSCs are known to be sufficient if the cosmological constant is zero. See [14].)

Implications of LSCs in quantum gravity were studied by Professor Moncrief. He found that the physical states in linearized quantum gravity in a background spacetime with compact Cauchy surfaces are required by the LSCs to be invariant under the continuous background isometries (see [11, 12]). Two such spacetimes with nontrivial isometries are static flat spacetime with toroidal topology mentioned before and de Sitter spacetime. In the former spacetime the invariance requirement on the physical states can be handled in analogy with the requirement of diffeomorphism invariance—i.e., the Hamiltonian and momentum constraints—in full quantum gravity (see [11] and [9]). In this article, we discuss how to incorporate the invariance requirement coming from LSCs in the latter spacetime (see [6, 7, 8]).

If one were to impose the invariance under the de Sitter group $SO(4,1)$ in the Fock space of gravitons in de Sitter spacetime, the vacuum would be the only physical state. Moncrief conjectured (see [5]) that there are de Sitter-invariant states with infinite norm, and that they can be normalized by factoring out infinity from the inner product. We will show that this conjecture is true. In section 2 the case of flat spacetime with toroidal topology is reviewed and the method used for de Sitter spacetime is introduced. In section 3 a Hilbert space of de Sitter-invariant states is constructed. The Roman letters $a, b, c, \ldots$ are used for space indices and the Greek letters $\mu, \nu, \ldots$ for spacetime indices throughout this article.

## 2.   Flat spacetime with toroidal spatial topology

Brill and Deser found that solutions to the linearized Einstein equations in static flat spacetime with topology $T^3$ do not extend to exact solutions unless they satisfy four quadratic constraints (LSCs) corresponding to the four Killing fields. Moncrief studied these constraints in the context of quantum gravity as an example of his general observation mentioned in the introduction. In this section this example is reviewed in order to motivate the method used for de Sitter spacetime.

In the Hamiltonian formulation of general relativity (see [1]) the line element is written as

$$ds^2 = -(N^2 - N^a N_a)dt^2 + N_a dx^a dt + g_{ab}dx^a dx^b. \tag{1}$$

We consider the linear approximation about the flat background metric with the points $(x^1, x^2, x^3)$ and $(x^1 + L^1, x^2 + L^2, x^3 + L^3)$ identified. We write the 3-metric as $g_{ab} = g^\flat_{ab} + h_{ab}$, where $g^\flat_{ab}$ is the flat Euclidean metric, and the conjugate momentum as $\pi^{ab} = 0 + p^{ab}$, and use linearized Hamilton's equations for $h_{ab}$ and $p^{ab}$ with $N = 1$ and $N^a = 0$. After imposing the linearized Hamiltonian and momentum constraints, one can let

$$(h_{ab}, p_{ab}) = (\bar{h}_{ab} + h^{TT}_{ab}, \bar{p}_{ab} + p^{TT}_{ab}), \tag{2}$$

where the indices are raised and lowered by the flat metric $g^{i}_{ab}$. The $(\bar{h}_{ab}, \bar{p}_{ab})$ are the spatially-constant part of the initial data on the Cauchy surface and the $(h^{TT}_{ab}, p^{TT}_{ab})$ are the transverse-traceless part with nonzero wave numbers. The Hamiltonian for the linearized theory is

$$H = \int d^3x \left[ \bar{p}^{ab}\bar{p}_{ab} - \frac{1}{2}\bar{p}^2 + p^{abTT}p^{TT}_{ab} + \frac{1}{4}\partial_c h^{TT}_{ab}\partial^c h^{abTT} \right] . \tag{3}$$

Solutions to Hamilton's equations with this Hamiltonian need to satisfy the following LSCs in order that they extend to exact solutions:

$$H = 0, \tag{4}$$

$$\int d^3x\, p^{abTT}\mathcal{L}_{X^{(i)}}h^{TT}_{ab} = 0, \tag{5}$$

where $X^{(i)}$ $(i = 1, 2, 3)$ are the Killing fields which generate the space translations, i.e., $\mathcal{L}_{X^{(i)}}h^{TT}_{ab} = \partial_i h^{TT}_{ab}$.

In the quantum theory, the wavefunctions can be taken to be functionals of $\bar{h}_{ab}$ and $h^{TT}_{ab}$. The LSCs (4) and (5) can be taken into account by requiring that the corresponding operators annihilate the physical states. The condition (5) then requires that the wavefunctions be invariant under space translations. (Note the analogy with the situation in full quantum gravity where the momentum constraint requires the invariance of the wavefunctions under arbitrary space diffeomorphisms.) This amounts to the requirement that the total momentum of the gravitons be zero.

The condition (4) is a little more involved. Define

$$\bar{h}_{ab} = \frac{1}{\sqrt{6V}}\delta_{ab}c + \frac{1}{\sqrt{V}}\sum_{i=1}^{5} T^{(i)}_{ab}c_i , \tag{6}$$

where $V = L^1L^2L^3$ is the volume of the background $T^3$ space, and where the $T^{(i)}_{ab}$ are constant traceless tensors satisfying $T^{(i)}_{ab}T^{(j)ab} = \delta^{ij}$. Similarly, let

$$\bar{p}_{ab} = \sqrt{\frac{2}{3V}}\delta_{ab}c_P + \frac{1}{\sqrt{V}}\sum_{i=1}^{5} T^{(i)}_{ab}c_{Pi} . \tag{7}$$

Then, $c$ and $-c_P$ as well as $c_i$ and $c_{Pi}$ are canonically conjugate to each other. Denote the sum of the third and fourth terms in (3)—i.e., the total graviton energy—by $E_T$. Then,

$$H = -c_P^2 + \sum_{i=1}^{5} c_{Pi}^2 + E_T . \tag{8}$$

When this system is quantized, the wavefunctions can be chosen to be eigenfunctions of the operator $E_T$ without loss of generality. Hence $E_T$ can be treated as a constant c-number. (Note, however, that this operator involves infinity, which needs to be subtracted.) By letting $c_P = i(\partial/\partial c)$ and $c_{P_i} = -i(\partial/\partial c_i)$, we obtain the quantum version of the condition (4):

$$\left(\frac{\partial^2}{\partial c^2} - \sum_{i=1}^{5}\frac{\partial^2}{\partial c_i^2} + E_T\right)\Psi(c, c_i) = 0, \tag{9}$$

where we have suppressed the variables $h_{ab}^{TT}$. This equation is nothing but the Klein-Gordon equation for a free scalar field in six dimensions with mass $\sqrt{E_T}$. Thus, one can introduce the Klein-Gordon inner product:

$$(\Psi^{(1)}|\Psi^{(2)})_{KG} \equiv i\int\prod_{i=1}^{5}dc_i\left[\Psi^{(1)*}\frac{\partial}{\partial c}\Psi^{(2)} - \frac{\partial}{\partial c}\Psi^{(1)*}\cdot\Psi^{(2)}\right]. \tag{10}$$

This integral is independent of $c$ due to the equation (9). To make this inner product positive-definite, one restricts the states to the 'positive-frequency' solutions, i.e., superpositions of eingenfunctions of $c_P$ with positive eigenvalues. This procedure is analogous to that advocated by Wheeler, Misner and others in full quantum gravity (see, e.g., [10]).

This method of obtaining a Hilbert space of quantum states satisfying the LSCs cannot be applied as it stands to the de Sitter case because the LSCs do not take the form of Klein-Gordon equations. However, one can re-express this method in such a way that it can be applied to the de Sitter case as well. Consider first the Hilbert space of states with fixed $E_T$ disregarding the condition (4). Its elements are the square-integrable functions of the variables $c$ and $c_i$. Let $\psi(c, c_i)$ be an element in this Hilbert space. The time evolution of this state is governed by the Hamiltonian (3). Then, a function satisfying (4) can be obtained by smearing it over the time-translation group:

$$\Psi(c, c_i) = \int_{-\infty}^{+\infty}d\tau e^{-iH\tau}\psi(c, c_i). \tag{11}$$

This wavefunction is not in the original Hilbert space, i.e., it is not square-integrable. However, one can introduce a finite inner product as follows. Let $\Psi^{(i)}$ ($i = 1, 2$) be the wavefunctions obtained using eq. (11), from the wavefunctions $\psi^{(i)}$ in the original Hilbert space of square-integrable functions. Then define

$$(\Psi^{(1)}|\Psi^{(2)}) \equiv \int dc\prod_{i=1}^{5}dc_i\int_{-\infty}^{+\infty}d\tau\psi^{(1)*}e^{-iH\tau}\psi^{(2)}. \tag{12}$$

By working with the Fourier-decomposed expressions of the wavefunctions, one can show that

$$(\Psi^{(1)}|\Psi^{(2)}) = (\Psi^{(1)+}|\Psi^{(2)+})_{KG} - (\Psi^{(1)-}|\Psi^{(2)-})_{KG}, \tag{13}$$

where $\Psi^{(i)+}$ and $\Psi^{(i)-}$ are the positive- and negative-frequency parts, respectively, of $\Psi^{(i)}$. Since these two parts do not interact with each other, the two inner products $(\Psi^{(1)}|\Psi^{(2)})$ and $(\Psi^{(1)}|\Psi^{(2)})_{KG}$ are essentially equivalent.

This 'smearing method' can be formulated in other spacetimes where nontrivial LSCs are present. One first constructs a Hilbert space of states $|\psi^{(i)}\rangle$ disregarding the LSCs. Then one smears the states $|\psi^{(i)}\rangle$ over the continuous isometry group $G$:

$$|\Psi^{(i)}\rangle \equiv \int dc\, U(c)|\psi^{(i)}\rangle, \tag{14}$$

where $U(c)$ is the unitary operator corresponding to the element $c \in G$, and where $dc$ is the invariant measure on $G$, which is assumed to exist. If the states $|\Psi^{(i)}\rangle$ are well-defined objects—i.e., if its projections to the basis states of the original Hilbert space are finite—as in the case of flat spacetime with spatial topology $T^3$, then they are invariant under the continuous isometries, and satisfy the quantum LSCs as a result. (To show this rigorously, one needs to justify the formula $U(b)\int dc\, U(c)|\psi\rangle = \int dc\, U(bc)|\psi\rangle$. This can be done for the de Sitter case.) The new inner product is defined by

$$(\Psi^{(i)}|\Psi^{(j)}) \equiv \int dc\, \langle\psi^{(i)}|U(c)|\psi^{(j)}\rangle = \langle\psi^{(i)}|\Psi^{(j)}\rangle = \langle\Psi^{(i)}|\psi^{(j)}\rangle. \tag{15}$$

We expect this to be positive-definite in general because it is related to the formally positive-definite inner product among $|\Psi^{(i)}\rangle$ by

$$\langle\Psi^{(i)}|\Psi^{(j)}\rangle = \left(\int dc\right)(\Psi^{(i)}|\Psi^{(j)}), \tag{16}$$

where $\int dc$ is the volume of the group $G$. However, it is not clear if this ill-defined formula can be turned into a *proof* of the positive-definiteness.

In the next section the 'smearing method' will be applied to linearized gravity in de Sitter spacetime. In particular, it will be shown that the invariant states are well-defined and that the inner product (15) is indeed positive-definite in this case.

## 3.   Hilbert space of de Sitter-invariant states

The line element of de Sitter spacetime can be written as

$$ds^2 = -dt^2 + \cosh^2 t\, dl_3^2, \tag{17}$$

where the cosmological constant is $\Lambda = 3$. We denote the line element of the unit $N$-sphere ($S^N$) by $dl_N^2$. The physical graviton modes for the usual 'Euclidean vacuum' (see [4]) are

$$h_{ab}^{(L\sigma)} = k_L(t)Y_{ab}^{(L\sigma)} \tag{18}$$

with $h_{\mu 0}^{(L\sigma)} = 0$, where the $Y_{ab}^{(L\sigma)}$ are the normalized transverse-traceless eigentensors of the Laplace-Beltrami operator $D_a D^a$ on $S^3$ with eigenvalues $-L(L+2)+2$ ($L =$

$2, 3, \ldots, \infty$). The label $\sigma$ represents the other quantum numbers. The function $k_L(t)$ is given by

$$k_L(t) = \left[ \frac{L+1}{L(L+2)} \right]^{1/2} \cosh t \left( 1 + \frac{i \sinh t}{L+1} \right) \left( \frac{1 - i \sinh t}{\cosh t} \right)^{L+1}. \qquad (19)$$

The modes $h_{\mu\nu}^{(L\sigma)}$ satisfy $\nabla^\alpha h_{\alpha\mu} = h^\alpha{}_\alpha = 0$ and $(\nabla^\alpha \nabla_\alpha - 2)h_{\mu\nu} = 0$. Define the equivalence relation $h_{\mu\nu} \approx h_{\mu\nu} + \nabla_\mu \Lambda_\nu + \nabla_\nu \Lambda_\mu$ for the solutions to these equations. Then the equivalence classes with the representative elements $h_{\mu\nu}^{(L\sigma)}$ form an orthonormal basis of a unitary representation of the de Sitter group $SO(4,1)$ with the Klein-Gordon inner product

$$\langle h^{(1)}, h^{(2)} \rangle = i \int d\Sigma^\alpha \left( h_{\mu\nu}^{(1)*} \nabla_\alpha h^{(2)\mu\nu} - \nabla_\alpha h_{\mu\nu}^{(1)*} \cdot h^{(2)\mu\nu} \right). \qquad (20)$$

This inner product is well-defined because the pure-gauge modes are orthogonal to all modes including themselves.

Now, we will write down the 'smearing formula' (14) for the de Sitter group $SO(4,1)$. Let $dl_3^2 = d\chi^2 + \sin^2 \chi \, dl_2^2$ in (17). Then, one of the boost Killing fields, which can be regarded as a generator of the de Sitter group, is

$$B = \cos \chi \frac{\partial}{\partial t} - \tanh t \sin \chi \frac{\partial}{\partial \chi}. \qquad (21)$$

Any element $c \in SO(4,1)$ can be written as $c = hx$, where $h$ is an element of $SO(4)$ and where $x$ is a representative element of the right coset $SO(4,1)/SO(4)$, which is the four-dimensional hyperbolic space $H^4$. The element $x$, in turn, can be written as $x = e^{\beta B} a$, where $a$ is an element of $SO(4)$. Then the invariant measure $dc$ is the product of the measure $dx = da d\beta \sinh^3 \beta$ on $H^4$ (where $da$ is the measure on $S^3$) and the measure $dh$ on $SO(4)$. Thus, we can write

$$|\Psi\rangle \propto \int dh \, U(h) \int_0^\infty d\beta \sinh^3 \beta \, e^{\beta \hat{B}} \int da \, U(a) |\psi\rangle, \qquad (22)$$

where $e^{\beta \hat{B}} \equiv U(e^{\beta B})$. Since the non-$SO(4)$-invariant part of $|\psi\rangle$ is annihilated in (22), the state $|\psi\rangle$ can be assumed to be $SO(4)$ invariant without loss of generality. Note also that $P_0 \equiv \int dh U(h)$ is proportional to the projection operator onto the $SO(4)$-invariant subspace. Thus, by introducing a suitable normalization factor, the invariant state can be written as

$$|\Psi\rangle = P_0 \int_0^\infty d\beta \sinh^3 \beta \, e^{\beta \hat{B}} |\psi\rangle. \qquad (23)$$

According to this formula, it is necessary for the matrix elements of $e^{\beta \hat{B}}$ to fall off faster than $e^{-3\beta}$ for large $\beta$ in order to have the invariant states $|\Psi\rangle$ that are linear combinations of basis states of the original Fock space of gravitons with finite coefficients. It is known that the unitary representation formed by the one-graviton states is a direct sum of two irreducible representations belonging to the so-called discrete series (see [5]; the discrete series is referred to as the exceptional series there). The matrix elements of the representations in the discrete series are, by definition, square-integrable as functions of the group element (see [16], p. 349). Thus, the matrix elements of $e^{\beta \hat{B}}$ for one-particle states fall off faster than $e^{-(3/2)\beta}$ because the measure grows like $e^{3\beta}$. (In fact, they fall off at least as fast as $e^{-3\beta}$.) Since $N$-particle states are tensor products of one-particle states when considered as representations of $SO(4,1)$, the matrix elements of $e^{\beta \hat{B}}$ for $N$-particle states fall off faster than $e^{-(3N/2)\beta}$. Therefore, if $|\psi\rangle$ is an $N$-particle state with $N \geq 2$, then the coefficients in the basis-state expansion of the invariant state $|\Psi\rangle$ are finite and, hence, the state $|\Psi\rangle$ is well-defined. [Note in this context that one does not obtain a well-defined state by smearing the de Sitter-invariant vacuum state $|0\rangle$. Hence, there is no vaccuum (0-particle) state in the Hilbert space we are constructing here. However, the state $|0\rangle$ could be added to the Hilbert space *by hand* with the norm $\langle 0|0\rangle = 1$.]

Next, let us show that the inner product (15) is indeed positive-definite. If $|\psi^{(1)}\rangle$ and $|\psi^{(2)}\rangle$ are $SO(4)$ invariant, then the inner product of the de Sitter-invariant states $|\Psi^{(1)}\rangle$ and $|\Psi^{(2)}\rangle$ obtained through (23) is

$$(\Psi^{(1)}|\Psi^{(2)}) = \int_0^\infty d\beta \sinh^3 \beta \, \langle \psi^{(1)}|e^{\beta \hat{B}}|\psi^{(2)}\rangle. \tag{24}$$

We will first evaluate this inner product explicitly for certain states and show that it is positive-definite for those states. Then we will show that they form a basis of the Hilbert space of the de Sitter-invariant states of the form (23).

The boost generator $B$ changes the 'angular momentum' $L$ by 1:

$$\mathcal{L}_B h_{\mu\nu}^{(L\sigma)} = C(L,\sigma)h_{\mu\nu}^{(L+1,\sigma)} - C(L-1,\sigma)^* h_{\mu\nu}^{(L-1,\sigma)} + \text{pure-gauge modes}, \tag{25}$$

where the coefficients $C(L,\sigma)$ can be found in [8]. Then, the boost operator in the second quantized theory is $\hat{B} = \hat{B}_+ - \hat{B}_-$, where

$$\hat{B}_+ \equiv \sum_{L\sigma} C(L,\sigma)a^\dagger(L+1,\sigma)a(L,\sigma) \tag{26}$$

and $\hat{B}_- \equiv \hat{B}_+^\dagger$. The $a(L,\sigma)$ and $a^\dagger(L,\sigma)$ are the annihilation and creation operators, respectively, for the mode $h_{\mu\nu}^{(L\sigma)}$. Define

$$\hat{J} \equiv \sum_{L\sigma}(L+1)a^\dagger(L,\sigma)a(L,\sigma). \tag{27}$$

The commutators of $\hat{B}_\pm$ and $\hat{J}$ turn out to be

$$[\hat{J}, \hat{B}_\pm] = \pm\hat{B}_\pm, \qquad [\hat{B}_-, \hat{B}_+] = \frac{1}{2}\hat{J}. \tag{28}$$

These commutators imply that $\hat{B}$, $\hat{C} \equiv -i(\hat{B}_+ + \hat{B}_-)$ and $i\hat{J}$ form an $SL(2, R)$ algebra with real coefficients. Note that this algebra is not a subalgebra of the de Sitter algebra under consideration because the operators $\hat{C}$ and $i\hat{J}$ are not elements of the latter algebra.

The $SL(2, R)$ algebra given by (28) enables us to compute the matrix elements of $e^{\beta\hat{B}}$ for the (multi-particle) states $|\psi(m, \rho)\rangle$ with the properties

$$\hat{B}_-|\psi(m, \rho)\rangle = 0, \tag{29}$$

$$\hat{J}|\psi(m, \rho)\rangle = m|\psi(m, \rho)\rangle, \tag{30}$$

where the label $\rho$ represents the quantum numbers other than $m$. (See [15] for a detailed discussion of the unitary representations of $SL(2, R)$.) We choose the labels $\rho$ so that $\langle\psi(m, \rho)|\psi(m, \rho')\rangle = \delta_{\rho\rho'}$. By differentiating the function $f(\beta) \equiv \langle\psi(m, \rho)|e^{\beta\hat{B}}|\psi(m', \rho')\rangle$ and using eq. (29), we obtain

$$f'(\beta) = -i\langle\psi(m, \rho)|\hat{C}e^{\beta\hat{B}}|\psi(m', \rho')\rangle = i\langle\psi(m, \rho)|e^{\beta\hat{B}}\hat{C}|\psi(m', \rho')\rangle. \tag{31}$$

Using the Baker-Hausdorff lemma

$$e^{\beta\hat{B}}\hat{C}e^{-\beta\hat{B}} = \sum_{n=0}^{\infty} \frac{\beta^n}{n!}\underbrace{[\hat{B}, [\hat{B}, [\cdots[\hat{B}, \hat{C}]\cdots]]]}_{n}, \tag{32}$$

and the commutators $[\hat{B}, \hat{C}] = i\hat{J}$ and $[\hat{B}, \hat{J}] = -i\hat{C}$, we find

$$e^{\beta\hat{B}}\hat{C} = \hat{C}e^{\beta\hat{B}}\cosh\beta + i\hat{J}e^{\beta\hat{B}}\sinh\beta. \tag{33}$$

Substituting this in (31) and using eq. (30), we obtain a differential equation satisfied by $f(\beta)$:

$$f'(\beta) = -m\tanh(\beta/2)f(\beta). \tag{34}$$

This equation and the condition $f(0) = \delta_{mm'}\delta_{\rho\rho'}$ imply that

$$\langle\psi(m, \rho)|e^{\beta\hat{B}}|\psi(m', \rho')\rangle = \frac{1}{\cosh^{2m}(\beta/2)}\delta_{mm'}\delta_{\rho\rho'}. \tag{35}$$

By substituting (35) in (24), we find that the de Sitter-invariant states

$$|\Psi(m, \rho)\rangle = \left[\frac{(m-2)(m-3)}{8}\right]^{1/2} P_0 \int_0^{\infty} d\beta \sinh^3\beta \, e^{\beta\hat{B}}|\psi(m, \rho)\rangle \tag{36}$$

form an orthonormal set. Thus, the inner product (24) is positive-definite for these states. Also, by moving $\hat{B}_-$'s to the right in eq. (36) using the commutators given by (28), one can show that $|\psi(m, \rho)\rangle$ is the lowest eigenstate of $\hat{J}$ in the eigenstate expansion of the invariant state $|\Psi(m, \rho)\rangle$.

Next, we will show that the states $|\Psi(m,\rho)\rangle$ form a basis for the Hilbert space of the invariant states of the form (23). It is sufficient to show that the invariant state $|\Psi\rangle$ obtained from an eigenstate $|\psi\rangle$ of $\hat{J}$ with eigenvalue $m_1$ is a linear combination of the states $|\Psi(m,\rho)\rangle$. Let the minimum eigenvalue of $\hat{J}$ in the eigenstate expansion of $|\Psi\rangle$ be $m_0$. Then one can write

$$|\Psi\rangle = |\Psi^{(1)}\rangle + \sum_\rho c(m_0,\rho)|\Psi(m_0,\rho)\rangle, \tag{37}$$

where the $c(m_0,\rho)$ are constants and where the minimum eigenvalue of $\hat{J}$ in the expansion of the invariant state $|\Psi^{(1)}\rangle$ is at least $m_0 + 1$. This is because the part of $|\Psi\rangle$ with $\hat{J} = m_0$ must be annihilated by $\hat{B}_-$. (Otherwise, the state $|\Psi\rangle$ would not be annihilated by $\hat{B}$.) Applying the decomposition (37) to $|\Psi^{(1)}\rangle$, one has

$$|\Psi^{(1)}\rangle = |\Psi^{(2)}\rangle + \sum_\rho c(m_0 + 1,\rho)|\Psi(m_0 + 1,\rho)\rangle, \tag{38}$$

where the minimum value of $\hat{J}$ in the invariant state $|\Psi^{(2)}\rangle$ is at least $m_0 + 2$. Continuing in this manner, one obtains

$$|\Psi\rangle = |\Psi'\rangle + \sum_{m=m_0}^{m_1} \sum_\rho c(m,\rho)|\Psi(m,\rho)\rangle, \tag{39}$$

where the lowest value of $\hat{J}$ in $|\Psi'\rangle$ is at least $m_1 + 1$. The invariant state $|\Psi'\rangle$ is obtained by smearing the following state:

$$|\psi'\rangle = |\psi\rangle - \sum_{m=m_0}^{m_1} \sum_\rho [(m-2)(m-3)/8]^{1/2} c(m,\rho)|\psi(m,\rho)\rangle. \tag{40}$$

To show that $|\Psi'\rangle = 0$, we assume that $|\Psi'\rangle \neq 0$ and will derive a contradiction. First, note that $(\Psi'|\Psi') = \langle\psi'|\Psi'\rangle = 0$ because the maximum value of $\hat{J}$ in $|\psi'\rangle$ is less than or equal to $m_1$. Let the minimum eigenvalue of $\hat{J}$ in $|\Psi'\rangle$ be $m_1'$ ($> m_1$). Then one can write

$$|\Psi'\rangle = |\Psi''\rangle + \sum_\rho c(m_1',\rho)|\Psi(m_1',\rho)\rangle, \tag{41}$$

where the minimum value of $\hat{J}$ in the invariant state $|\Psi''\rangle$ is at least $m_1' + 1$. The invariant states $|\Psi''\rangle$ and $|\Psi(m_1',\rho)\rangle$ are orthogonal to one another with respect to the inner product (24) because $\langle\psi(m_1',\rho)|\Psi''\rangle = 0$. On the other hand, the states $|\Psi'\rangle$ and $|\Psi''\rangle$ are orthogonal to themselves. Then, by evaluating the norm of (41), we have $|c(m_1',\rho)|^2 = 0$, which is a contradiction. Hence $|\Psi'\rangle = 0$.

Thus, we have constructed a basis $\{|\Psi(m,\rho)\rangle\}$ of the Hilbert space of the de Sitter-invariant states of the form (23). The inner product (24) is positive-definite and is given by $\langle\Psi(m,\rho)|\Psi(m',\rho')\rangle = \delta_{mm'}\delta_{\rho\rho'}$.

It will be interesting to see if the method proposed in this article can be applied to other spatially-compact spacetimes with boost-like isometries (see [13]). In particular, it will be nice to prove that the inner product (15) is positive-definite without using specific properties of the isometry group. Finally, it will be very interesting to incorporate nonlinear effects (i.e., interactions) and construct (tree-level) perturbative quantum gravity in de Sitter spacetime using the Hilbert space presented in this article to see if one obtains a physically acceptable theory.

## Acknowledgement

Discussions with Dr. Roberto Camporesi were very helpful in understanding the unitary representations of the de Sitter group.

## References

[1] Arnowitt R., Deser S., Misner C.W., *The dynamics of general relativity. Gravitation: An introduction to current research*, ed. L. Witten, Wiley, 1962, p.p. 227-265.

[2] Brill D., *Isolated solutions in general relativity. University of Maryland Technical Report No. 71-076 (1971).*

[3] Brill D., Deser S., *Instability of closed spaces in general relativity. Commun. Math. Phys., Vol. 32 (1973), pp. 291-304.*

[4] Gibbons G., Hawking S.W., *Cosmological event horizons, thermodynamics, and particle creation. Phys. Rev. D, Vol. 15 (1977), pp. 2738-2751.*

[5] Higuchi A., *Quantum fields of nonzero spin in de Sitter spacetime, PhD thesis. Yale University, 1987, unpublished.*

[6] Higuchi A., *Linearization instabilities of de Sitter spacetime: I. Class. Quantum Grav., Vol. 8 (1991), pp. 1961-1981.*

[7] Higuchi A., *Linearization instabilities of de Sitter spacetime: II. Class. Quantum Grav., Vol. 8 (1991), pp. 1983-2004.*

[8] Higuchi A., *Linearized gravity in de Sitter spacetime as a representation of SO(4,1). Class. Quantum Grav., Vol. 8 (1991), pp. 2005-2021.*

[9] Higuchi A., *Linearized quantum gravity in flat space with toroidal topology. Class. Quantum Grav., Vol. 8 (1991), pp. 2023-2034.*

[10] Kuchař K., *Canonical methods of quantization. Quantum gravity 2, ed. R. Penrose, D.W. Sciama, Oxford, 1981, p.p. 329-376.*

[11] Moncrief V., *Invariant states and quantized gravitational perturbations.* Phys. Rev. D, Vol. 18 (1978), pp. 983-989.

[12] Moncrief V., *Quantum linearization instabilities.* Gen. Rel. Grav., Vol. 10 (1979), pp. 93-97.

[13] Moncrief V., *Boost symmetries in spatially compact spacetimes with a cosmological constant.* Yale University preprint, 1992.

[14] Moncrief V., *Recent advances and open problems in linearization stability.* This volume.

[15] Sugiura M., *Unitary representations and harmonic analysis.* Wiley, 1975.

[16] Warner G., *Harmonic analysis on semi-simple Lie groups I.* Springer-Verlag, 1972.

# What is the True Description of Charged Black Holes?

## GARY T. HOROWITZ

University of California, Santa Barbara

**ABSTRACT**

If string theory describes nature, then charged black holes are not described by the Reissner-Nordström solution. This solution must be modified to include a massive dilaton. In the limit of vanishing dilaton mass, the new solution can be found by a generalization of the Harrison transformation for the Einstein-Maxwell equations. These two solution generating transformations and the resulting black holes are compared. It is shown that the extremal black hole with massless dilaton can be viewed as the "square root" of the extremal Reissner-Nordström solution. When the dilaton mass is included, extremal black holes are repulsive, and it is energetically favorable for them to bifurcate into smaller holes.

## 1 INTRODUCTION

It is a pleasure to honor Dieter Brill on the occasion of his sixtieth birthday. Over the years, Dieter has worked on many aspects of general relativity. But two of his recent interests are negative energy (in higher dimensional theories) (Brill and Pfister 1989; Brill and Horowitz 1991), and the possibility that extremal charged black holes can quantum mechanically bifurcate (Brill 1992). I would like to describe some recent work which touches on both of these areas.

For many years, it has been widely believed that static charged black holes in nature are accurately described by the Reissner-Nordström solution. This is the result of two powerful theorems: the uniqueness theorem (Israel 1968) which proves that the only static black hole solution to the Einstein-Maxwell equations is Reissner-Nordström, and the no-hair theorem (Chase 1970; Bekenstein 1972) which shows that if one adds simple additional matter fields to the theory, the only static black hole is still Reissner-Nordström. The extra fields either fall into the hole or radiate out to infinity. Although this solution seems physically reasonable in most respects, its extremal limit has two puzzling features. The first is that the event horizon is infinitely far away along a spacelike geodesic, but only a finite distance away along a timelike or null geodesic. The second is that the extremal black hole has zero Hawking temperature,

157

but nonzero entropy as measured by its surface area.

Recently, a slightly different picture of charged black holes has emerged. This is the result of studying gravitational consequences of string theory. A new class of charged black hole solutions have been found which do not have the puzzling features of Reissner-Nordström. (For recent reviews, see Horowitz 1992 and Sen 1992.) In the extremal limit, the area of the event horizon of these new black holes goes to zero, which is consistent with zero entropy. From another standpoint (for a magnetically charged black hole) the horizon moves off to infinity in all directions: timelike, space-like, and null. These solutions should be of particular interest to Dieter Brill since we will see that they are even more likely to bifurcate than the analogous black holes in the Einstein-Maxwell theory. In addition, although the total energy remains positive for these four dimensional holes, we will see that some of them resemble negative energy objects in that they are repulsive.

In addition to the metric $g_{\mu\nu}$ and Maxwell field $F_{\mu\nu}$, string theory predicts the existence of a scalar field called the dilaton. The dilaton has an exponential coupling to $F^2$ which implies that it cannot remain zero when $F^2 \neq 0$. This is the key difference from the previous no-hair theorems and shows that the charged black holes will differ from the Reissner-Nordström solution. Although the dilaton is massless while supersymmetry is unbroken, it is expected to become massive at low energies. The breaking of supersymmetry in string theory is a nonperturbative effect which is not well understood at this time. So we cannot yet calculate reliably the dilaton potential from first principles. However the qualitative behavior of the solutions can be obtained by considering the simplest possibility $m^2\phi^2$. We are thus led to consider the low energy action:

$$S = \int d^4x\sqrt{-g}\left[R - 2(\nabla\phi)^2 - 2m^2\phi^2 - e^{-2\phi}F^2\right] \tag{1}$$

with equations of motion

$$\nabla_\mu(e^{-2\phi}F^{\mu\nu}) = 0, \tag{2a}$$

$$\nabla^2\phi - m^2\phi = -\tfrac{1}{2}e^{-2\phi}F^2, \tag{2b}$$

$$G_{\mu\nu} = 2\nabla_\mu\phi\nabla_\nu\phi + 2e^{-2\phi}F_{\mu\rho}F_\nu{}^\rho$$
$$- g_{\mu\nu}[(\nabla\phi)^2 + m^2\phi^2 + \tfrac{1}{2}e^{-2\phi}F^2] \tag{2c}$$

When $F_{\mu\nu} = 0$, these equations reduce to those of the Einstein-massive scalar field theory. The no-hair theorems show that the only black hole solutions are Schwarzschild with $\phi = 0$. However, as we just remarked, when the charge is nonzero, the dilaton will not be constant, and will alter the geometry. It does not appear possible to write the exact solution to (2) describing a static, spherically symmetric, charged black hole in closed form. But we will see that by combining general arguments, approximation methods, and numerical results, we can obtain a fairly complete understanding of their properties.

## 2 SOLUTIONS WITH A MASSLESS DILATON

Let us begin by considering the limit where the dilaton mass is negligible. It will be shown in the next section that this is appropriate for small black holes. Even with $m = 0$, the exponential coupling of the dilaton to the Maxwell field appears to prevent a simple closed form solution. However exact solutions can be found which are, in some respects, even simpler than the Reissner-Nordström solution (Gibbons 1982; Gibbons and Maeda 1988; Garfinkle, Horowitz and Strominger 1991). In the Einstein-Maxwell theory there is a transformation (Harrison 1968) which maps stationary vacuum solutions into stationary charged solutions. It turns out that there is a similar transformation for the theory (1) with $m = 0$ (Hassan and Sen 1992; Maharana and Schwarz 1992). In this section, we compare these two transformations and the resulting charged black holes. For simplicity, we will consider only static solutions.

We start with the familiar Einstein-Maxwell theory. Suppose

$$ds^2 = -\lambda dt^2 + h_{ij} dx^i dx^j \tag{3}$$

is a static vacuum solution. The Harrison transformation begins by rewriting this metric as

$$ds^2 = -\lambda dt^2 + \lambda^{-1} \gamma_{ij} dx^i dx^j \tag{4}$$

where $\gamma_{ij} = \lambda h_{ij}$. The charged solution is obtained by keeping $\gamma_{ij}$ unchanged and setting

$$\tilde{\lambda} = \frac{\lambda}{(\cosh^2 \alpha - \lambda \sinh^2 \alpha)^2} \tag{5a}$$

$$A_t = \frac{(\lambda - 1)\sinh 2\alpha}{2(\cosh^2 \alpha - \lambda \sinh^2 \alpha)} \tag{5b}$$

where $\alpha$ is a free parameter. Clearly, $\alpha = 0$ leaves the solution unchanged. Since the factor in the denominator appears frequently, it is convenient to introduce the notation

$$e^{-2\phi} \equiv \cosh^2 \alpha - \lambda \sinh^2 \alpha \tag{6}$$

Notice that $\phi$, at this point, has nothing to do with the dilaton. We are considering solutions to the Einstein-Maxwell equations and $\phi$ is simply defined by (6). Equation (5a) can now be rewritten $\tilde{\lambda} = \lambda e^{4\phi}$ and the new metric can be expressed

$$ds^2 = -\lambda e^{4\phi} dt^2 + e^{-4\phi} h_{ij} dx^i dx^j \tag{7}$$

Let us apply this transformation to the Schwarzschild solution:

$$\lambda = 1 - \frac{2M_0}{\rho}, \qquad h_{ij}\,dx^i dx^j = \left(1 - \frac{2M_0}{\rho}\right)^{-1} d\rho^2 + \rho^2 d\Omega, \qquad (8)$$

In this case, $e^{-2\phi} = 1 + 2M_0 \sinh^2 \alpha / \rho$. So the new metric becomes

$$ds^2 = -\left(1 - \frac{2M_0}{\rho}\right)\left(1 + \frac{2M_0 \sinh^2 \alpha}{\rho}\right)^{-2} dt^2 +$$
$$\left(1 + \frac{2M_0 \sinh^2 \alpha}{\rho}\right)^2 \left[\left(1 - \frac{2M_0}{\rho}\right)^{-1} d\rho^2 + \rho^2 d\Omega\right] \qquad (9)$$

This can be simplified by introducing a new radial coordinate $r = \rho + 2M_0 \sinh^2 \alpha$. Then

$$\rho(\rho - 2M_0) = (r - 2M_0 \sinh^2 \alpha)(r - 2M_0 \cosh^2 \alpha) \equiv (r - r_-)(r - r_+) \qquad (10)$$

where we have defined $r_+ \equiv 2M_0 \cosh^2 \alpha$ and $r_- \equiv 2M_0 \sinh^2 \alpha$. The metric (9) now becomes:

$$ds^2 = -\frac{(r - r_+)(r - r_-)}{r^2} dt^2 + \frac{r^2}{(r - r_+)(r - r_-)} dr^2 + r^2 d\Omega \qquad (11)$$

This is the familiar form of the Reissner-Nordström solution with the event horizon at $r = r_+$ and the inner horizon at $r = r_-$. Notice that the original Schwarzschild singularity $\rho = 0$ has become the inner horizon, while the Reissner-Nordström singularity $r = 0$ corresponds to a negative value of $\rho$ in the original Schwarzschild metric. The Maxwell field is

$$F_{rt} = \frac{M_0 \sinh 2\alpha}{r^2} \qquad (12)$$

which shows that the total charge is $Q = M_0 \sinh 2\alpha$. The total mass is $M = (r_+ + r_-)/2 = M_0 \cosh 2\alpha$. The extremal limit is obtained by taking $\alpha \to \infty$, $M_0 \to 0$ keeping $M_0 e^{2\alpha}$ finite. In this limit, the two horizons coincide $r_+ = r_-$. The event horizon becomes degenerate, with nonzero area but vanishing Hawking temperature.

Now we turn to the theory with a massless dilaton. There is a solution generating transformation directly analogous to the Harrison transformation. It is usually discussed in terms of the conformally rescaled metric $\hat{g}_{\mu\nu} = e^{2\phi} g_{\mu\nu}$ (where $\phi$ is now the dilaton) which is called the string metric. This is the metric that a string directly couples to. To avoid confusion, the metric we have been using until now will be called the Einstein metric (since it has the standard Einstein action). If one starts with a static vacuum solution with (string) metric of the form $\widehat{ds}^2 = -\lambda dt^2 + h_{ij} dx^i dx^j$ and $\phi = 0$, then a new solution is obtained by keeping $h_{ij}$ unchanged, and setting

$$\tilde{\lambda} = \frac{\lambda}{(\cosh^2 \alpha - \lambda \sinh^2 \alpha)^2} \tag{13a}$$

$$A_t = \frac{(\lambda - 1) \sinh 2\alpha}{2\sqrt{2}(\cosh^2 \alpha - \lambda \sinh^2 \alpha)} \tag{13b}$$

$$e^{-2\phi} = \cosh^2 \alpha - \lambda \sinh^2 \alpha \tag{13c}$$

The similarity to (5) is remarkable. The change in $\lambda$ is identical to the Harrison transformation, and the expression for the vector potential $A_t$ differs only by a factor of $\sqrt{2}$. In addition, the solution for the dilaton is exactly what we called $\phi$ in (6). (This is why we chose that definition for $\phi$.) To compare with the Reissner-Nordström solution, we should rescale back to the Einstein metric:

$$ds^2 = -\tilde{\lambda} e^{-2\phi} dt^2 + e^{-2\phi} h_{ij} dx^i dx^j \tag{14}$$

In terms of the original $\lambda$ we get

$$ds^2 = -\lambda e^{2\phi} dt^2 + e^{-2\phi} h_{ij} dx^i dx^j \tag{15}$$

If one now compares (15) with the solution to the Einstein-Maxwell theory (7), one sees that *the only difference in the solutions with and without a massless dilaton is a factor of two in the exponent*.

Applying this transformation to the Schwarzschild solution (8), we obtain the metric

$$ds^2 = -\left(1 - \frac{2M_0}{\rho}\right)\left(1 + \frac{2M_0 \sinh^2 \alpha}{\rho}\right)^{-1} dt^2 +$$
$$\left(1 + \frac{2M_0 \sinh^2 \alpha}{\rho}\right)\left[\left(1 - \frac{2M_0}{\rho}\right)^{-1} d\rho^2 + \rho^2 d\Omega\right] \tag{16}$$

together with the Maxwell field and dilaton

$$A_t = -\frac{M_0 \sinh 2\alpha}{\sqrt{2}(\rho + 2M_0 \sinh^2 \alpha)} \tag{17a}$$

$$e^{-2\phi} = 1 + \frac{2M_0 \sinh^2 \alpha}{\rho} \tag{17b}$$

The factor of two difference in the exponent has the following important consequence: The surface $\rho = 0$ in the metric (16) still has zero area and does not become an inner horizon. The total mass is now $M = M_0 \cosh^2 \alpha$ and the charge is $Q =$

$M_0 \sinh 2\alpha/\sqrt{2}$. Introducing the same radial coordinate as before, $r = \rho + 2M_0 \sinh^2 \alpha$, the solution takes the remarkably simple form

$$ds^2 = -\left(1 - \frac{2M}{r}\right) dt^2 + \left(1 - \frac{2M}{r}\right)^{-1} dr^2 + r\left(r - \frac{Q^2}{M}\right) d\Omega$$

$$F_{rt} = \frac{Q}{r^2} \qquad e^{2\phi} = 1 - \frac{Q^2}{Mr} \tag{18}$$

Note that the metric in the $r - t$ plane is identical to Schwarzschild! (Although the mass is different from the one we started with.) There is an event horizon at $r = 2M$ and no inner horizon. The only difference from Schwarzschild is that the area of the spheres is reduced by an amount depending on the charge. This area goes to zero when $r = Q^2/M$ ($\rho = 0$) resulting in a curvature singularity.

Clearly, the solution (18) describes a black hole only when the singularity is inside the event horizon i.e. $Q^2 < 2M^2$. Linearized perturbations about this solution have been studied (Holzhey and Wilczek 1992) and remain well behaved outside the horizon: the black hole is stable. In the extremal limit, $Q^2 = 2M^2$, the event horizon shrinks down to zero area and becomes singular. The resulting spacetime describes neither a black hole (with a spacelike singularity), nor a conventional naked singularity (which is timelike). Since the causal structure in the $r - t$ plane is independent of $Q$, the extremal black hole has a null singularity. Its Penrose diagram is identical to the region $r > 2M$ of Schwarzschild, with both the future and past horizons, $r = 2M$, replaced by singularities.

The fact that the area of the event horizon goes to zero in the extremal limit has two important consequences. First, it follows immediately from the area theorem that there is no classical process by which a nearly extremal black hole can become extremal (Garfinkle 1992). This is also true for the Reissner-Nordström solution, but the proof is more involved. Second, if one interprets the area as a measure of the entropy of a black hole through the Hawking-Bekenstein formula, $S = A/4$, then the extremal black holes have zero entropy. This is what one might expect for a ground state. The nonzero area of the extremal Reissner-Nordström solution has always been puzzling from this standpoint, since it seems to indicate a highly degenerate ground state. Although this black hole resolves one puzzle, it creates a new one. For a static, spherically symmetric black hole, the Hawking temperature can be found by analytically continuing to imaginary time and computing what periodicity is necessary to avoid a conical singularity. Since this only involves the $r - t$ part of the metric which is identical to Schwarzschild for the solution (18), the Hawking temperature is also identical: $T = 1/8\pi M$, independent of $Q$! In particular, the Hawking temperature does not go to zero as one approaches the extremal limit.

This leads one to worry that the black hole might continue to evaporate past the extremal limit to form a true naked singularity. However, the backreaction clearly becomes important as one approaches the extremal limit, and a complete calculation is necessary to understand the evolution. (For a review of recent work in this area, see Harvey and Strominger 1992.)

In terms of the original Schwarzschild mass $M_0$ and transformation parameter $\alpha$, the extremal limit again corresponds to $M_0 \to 0, \alpha \to \infty$ keeping $M_0 e^{2\alpha}$ fixed. In this limit, $\lambda \to 1$ and $h_{ij} \to \delta_{ij}$. So the extremal black hole metrics (7) and (15) are determined entirely by the function $e^{2\phi}$. Since the solutions with and without a dilaton differ only by a factor of two in the exponent, *one can view the extremal black hole with dilaton as the square root of the extremal Reissner-Nordström solution.* It is not clear what the physical interpretation of this is. On the one hand, the statement is clearly coordinate dependent. It holds in isotropic coordinates where the spatial metric is manifestly conformally flat. On the other hand, the statement applies not just to single black holes but extends to multi-black hole solutions as well. In the Einstein-Maxwell theory, since there is no force between extremal black holes (with the same sign of the charge), there exist static, multi-black hole solutions. These are the Majumdar-Papapetrou solutions (Majumdar 1947; Papapetrou 1947; Hartle and Hawking 1972) and take the form

$$ds^2 = -e^{4\phi}dt^2 + e^{-4\phi}d\vec{x} \cdot d\vec{x} \tag{19}$$

with

$$e^{-2\phi} = 1 + \Sigma_i \frac{2M_i}{|\vec{x} - \vec{x}_i|} \tag{20}$$

In exact analogy, the extremal black holes with a massless dilaton also have no force between them. The fact that $Q^2 = 2M^2$ rather than $Q^2 = M^2$ in the extremal limit means that there is a stronger electrostatic repulsion which exactly cancels the additional attractive force due to the dilaton. The multi-black hole solution turns out to be simply the "square root" of (19):

$$ds^2 = -e^{2\phi}dt^2 + e^{-2\phi}d\vec{x} \cdot d\vec{x} \tag{21}$$

with $e^{-2\phi}$ again given by (20).

The solution generating transformations (5) and (13) produce solutions with electric charge. But one can easily obtain solutions with magnetic charge by applying a duality transformation. When the dilaton is present, the duality transformation corresponds to replacing $F_{\mu\nu}$ and $\phi$ with $\tilde{F}_{\mu\nu} \equiv \frac{1}{2}e^{-2\phi}\epsilon_{\mu\nu}{}^{\lambda\rho}F_{\lambda\rho}$ and $\tilde{\phi} = -\phi$. This leaves the stress energy tensor invariant and hence does not change the Einstein metric. But since the dilaton changes sign, it does change the string metric. For electrically charged black holes, the string metric is obtained by multiplying (21) by $e^{2\phi}$. The result is that the spatial metric is completely flat and $g_{tt}$ is identical to the Majumdar-Papapetrou solutions. On the other hand, since $\phi$ changes sign under duality, the string metric describing several magnetically charged black holes is obtained by multiplying (21) by $e^{-2\phi}$ where $\phi$ is still given by (20). The resulting metric has $g_{tt} = -1$ and a spatial metric which is identical to the Majumdar-Papapetrou solution. In other words, the string metric describing several extremal black holes is obtained from the analogous Einstein-Maxwell solution by flattening either the time or space part of the metric (depending on the type of charge) leaving the rest unchanged.

The magnetically charged case is of particular interest. Recall that in the spatial part of the Reissner-Nordström metric, the horizon moves off to infinity as one approaches the extremal limit. The extremal geometry resembles an infinite throat attached to an asymptotically flat region. Since $g_{tt} = -1$ for the string solution, the horizon is now infinitely far away in timelike and null directions as well as space-like. It has been suggested that this type of geometry may play an important role in explaining what happens to the information that falls into a black hole, after it evaporates (Banks, Dabholkar, Douglas and O'Loughlin 1992; Banks and O'Loughlin 1992; Banks, O'Loughlin, and Strominger 1992).

## 3 SOLUTIONS WITH A MASSIVE DILATON

We now wish to include the effects of the dilaton mass. (The following results were obtained in collaboration with J. Horne (Horne and Horowitz 1992). For another discussion of black holes with a massive dilaton see Gregory and Harvey 1992.) Unfortunately, exact black hole solutions with a massive dilaton do not seem to be expressible in closed form. Given the simple form of the Reissner-Nordström (11) and massless dilaton (18) solutions, we will assume a metric of the form

$$ds^2 = -\lambda dt^2 + \lambda^{-1} dr^2 + R^2 d\Omega \tag{22}$$

where $\lambda$ and $R$ are functions of $r$ only.

We first consider the asymptotic form of the solution. We are interested in solutions that are asymptotically flat, which requires that $\phi \to 0$ at infinity. For large $r$, the right hand side of the dilaton equation (2b) behaves like $Q^2/r^4$, and the derivative term becomes negligible. The dilaton thus falls off like $|\phi| \sim Q^2/m^2 r^4$. Recall that a massless scalar field falls off as $1/r$ while a massive field with localized sources falls off exponentially. Here we have a massive field with a source that falls off polynomially, which results in the unusual asymptotic form of $\phi$. Note that the limit $m \to 0$ keeping $r$ fixed is not well behaved. This is because $|\phi| \sim Q^2/m^2 r^4$ only in the asymptotic region where $r$ is large compared to the Compton wavelength of the dilaton, i.e. $rm \gg 1$.

Now consider the metric equation. At large distances, the Maxwell contribution to the stress tensor will be $O(1/r^4)$ while all terms involving the dilaton will fall off much faster. Thus, the asymptotic form of the field equation is identical to the Einstein-Maxwell theory, and the solution is just Reissner-Nordström. One can calculate the first order correction to the Reissner-Nordström solution by treating the dilaton terms as a perturbation and one finds

$$\lambda \sim 1 - \frac{2M}{r} + \frac{Q^2}{r^2} - \frac{Q^4}{5m^2r^6}, \tag{23a}$$

$$R \sim r\left(1 - \frac{2Q^4}{7m^4r^8}\right). \tag{23b}$$

Although the value of the dilaton mass is not known, one can place a lower limit from the fact that the $1/r^2$ force law has been confirmed down to scales of about 1 cm. This requires that the dilaton Compton wavelength be less than 1 cm or $m > 10^{-5}$eV. It follows that for a solar mass black hole, $Mm > 10^5$. In other words, the black hole is much larger than the Compton wavelength of the dilaton. (In geometrical units, the dilaton mass $m$ has dimensions of inverse length and is related to the inverse Compton wavelength. On the other hand, the black hole mass $M$ has dimensions of length and is related to the size of the black hole. Thus $Mm$ is dimensionless. Alternatively, one can view all quantities as dimensionless and measured in Planck units.) When $Mm \gg 1$, the deviation from Reissner-Nordström remains small until one is well inside the event horizon: At the horizon, $\delta\lambda < (Q/M)^4(Mm)^{-2} \ll 1$ and $\delta R < (Q/M)^4(Mm)^{-4} \ll 1$. So outside the horizon, the solution will be very similar to Reissner-Nordström. Since $\phi \sim Q^2/m^2r^4 < (Q/M)^2(Mm)^{-2} \ll 1$, the dilaton remains small everywhere outside the event horizon. For $Q \approx M$, the inner horizon of Reissner-Nordström is close to the event horizon and the corrections due to the dilaton will still remain small there. So unlike the massless dilaton solution, a large black hole coupled to a massive dilaton can have two horizons. This is what one should expect physically. When $Mm \gg 1$, the dilaton is essentially stuck in the bottom of its potential well and does not affect the solution significantly.

Even though the dilaton does not qualitatively affect the geometry, it has an important consequence near the extremal limit. Like Reissner-Nordström, the horizons will coalesce in the extremal limit. However, the condition for when this occurs is no longer exactly $Q^2 = M^2$. The new condition is found by asking when $\lambda$ has a double zero. If we substitute the unperturbed condition, $Q^2 = M^2$ and $r = M$, into the correction term in (23a) we find that to first order in the dilaton, $\lambda = 0$ is equivalent to

$$r^2 - 2Mr + \left(Q^2 - \frac{1}{5m^2}\right) = 0 \qquad (24)$$

This will have a double zero when

$$M^2 = Q^2 - \frac{1}{5m^2} \qquad (25)$$

So $M^2 < Q^2$ in the extremal limit, just like the massless dilaton solutions. But at infinity, we have seen that the solution always approaches the Reissner-Nordström solution, for which objects with $Q > M$ are repulsive. Since $Q^2$ is strictly greater than $M^2$ in the extremal limit, nonextremal black holes can also have a charge greater than their mass. We conclude that *nearly extremal and extremal black holes with a massive dilaton are repulsive.* Roughly speaking, this is a result of the fact that the presence of the dilaton near the horizon allows $Q^2 > M^2$ but the dilaton mass cuts off the attractive dilaton force at large separations. This appears to be the first example of gravitationally bound repulsive objects.

Since the extremal black hole has a degenerate horizon, its Hawking temperature vanishes. However, this does not guarantee that it is stable. Extremal black holes might quantum mechanically bifurcate. Brill considered this possibility in the context of the Einstein-Maxwell theory (Brill 1992). He found an instanton describing the splitting of an extremal Reissner-Nordström throat, and argued that there should exist a similar finite action instanton describing the splitting of the extremal Reissner-Nordström black hole. This has also been considered in the theory with a massless dilaton (Kallosh, Linde, Ortin, Peet and Van Proeyen 1992; Kallosh, Ortin and Peet 1992). However in both of these theories, the total mass $M$ of the black hole is proportional to its charge $Q$ in the extremal limit. Thus one black hole of charge $Q$ has the same mass as $n$ black holes of charge $Q/n$. The solutions are degenerate. This is not the case for the theory with a massive dilaton. It is now energetically favorable for an extremal black hole to bifurcate. From (25), the mass of a single extremal black hole of charge $Q$ is $M_1 = (Q^2 - 1/5m^2)^{\frac{1}{2}}$ while the mass for $n$ widely separated extremal black holes with charge $Q/n$ is

$$M_n = n\left[\left(\frac{Q}{n}\right)^2 - \frac{1}{5m^2}\right]^{\frac{1}{2}} = \left[Q^2 - \frac{n^2}{5m^2}\right]^{\frac{1}{2}} . \qquad (26)$$

Clearly, $M_n$ is a decreasing function of $n$. These black holes should be even more likely to bifurcate than the ones considered previously.

This expression for $M_n$ only applies to large black holes, since it is based on (25). (More precisely, it is valid when $mM_n/n \gg 1$.) However the conclusion is likely to hold more generally. This is because it follows from the fact that extremal black holes are repulsive, which in turn is a consequence of two general properties of the extremal solutions: they always approach Reissner-Nordström asymptotically and have $Q^2 > M^2$. Thus extremal black holes will probably continue to bifurcate until they have a single unit of charge. At this point the black hole is small compared to the Compton wavelength of the dilaton, $Mm \ll 1$. (Black holes of this type can also be obtained by starting with a large hole with small charge, and letting it evaporate. The extremal limit will not be reached until $Mm \ll 1$.) For these small black holes there is a region $M \ll r \ll 1/m$ which is far from the black hole but inside the dilaton Compton wavelength. One can solve the dilaton equation (2b) exactly for $r \gg M$ since the spacetime is essentially flat. The result is that for $M \ll r \ll 1/m$, $\phi$ behaves like a massless dilaton: $\phi \propto 1/r$. So one expects the black holes to behave qualitatively like the massless dilaton solution discussed in the previous section. Numerical calculations confirm that this is indeed the case (Horne and Horowitz 1992). In particular, there is no inner horizon and in the extremal limit, the event horizon shrinks to zero area and becomes singular. The string metric describing an extremal magnetically charged black hole has an infinite throat.

We have seen that black holes with a massive dilaton can be viewed as interpolating between the Reissner-Nordström solution (when their mass is large), and the massless dilaton solution (when their mass is small). One can say something about the transition region $Mm \approx 1$ by considering when a degenerate horizon can exist. If there is a radius $r_0$ for which $\lambda$ and $\lambda'$ both vanish, the field equations yield the following condition on the value of $\phi$ at $r_0$

$$\frac{e^{2\phi}}{\phi(1+\phi)^2} = Q^2 m^2 . \tag{27}$$

The left hand side has a minimum of $e^2/4$ when $\phi = 1$. So for $Q^2 m^2 < e^2/4$, there cannot exist a degenerate horizon. For $Q^2 m^2 > e^2/4$ there are two possible values of $\phi$, but the equations of motion also imply that $\lambda'' \propto (1 - \phi)$ so one needs $\phi < 1$ to have $\lambda'' > 0$ as required for an event horizon. At the critical value, $Q^2 m^2 = e^2/4$, $\lambda''$ also vanishes yielding a triple horizon. This suggests that there may be solutions with three distinct horizons. But at the moment, there is no direct evidence for this.

These intermediate size black holes can have an unusual property when described in terms of the string metric. One can show that the radius of the spheres of spherical symmetry in the Einstein metric, $R$, is monotonically increasing outside the horizon as expected. In the string metric, the radius of the spheres is $Re^\phi$. For an electrically charged hole, it turns out that $\phi$ is also increasing outside the horizon so the spheres indeed become larger as $r$ increases. However, for a magnetically charged hole, $\phi$ changes sign and is monotonically decreasing outside the horizon. So $Re^\phi$ need not be an increasing function of $r$. It certainly increases near infinity, and for the massless dilaton solution (18), it is monotonically increasing everywhere. However, for nonzero dilaton mass, one can show that it decreases outside the horizon for nearly extremal black holes with $Mm \approx 1$. This means that the spatial geometry contains a wormhole. Unlike the familiar wormhole in the maximally extended Schwarzschild solution, this wormhole is outside the horizon. It is static and transversable.

Combining this with our previous results, we see that the string metric describing an extremal, magnetically charged black hole has a rather exotic dependence on mass. It resembles the Reissner-Nordström metric with a degenerate horizon when $Mm \gg 1$. A wormhole forms outside the degenerate horizon when $Mm \approx 1$. Finally, the horizon and wormhole both disappear and are replaced by an infinite throat when $Mm \ll 1$.

## 4 CONCLUSIONS

We have discussed two related issues. On the mathematical side, we have seen that despite the apparent complication introduced by a massless dilaton, the theory shares some of the features of the Einstein-Maxwell theory. In particular, there is a solution generating transformation which is remarkably similar to the Harrison transformation. The resulting extremal black hole can be viewed as the "square root" of the extremal Reissner-Nordström solution. On the physical side, we have seen that the evolution of charged black holes predicted by string theory is quite different from standard lore. Black holes initially (when $Mm \gg 1$) are quite similar to the Reissner-Nordström solution. But as they evolve (through Hawking evaporation and possible quantum bifurcation) they reach a stage where $Mm \ll 1$. At this point, the black holes are similar to the massless dilaton solutions. For a dilaton mass of $m = 1 \text{TeV}$, the transition will occur at a black hole mass of $M = 10^{11} \text{gms}$ which is well above the Planck scale. This ensures that Planck scale corrections, which have been neglected throughout, will still be negligible.

What is the most likely endpoint of this evolution? Assuming the charge is not completely radiated away, one expects to be left with an extremal black hole with unit charge. (Although likely, this is still uncertain due to the fact that the Hawking temperature does not vanish as one approaches the extremal limit of these small black holes.) Since charge is quantized, this must be stable against further bifurcation. These remnants would be pointlike objects, with charge greater than their mass, which repel each other. In other words, they would appear very much like elementary particles. (See also Holzhey and Wilczek 1992.) In fact, by shifting the dilaton potential so that its minimum is at $\phi_0 \neq 0$, one can change the charge to mass ratio of the extremal black holes. There even exists a value of $\phi_0$ such that the extremal black hole of unit charge has the mass of an electron. In the past, one of the main objections to interpreting an electron as a black hole was the fact that its charge is much larger than its mass (in geometrical units) so that according to the Reissner-Nordström solution, it would be a naked singularity. The massive dilaton predicted by string theory removes this objection and opens up the possibility of a much closer connection between elementary particles and black holes.

## ACKNOWLEDGMENTS

It is a pleasure to thank my collaborators D. Garfinkle, A. Strominger, and especially J. Horne. This work was supported in part by NSF Grant PHY-9008502.

## REFERENCES

Banks T, Dabholkar A, Douglas M and O'Loughlin M (1992) "Are Horned Particles the Climax of Hawking Evaporation?", *Phys. Rev.* **D45** 3607

Banks T and O'Loughin M (1992) "Classical and Quantum Production of Cornucopions at Energies Below $10^{18}$ GeV", Rutgers preprint RU-92-14, hep-th/9206055

Banks T, O'Loughin M, and Strominger A (1992) "Black Hole Remnants and the Information Puzzle", Rutgers preprint RU-92-40, hep-th/9211030

Bekenstein J (1972) "Nonexistence of Baryon Number for Static Black Holes", *Phys. Rev.* **D5** 1239

Brill D (1992) "Splitting of an Extremal Reissner-Nordström Throat via Quantum Tunneling", *Phys. Rev.* **D46** 1560

Brill D and Horowitz G (1991) "Negative Energy in String Theory", *Phys. Lett.* **B262** 437

Brill D and Pfister H (1989) "States of Negative Total Energy in Kaluza-Klein Theory", *Phys. Lett.* **B228** 359

Chase J (1970) "Event Horizons in Static Scalar-Vacuum Spacetimes", *Commun. Math. Phys.* **19** 276

Garfinkle D (1992), private communication.

Garfinkle D, Horowitz G and Strominger A (1991) "Charged Black Holes in String Theory," *Phys. Rev.* **D43** 3140; (1992) **D45**, 3888(E)

Gibbons G (1982) "Antigravitating Black Hole Solitons with Scalar Hair in N=4 Supergravity", *Nucl. Phys.* **B207** 337

Gibbons G and Maeda K (1988) "Black Holes and Membranes in Higher-Dimensional Theories with Dilaton Fields", *Nucl. Phys.* **B298** 741

Gregory R and Harvey J (1992) "Black Holes with a Massive Dilaton", Enrico Fermi Preprint EFI-92-49, hep-th /9209070

Harrison B (1968) "New Solutions of the Einstein-Maxwell Equations from Old", *J. Math. Phys.* **9** 1744

Hartle J and Hawking S (1972) "Solutions of the Einstein-Maxwell Equations with Many Black Holes", Commun. Math. Phys. **26** 87

Harvey J and Strominger A (1992) "Quantum Aspects of Black Holes", to appear in the proceedings of the 1992 Trieste Spring School on String Theory and Quantum Gravity, hep-th/9209055

Hassan S and Sen A (1992) "Twisting Classical Solutions in Heterotic String Theory", *Nucl. Phys.* **B375** 103

Holzhey C and Wilczek F (1992) "Black Holes as Elementary Particles", *Nucl. Phys.* **B380** 447

Horne J and Horowitz G (1992) "Black Holes Coupled to a Massive Dilaton", *Nucl. Phys.* **B** to appear, hep-th/9210012

Horowitz G (1992) "The Dark Side of String Theory: Black Holes and Black Strings", to appear in the proceedings of the 1992 Trieste Spring School on String Theory and Quantum Gravity, hep-th/9210119

Israel W (1968) "Event Horizons in Static Electrovac Spacetimes", *Commun. Math. Phys.* **8** 245

Kallosh R, Linde A, Ortin T, Peet A and Van Proeyen A (1992) "Supersymmetry as a Cosmic Censor", Stanford preprint SU-ITP-92-13

Kallosh R, Ortin T and Peet A (1992) "Entropy and Action of Dilaton Black Holes", Stanford preprint SU-ITP-92-29, hep-th/9211015

Maharana J and Schwarz J (1992) "Noncompact Symmetries in String Theory", Caltech preprint CALT-68-1790, hep-th/9207016

Majumdar S (1947) *Phys. Rev.* **72** 930

Papapetrou A (1947) *Proc. Roy. Irish Acad.* **A51** 191

Sen A (1992) " Black Holes and Solitons in String Theory", Tata preprint TIFR-TH-92-57, hep-th/9210050

# Limits on the Adiabatic Index in Static Stellar Models

## LEE LINDBLOM AND A.K.M. MASOOD-UL-ALAM

Physics Department, Montana State University, Bozeman, Montana 59717

## 1 SUMMARY

Dieter Brill has made important contributions to the study of the positive mass theorem in general relativity theory (Brill 1959) and to the analysis of the conformal properties of asymptotically flat space-times (Cantor & Brill 1981). It is a pleasure to honor him on this occasion by applying these analytical tools to the study of static stellar models. We use the positive mass theorem and the conformal properties of these spacetimes to deduce an interesting constraint on the allowed values of the adiabatic index in the static stellar models of general relativity theory.

It is well known that non-singular polytropic stellar models in the Newtonian theory can exist only if the adiabatic index of the fluid is sufficiently large, $\gamma \geq 6/5$ (or equivalently if the polytropic index is sufficiently small, $n \leq 5$; see e.g., Chandrasekhar 1939). In this paper we derive two limits on the allowed values of the adiabatic index in the stellar models of general relativity theory. Our limits apply to static stellar models that have a finite radius and an equation of state which satisfies certain fairly weak smoothness assumptions. Our first bound on the adiabatic index guarantees that

$$\gamma \equiv \frac{\rho + p}{p}\frac{dp}{d\rho} > 1, \tag{1}$$

at some points in every neighborhood of the surface of such stars. This limit, although weaker than the traditional Newtonian limit, is simply a consequence of the regularity of the spacetime at the surface of these stars. This result makes no assumption about the high-density portion of the equation of state. Thus it applies to (essentially) every static stellar model in general relativity theory. Our second bound on the adiabatic index guarantees that

$$\gamma > \frac{6}{5}\left(1 + \frac{p}{\rho}\right)^2. \tag{2}$$

172

in some portion of every static stellar model with finite radius. This result is the direct general-relativistic analogue of the Newtonian limit $\gamma > 6/5$ for spherical polytropic stars with finite radius. Our result is considerably more general, however, because we do not assume that the equation of state is polytropic (i.e., of the form $p = \kappa \rho^{1+1/n}$ where $\kappa$ and $n$ are constants). Our result is also more general because it does not assume that the stellar model has spherical symmetry. Thus, our result compliments the work on the necessity of spherical symmetry in static general-relativistic stellar models by Masood-ul-Alam (1988) and by Beig & Simon (1991, 1992). Those proofs of spherical symmetry specifically exclude any equation of state which violates equation (2) at any point within the star. Thus, our second bound eliminates a large class of potential counterexamples to the spherical-symmetry conjecture. In particular, our result shows that no stellar model whose equation of state violates equation (2) everywhere can exist in general relativity theory.

## 2 THE SURFACE LIMIT

A stellar model in general relativity theory is an asymptotically-flat spacetime that satisfies Einstein's equation with a perfect-fluid source. A static (i.e., time independent and non-rotating) stellar model has a time-translation symmetry (by assumption) whose trajectories are hypersurface orthogonal. Thus, coordinates may be chosen in which the spacetime metric tensor has the representation

$$ds^2 = -V^2 dt^2 + g_{ab} dx^a dx^b, \qquad (3)$$

where $V$ and the three-dimensional spatial metric $g_{ab}$ are independent of $t$. The topology of the constant-$t$ hypersurfaces must be $\mathbf{R}^3$ (Lindblom & Brill 1980, Masood-ul-Alam 1987a). Einstein's equation for static stellar models reduces in this representation to the pair of equations

$$D^a D_a V = 4\pi V(\rho + 3p), \qquad (4)$$

$$R_{ab} = V^{-1} D_a D_b V + 4\pi(\rho - p) g_{ab}. \qquad (5)$$

The density and pressure of the fluid are denoted $\rho$ and $p$; and the density is assumed to be a given function of the pressure, $\rho = \rho(p)$ (the equation of state), which is non-negative and non-decreasing. The spatial covariant derivative compatible with $g_{ab}$ is denoted $D_a$, and its Ricci curvature is denoted $R_{ab}$. The Bianchi identity for the three-dimensional spatial geometry may be reduced to the form

$$D_a p = -V^{-1}(\rho + p) D_a V, \qquad (6)$$

with the use of equations (4)–(5).

The surface of a stellar model is the boundary between the interior fluid region where the pressure is positive, and the exterior region where the vacuum Einstein equation is satisfied. We limit our consideration here to stars whose surfaces occur at a finite radius (i.e., to stars which have non-trivial exterior vacuum regions). The pressure must vanish on the surface of the star in order to insure that the spacetime is non-singular there. We first prove that the ratio $p/\rho$ must also vanish on this surface: Consider a smooth curve $x^a(\lambda)$ that lies inside the star for $\lambda_0 \leq \lambda < \lambda_S$, while the point $x^a(\lambda_S)$ lies on the surface. Let $n^a = dx^a/d\lambda$ denote the tangent to this curve. We evaluate the integral of $(\rho + p)^{-1} n^a D_a p$ along this curve using equation (6):

$$\int_{p(\lambda_0)}^{p(\lambda)} \frac{d\hat{p}}{\rho(\hat{p}) + \hat{p}} = \int_{\lambda_0}^{\lambda} \frac{n^a D_a p \, d\lambda}{\rho + p} = -\int_{\lambda_0}^{\lambda} \frac{n^a D_a V \, d\lambda}{V} = \log\left[\frac{V(\lambda_0)}{V(\lambda)}\right]. \qquad (7)$$

In the limit $\lambda \to \lambda_S^-$ the expression on the right is well behaved, thus, the integral on the left must also be well behaved. Consequently the function,

$$h(p) \equiv \int_0^p \frac{d\hat{p}}{\rho(\hat{p}) + \hat{p}}, \qquad (8)$$

is well defined for all $p$ within any static stellar model in general relativity theory. The equation of state, $\rho(p)$, is a non-decreasing function; thus, we may estimate $h(p)$ as follows,

$$h(p) = \int_0^p \frac{d\hat{p}}{\rho(\hat{p}) + \hat{p}} \geq \int_0^p \frac{d\hat{p}}{\rho(p) + \hat{p}} = \log\left(1 + \frac{p}{\rho}\right). \qquad (9)$$

Since $\lim_{p \to 0+} h(p) = 0$, we conclude from equation (9) that

$$0 = \lim_{p \to 0+} e^{h(p)} - 1 \geq \lim_{p \to 0+} \frac{p}{\rho}. \qquad (10)$$

Thus, the limit of $p/\rho$ must vanish. So, we have established the desired result:

**Lemma.** Consider a static stellar model in general relativity theory whose surface occurs at a finite radius. Assume that the equation of state $\rho = \rho(p)$ is a positive and non-decreasing function in some open neighborhood of the surface of the star: i.e., for pressures $0 < p < \epsilon$. Then $\lim_{p \to 0+}(p/\rho) = 0$.

We now turn to the derivation of our first limit on the adiabatic index,

$$\gamma(p) \equiv \frac{\rho + p}{p} \frac{dp}{d\rho}. \qquad (11)$$

We assume that the equation of state, $\rho(p)$, is a positive and non-decreasing $C^1$ function in the open interval, $0 < p < \epsilon$ (for some $\epsilon > 0$). Thus, $1/\gamma(p)$ is a continuous and non-negative function there. Let $\Gamma(\epsilon)$ be the least upper bound of $\gamma(p)$ for $0 < p < \epsilon$: i.e., let $\Gamma(\epsilon) = \sup_{0<p<\epsilon} \gamma(p)$. If $\Gamma(\epsilon)$ is not finite, then there exist points in the open interval $0 < p < \epsilon$ where $\gamma > 1$, trivially. We turn our attention, therefore, to the case where $\Gamma(\epsilon)$ is finite. From the definition of the adiabatic index, we obtain the following inequality

$$\Gamma(\epsilon) \geq \gamma = \frac{\rho + p}{p} \frac{dp}{d\rho} \geq \frac{\rho}{p} \frac{dp}{d\rho}. \tag{12}$$

This inequality may be integrated to obtain an upper bound for the density function on the open interval $0 < p < \epsilon$:

$$\rho \leq \rho(\epsilon) \left( \frac{p}{\epsilon} \right)^{1/\Gamma(\epsilon)}. \tag{13}$$

This upper bound and the Lemma may then be used to obtain the following condition on $\Gamma(\epsilon)$,

$$\lim_{p \to 0+} \frac{p}{\rho} \geq \lim_{p \to 0+} \frac{\epsilon}{\rho(\epsilon)} \left( \frac{p}{\epsilon} \right)^{1-1/\Gamma(\epsilon)}. \tag{14}$$

The limit on the right vanishes only if $\Gamma(\epsilon) > 1$. Since $\lim_{p \to 0+}(p/\rho) = 0$ as a consequence of the Lemma, it follows that $\Gamma(\epsilon) > 1$. Thus there exist points in the open interval $0 < p < \epsilon$ where $\gamma > 1$. The constant $\epsilon$ may be chosen to be arbitrarily small, so there must exist points in every open neighborhood of the surface of the star where $\gamma > 1$. In summary then our first limit on the adiabatic index is:

**Theorem 1.** Consider a static stellar model in general relativity theory whose surface occurs at a finite radius. Assume that the equation of state $\rho = \rho(p)$ is a positive and non-decreasing $C^1$ function of the pressure in some neighborhood of the surface of the star, $0 < p < \epsilon$. Then the adiabatic index $\gamma(p) > 1$ at some point in every open neighborhood of the surface of the star.

## 3 THE MAXIMUM LIMIT

Our method for deriving a limit on the maximum value of the adiabatic index is very similar to the technique developed by Masood-ul-Alam (1987b) for proving the necessity of spherical symmetry in certain static stellar models. We construct a particular conformal factor which is used to transform the spatial metric $g_{ab}$. This conformal factor scales the mass associated with the spatial geometry to zero while leaving the transformed scalar curvature non-negative—unless the adiabatic index of the fluid is sufficiently large. For stellar models composed of fluid with small adiabatic index, then, the positive mass theorem implies that the transformed geometry is flat. We show, however, that this is inconsistent with our assumption that the stellar model has a finite radius. Thus, we conclude that the maximum value of the adiabatic index must exceed a certain lower bound in every static stellar model with a finite radius.

The Bianchi identity, equation (6), for static fluid spacetimes determines a functional relationship between the fluid variables, $\rho$ and $p$, and the potential $V$. In particular, with the aid of equations (7)–(8), the density and pressure may be expressed as explicit functions of $V$: $p(V) = h^{-1}[\log(V_S/V)]$, and $\rho(V) = \rho[p(V)]$, where $V_S$ is the value of the potential on the surface of the star. We use these functions to define a conformal factor $\psi$ in the interior region of the star (where $0 < V \le V_S$):

$$\psi(V) = \frac{1}{2}(1 + V_S) \exp\left\{-\frac{V_S}{1 + V_S} \int_V^{V_S} \frac{\rho(\hat{V})\, d\hat{V}}{\hat{V}[\rho(\hat{V}) + 3p(\hat{V})]}\right\}. \tag{15}$$

If the interior of the star is composed of more than one connected region, then $\psi(V)$ is defined in each region by equation (15) with the constant $V_S$ taken to be the surface value of $V$ for that region. In the exterior of the star (e.g., where $V_S \le V \le 1$) we define the conformal factor to be,

$$\psi(V) = \frac{1}{2}(1 + V). \tag{16}$$

It is easy to see that this conformal factor is continuous across each component of the surface of the star by setting $V = V_S$ in equations (15) and (16). It is also straightforward to verify that $d\psi/dV$ is continuous at the surface of the star. The derivative of $\psi$ in the interior of the star satisfies the equation

$$\frac{d\psi}{dV} = \frac{V_S \psi}{V(1 + V_S)(1 + 3p/\rho)}. \tag{17}$$

Using the result of the Lemma, $\lim_{V \to V_S}(p/\rho) = 0$, it is easy to take the limit of the right side of equation (17) to verify that $d\psi/dV = 1/2$ at the surface of the star. The second derivative $d^2\psi/dV^2$ vanishes in the exterior of the star, while it is given by the expression

$$\frac{d^2\psi}{dV^2} = \frac{V_S \psi}{V^2(1 + V_S)(1 + 3p/\rho)^2}\left[\frac{2 + 3V_S}{1 + V_S} - \frac{3}{\gamma}\left(1 + \frac{p}{\rho}\right)^2\right], \tag{18}$$

in the interior of the star. The function $d^2\psi/dV^2$ is continuous and has a finite upper bound for $V < V_S$ whenever $1/\gamma$ is bounded. Thus, $\psi$ is $C^{1,1}$ (i.e, its first derivative satisfies a Lipshitz condition) even across the surface of the star as long as $1/\gamma$ is bounded. Note that $1/\gamma(p)$ is finite for any $p > 0$ as a consequence of our assumption that $\rho(p)$ is a $C^1$ function. Thus, our assumption that $1/\gamma$ is bounded is a restriction on the behavior of the equation of state only near $p = 0$. The boundedness of $1/\gamma$ follows automatically for equations of state which are smooth enough that the limit of $\gamma(p)$ is well defined (in the sense that $\lim \inf \gamma = \lim \sup \gamma$) as $p \to 0^+$. In this case Theorem 1 implies that $\lim \inf \gamma \geq 1$, and so $1/\gamma$ is bounded as $p \to 0^+$.

We use the function $\psi$ given in equations (15) and (16) to define a conformally transformed spatial metric tensor $\bar{g}_{ab}$:

$$\bar{g}_{ab} = \psi^4 g_{ab}. \tag{19}$$

This transformed metric has two important properties: a) the mass associated with $\bar{g}_{ab}$ is zero; and b) the scalar curvature associated with $\bar{g}_{ab}$ is non-negative unless the adiabatic index is too large. The first of these properties can be deduced by examining the asymptotic boundary conditions on the stellar model. In general relativity theory a stellar model is a non-singular asymptotically-flat solution to equations (4)–(5). The appropriate asymptotic forms for $V$ and $g_{ab}$ are therefore,

$$V = 1 - \frac{M}{r} + O(r^{-2}), \tag{20}$$

$$g_{ab} = \left(1 + \frac{2M}{r}\right)\delta_{ab} + O(r^{-2}), \tag{21}$$

where the constant $M$ is the mass of the star, $\delta_{ab}$ is the flat Euclidean metric, and $r$ is a spherical radial coordinate associated with the metric $\delta_{ab}$. These asymptotic forms imply that the conformal factor defined in equation (16) has the asymptotic form

$$\psi = 1 - \frac{M}{2r} + O(r^{-2}). \tag{22}$$

It follows, then, that the asymptotic form of the transformed metric is

$$\bar{g}_{ab} = \delta_{ab} + O(r^{-2}). \tag{23}$$

Thus, the mass associated with $\bar{g}_{ab}$ is zero. The second important property of $\bar{g}_{ab}$, the sign of its scalar curvature, is easily deduced. The general expression for the scalar curvature $\bar{R}$ associated with the metric $\bar{g}_{ab}$ is

$$\bar{R} = 8\psi^{-5}\left(2\pi\rho\psi - D^a D_a \psi\right). \tag{24}$$

The term $D^a D_a \psi$ can be evaluated by using equations (4) and (15)–(18). The result can be expressed in the form

$$\psi^4(1+V_S)\bar{R} = 16\pi\rho(1-V_S) + \frac{8\rho^2 V_S D^a V D_a V}{\gamma V^2(\rho+3p)^2}\left[3\left(1+\frac{p}{\rho}\right)^2 - \gamma\frac{2+3V_S}{1+V_S}\right] \quad (25)$$

in the interior of each connected component of the star, while $\bar{R}$ vanishes in the exterior of the star. The potential on the surface of the star is strictly less than one, $V_S < 1$, in any stellar model that has a finite radius. Thus, the conformally transformed scalar curvature is non-negative as long as the condition,

$$\gamma \le 3\left(1+\frac{p}{\rho}\right)^2 \frac{1+V_S}{2+3V_S}, \quad (26)$$

is satisfied by the adiabatic index.

The positive mass theorem (Schoen & Yau 1979, Parker & Taubes 1982) implies that a $C^{1,1}$ asymptotically-flat Riemannian three-metric is flat if its scalar curvature is non-negative and its mass is zero. The needed smoothness of $\bar{g}_{ab}$ follows from the smoothness of $\psi$ established above. Thus, the spatial metric $\bar{g}_{ab}$ is flat if the adiabatic index satisfies equation (26). But, if $\bar{g}_{ab}$ is flat, then the scalar curvature $\bar{R}$ vanishes. This implies, from equation (25), that $V_S = 1$ and that equality must hold in equation (26). But $V_S < 1$ in any star with a finite radius, and so we conclude that $\bar{R}$ must be negative somewhere in such stars. Thus equation (26) must be violated. Consequently, at some point in every static star with a finite radius the adiabatic index must satisfy

$$\gamma > 3\left(1+\frac{p}{\rho}\right)^2 \frac{1+V_S}{2+3V_S} > \frac{6}{5}\left(1+\frac{p}{\rho}\right)^2 \ge \frac{6}{5}. \quad (27)$$

In summary, then, we have established the following lower bound on the maximum value of the adiabatic index:

**Theorem 2.** Consider a static stellar model in general relativity theory whose surface occurs at a finite radius. Assume that the equation of state $\rho = \rho(p)$ is a positive and non-decreasing $C^1$ function of the pressure. Assume that $1/\gamma(p)$ is bounded as $p \to 0^+$. Then the adiabatic index must satisfy the inequality

$$\gamma > \frac{6}{5}\left(1+\frac{p}{\rho}\right)^2 \ge \frac{6}{5}, \quad (28)$$

at some point within the star.

We point out that the extreme case in the argument that leads to Theorem 2—when equality holds in equation (26) and $V_S = 1$—is not vacuous. This case constrains the equation of state of the fluid to be the one used by Buchdahl (1964) to construct a family of non-singular but infinite-radius stellar models. The existence of these asymptotically flat, albeit infinite radius, stellar models suggests that our limit on the maximum value of the adiabatic index is the strongest possible limit of this kind.

## 4 CONCLUDING REMARKS

It is instructive to examine the implications of Theorem 2 for stellar models constructed from fluid that has a polytropic equation of state:

$$p = \kappa \rho^{1+1/n}, \tag{29}$$

where $\kappa$ and $n$ are constants. The constraint on the adiabatic index, equation (28), is equivalent for polytropes to a constraint on the polytropic index $n$:

$$n < 5(1 + 6p/\rho)^{-1} \leq 5. \tag{30}$$

Thus, $n < 5$ for general-relativistic stellar models that have a polytropic equation of state. This shows that equation (28) is the direct general-relativistic analogue of the familiar Newtonian limit on the polytropic index $n < 5$ (see e.g., Chandrasekhar 1939).

We have chosen to express the limit on the adiabatic index, equation (28), in a form which involves only the equation of state without making any reference to the macroscopic properties of the particular stellar model. In fact, a somewhat stronger limit was obtained in equation (27). That limit on the adiabatic index can be expressed in another form that involves the surface value of the potential $V_S$ in a simple way:

$$\gamma > 3\frac{1 + V_S}{2 + 3V_S} > \frac{6}{5}. \tag{31}$$

Thus, for a spherical star of mass $M$ and radius $R$ the adiabatic index must satisfy the inequality

$$\gamma > 3\frac{1 + (1 - 2M/R)^{1/2}}{2 + 3(1 - 2M/R)^{1/2}} > \frac{6}{5}, \tag{32}$$

at some point within the star. For polytropic stars this condition is equivalent to the bound

$$n < 2 + 3V_S = 2 + 3(1 - 2M/R)^{1/2} < 5, \tag{33}$$

on the polytropic index.

The limits on the adiabatic index that we derive in this paper are necessary conditions for the existence of a static stellar model in general relativity theory. Our limits are weaker, therefore, than the bounds needed to guarantee the stability of these models. Glass and Harpaz (1983) have shown, for example, that the adiabatic index must exceed 4/3 in a stable relativistic polytrope by an amount that depends on the ratio $p/\rho$ at the center of the star. A similar limit on a suitably averaged value of the adiabatic index, $\bar{\gamma} > 4/3$, is necessary for the stability of nearly Newtonian stars with any equation of state (see e.g., Misner, Thorne & Wheeler 1973).

The bound on the adiabatic index derived in Theorem 2 does not assume that the stellar model has spherical symmetry. Thus, our result compliments the work on the necessity of spherical symmetry in static general-relativistic stellar models by Masood-ul-Alam (1988) and by Beig & Simon (1991, 1992). Those proofs of spherical symmetry apply only to equations of state for which the adiabatic index satisfies the inequality (28) at every point. Thus, Theorem 2 eliminates a large class of potential counterexamples to the spherical-symmetry conjecture. In particular, our result shows that no stellar model whose equation of state violates equation (28) everywhere can exist in general relativity theory. Of course many equations of state exist, including many realistic models of neutron-star matter, which satisfy equation (28) for some values of the pressure and violate it for others.

This research was supported by the grant PHY-9019753 from the National Science Foundation, and grant NAGW-2951 from the National Aeronautics and Space Administration.

## REFERENCES

Beig, R. & Simon, W. 1991, *Lett. Math. Phys.*, **21**, 245–250.

Beig, R. & Simon, W. 1992, *Commun. Math. Phys.* **144**, 373–390.

Brill, D.R. 1959, *Ann. Phys.* **7**, 466–483.

Buchdahl, H.A. 1964, *Ap. J.* **140**, 1512–1516.

Cantor, M., & Brill, D.R. 1981, *Comp. Math.* **43**, 317–330.

Chandrasekhar, S. 1939, *An Introduction to the Study of Stellar Structure*, (Chicago: University of Chicago Press).

Lindblom, L. & Brill, D.R. 1980, in *Essays in General Relativity*, pp. 13–19, ed. by F.J. Tipler, (New York: Academic Press).

Masood-ul-Alam, A.K.M. 1987a, *Commun. Math. Phys.* **108**, 193–211.

Masood-ul-Alam, A.K.M. 1987b, *Class. Quantum Grav.* **4**, 625–633.

Masood-ul-Alam, A.K.M. 1988, *Class. Quantum Grav.* **5**, 409–421.

Misner, C.W., Thorne, K.S. & Wheeler, J.A. 1973, *Gravitation*, (San Francisco: W.H. Freeman and Co.).

Parker, T. & Taubes, C. 1982, *Commun. Math. Phys.* **84**, 223–238.

Schoen, R. & Yau, S.T. 1979, *Commun. Math. Phys.* **65**, 45–76.

# On the Relativity of Rotation

## BAHRAM MASHHOON

Department of Physics and Astronomy
University of Missouri-Columbia
Columbia, Missouri 65211, USA

### ABSTRACT

The problem of the origin of rotational inertia is examined within the framework of the relativistic theory of gravitation. It is argued that gravitomagnetic effects cannot be interpreted in terms of the relativity of rotation. Absolute and relative motion are discussed on the basis of the hypothesis that these are complementary classical manifestations of movement.

What is the origin of inertia? For instance, with respect to what does the Earth rotate around its axis? The rotation of a body does not generate any basic new gravitational effect in the Newtonian theory. This is not the case, however, in Einstein's theory of gravitation. The striking analogy between Newton's law of gravitation and Coulomb's law of electricity has led to a description of Newtonian gravity in terms of a gravitoelectric field. Any theory that combines Newtonian gravity with Lorentz invariance is expected to contain a gravitomagnetic field in some form; in general relativity, the gravitomagnetic field is usually caused by the angular momentum of the source of the gravitational field. The first gravitomagnetic effects were described by de Sitter[1] soon after Einstein's fundamental work on general relativity. The question of relativity of rotation was also discussed by de Sitter following his investigation of the astronomical consequences of Einstein's relativistic theory of gravitation; de Sitter concluded that the problem of inertia did not have a solution in the general theory of relativity.[1] There is no direct experimental evidence for the existence of the gravitomagnetic field at present; however, various proposals exist for measuring the gravitomagnetic field of the Earth. The most significant proposal has developed into the Gravity Probe B which will involve a telescope and a superconducting gyroscope on board a drag-free satellite in a polar orbit about the Earth.[2,3]

In connection with the origin of rotational inertia, an attempt is made here to describe the physical origin of the gravitomagnetic field in terms of relativity of rotation within the framework of the standard relativistic theory of gravitation that is consistent with all of the observational data available at present, namely, Einstein's general theory of relativity. The problem may be described as follows: Imagine an idealized situation involving an isolated astronomical body (henceforth called "Earth") whose center-of-mass is at rest in an appropriate system of coordinates. If the Earth is static, there is only a gravitoelectric field which consists of the Newtonian part together with the post-Newtonian relativistic corrections. Now suppose that the Earth rotates rigidly with constant frequency $\omega_\oplus$ around its axis, then a gravitomagnetic field exists in addition to the gravitoelectric field. However, a coordinate system may be chosen in which the rotating Earth is completely at rest; this comoving system is in fact the fundamental astronomical coordinate system that has been employed since time immemorial. In this system the Earth appears to be the center of the universe since the heavens revolve around it. According to general relativity, the gravitomagnetic field of the Earth is still present in this coordinate system; in fact, an experiment has been proposed to measure this field by a Foucault pendulum located at the south pole.[4] With respect to comoving observers at rest on the Earth, the gravitomagnetic field should be interpreted as the induced field of the moving heavens if rotation is relative. Is such an interpretation permitted by the general theory of relativity? If not, then the rotation of the Earth must be considered to be absolute according to general relativity. The gravitomagnetic field would thus be a manifestation of the absolute rotation of the Earth. It is argued in this work that the rotation of the Earth is absolute; in general, the translational and rotational accelerations of a physical system are absolute.[5] Mach's fundamental insight[6] is interpreted instead in light of the quantum theory.[7] Neither relative motion nor absolute motion can provide a completely satisfactory description of physical phenomena. Movement of a physical system has complementary classical manifestations in relative and absolute motion.

It is a great pleasure for me to dedicate this paper to Dieter R. Brill on the occasion of his sixtieth birthday.

## 1. INTRODUCTION

The description of observers and their motion in spacetime is of basic importance for a proper discussion of the notion of relativity of arbitrary motion. The path of an observer, i.e., a classical measuring device, in space and time is characterized by its worldline $x^\mu = x^\mu(\tau)$, where $\tau$ is the proper time along the path. The idealization of replacing an extended classical system by a geometrical line is necessary for theoretical purposes; however, this does not appear to constitute a fundamental physical limitation.[8,9] To perform physical measurements over an extended period of time, an observer must define a local spatial coordinate system at each instant of time; this requirement implies that a tetrad frame $\lambda^\mu_{(\alpha)}$ must be carried along the path of the observer such that $\lambda_{(0)}^\mu$ is a unit vector tangent to the path and constitutes the local temporal axis and $\lambda_{(i)}^\mu$ are unit vectors that constitute the local spatial axes.[10] The motion of the orthonormal frame $\lambda_{(\alpha)}^\mu$ along the path may be described by an antisymmetric Lorentz tensor $\phi_{\alpha\beta}$,

$$\frac{D\lambda_{(\alpha)}^\mu}{D\tau} = \phi_\alpha{}^\beta \lambda_{(\beta)}^\mu \quad , \tag{1}$$

which describes the deviation of the frame from parallel transport along the path. The tensor $\phi_{\alpha\beta}$ can be decomposed in a natural way in terms of two local Cartesian vectors $\mathbf{g}$ and $\boldsymbol{\omega}$, $\phi_{\alpha\beta} \to (\mathbf{g}, \boldsymbol{\omega})$, which represent, respectively, the local translational acceleration of the observer and the rotational frequency of the spatial frame with respect to a nonrotating (i.e., Fermi-Walker transported) triad. This interpretation of the "electric" and "magnetic" parts of the antisymmetric tensor $\phi_{\alpha\beta}$ is based on the following considerations. Let $\lambda_{(0)}^\mu = dx^\mu/d\tau$ be the velocity vector, then

$$\frac{d^2 x^\mu}{d\tau^2} + \Gamma_{\nu\rho}^\mu \frac{dx^\nu}{d\tau} \frac{dx^\rho}{d\tau} = A^\mu \quad , \tag{2}$$

where $A^\mu$ is the acceleration vector of the observer, and $A_{(\alpha)} = A_\mu \lambda_{(\alpha)}^\mu = (0, -\mathbf{g})$ since $\lambda_{(0)}^\mu$ is a unit timelike vector and hence $A^\mu$, $\lambda_{(0)}^\mu A_\mu = 0$, is spacelike. Thus $\phi_{0i} = -g_i$ and $g^2 = A_\mu A^\mu$ is the square of the magnitude of the acceleration vector. The signature of the spacetime metric is $+2$ throughout. While the "electric" part of $\phi_{\alpha\beta}$ determines the translational acceleration of the observer, the "magnetic" part determines its rotational frequency. Imagine that the observer carries along its path a triad of ideal gyroscopes that establish a local nonrotating spatial frame. Let this orthonormal tetrad system be denoted by $\lambda_{(\alpha)}^\mu$, where $\lambda_{(0)}^\mu = \lambda_{(0)}^\mu$ and $D\lambda_{(i)}^\mu/D\tau$ is purely tangential, i.e.

$$\frac{D\lambda_{(\alpha)}^\mu}{D\tau} = \lambda_{(0)}^\mu A_\nu \lambda_{(\alpha)}^\nu - A^\mu \eta_{0\alpha} \quad . \tag{3}$$

Any other orthonormal tetrad frame is related to the nonrotating system by a simple rotation of the spatial triad, i.e., $\lambda_{(i)}^\mu = R_{ij} \lambda_{(j)}^\mu$, where

$$\frac{dR_{ij}}{d\tau} = \phi_{ik} R_{kj} \quad , \tag{4}$$

so that the "magnetic" part of $\phi_{\alpha\beta}$ determines the rate of rotation of the actual frame with respect to the nonrotating system. The frequency of rotation is given by $\omega_i = \frac{1}{2}\epsilon_{ijk}\phi_{jk}$.

It is important to recognize that $\mathbf{g}$ and $\boldsymbol{\omega}$ are scalars, i.e., they are totally independent of the underlying coordinate system employed in spacetime. The observer may perceive itself to be at rest and interpret its measurements in terms of the "acceleration"

and "rotation" of the rest of the masses in the universe toward it; however, this interpretation will not change the fact that the observer has at each instant an absolute
translational acceleration $\mathbf{g}(\tau)$ and an absolute rotation of frequency $\boldsymbol{\omega}(\tau)$. The absolute character of translational and rotational accelerations may be expressed in terms
of the existence of absolute acceleration lengths that may be associated with the observer. In fact, all actual observers are endowed with acceleration lengths; however,
it is important to imagine locally inertial observers that have infinite acceleration
lengths. Such ideal observers follow geodesics and refer their measurements to locally
nonrotating spatial frames. For an accelerated observer referring its measurements to
a nonrotating frame, the proper (translational) acceleration length is $c^2/g(\tau)$ while
for a geodesic observer that refers its measurements to a rotating frame the proper
(rotational) acceleration length is $c/\omega(\tau)$. In general, the *proper* acceleration scales
may be defined[11] in terms of the Lorentz invariants of $\phi_{\alpha\beta}$, i.e., $\phi_{\alpha\beta}\phi^{\alpha\beta} = 2(-g^2 + \omega^2)$
and $\phi^*_{\alpha\beta}\phi^{\alpha\beta} = 4\mathbf{g} \cdot \boldsymbol{\omega}$, where $\phi^*_{\alpha\beta}$ is the dual of $\phi_{\alpha\beta}$.

The theoretical origin of these absolute (translational and rotational) accelerations
can be traced back to the fact that in Newtonian mechanics an observer can decide
by purely local measurements whether it is noninertial. That is, the local laws of
motion already embody the fundamental Newtonian notions of absolute space and
time. Therefore, it should not be surprising that the relativistic generalization of
the Newtonian equation of motion still contains the notion of absolute accelerated
motion. It should be clear from this analysis that general relativity simply inherits
the essential aspects of Newtonian theory's problem of the origin of inertia.[5]

The determination of spacetime metric is the basic theoretical requirement underlying
the construction of an orthonormal tetrad associated with an ideal test observer. The
metric tensor $g_{\mu\nu}$ satisfies the gravitational field equations

$$R_{\mu\nu} - \frac{1}{2}g_{\mu\nu}R = \frac{8\pi G}{c^4}T_{\mu\nu} \quad , \tag{5}$$

which are nonlinear partial differential equations for the metric tensor once the
energy-momentum tensor of matter, $T_{\mu\nu}$, is specified. To obtain a definite solution,
boundary conditions are needed. It appears natural to impose boundary conditions
on $g_{\mu\nu}$ at infinity. However, it is likely that the actual physical conditions at infinity
will never be known.[1] Therefore, the imposition of boundary conditions amounts to
restrictions on the solutions of the field equations in order to choose from all possible
solutions only those that have actual physical relevance. The main experimental suc-
cesses of general relativity are related to those solutions of the field equations that
are asymptotically flat. This naturally corresponds to the notion that spacetime is
Minkowskian; however, where mass-energy exists, the spacetime curvature deviates
from zero in a certain proportion to the accumulation of mass-energy. In this ap-

proach, the absolute space and time of Newton are thus replaced by Minkowski's absolute spacetime; however, no attempt is made to account for this underlying continuum.

To avoid the introduction of a global Minkowski spacetime, the hypothesis has been introduced that the universe is spatially closed.[12,13] Spatial closure depends upon the existence of sufficient mass-energy in the world. Observations do not compel us to choose a spatially closed cosmos at the present time; however, the adoption of spatial closure has led to elegant analyses of the dynamics of the gravitational field.[14,15] It can be shown that the idea of relativity of arbitrary motion is at variance with general relativity regardless of the boundary conditions. In fact, it may eventually turn out that even a spatially closed world is also compatible with the existence of an underlying Minkowski spacetime (cf. section 4).

The basic physical steps leading from the Newtonian theory to the relativistic theory of gravitation have been delineated in a recent work[16] on the basis of the theory of measurement. The analysis of measurements provides a firm foundation for the discussion of the relativity of arbitrary motion in the theory of general relativity. The structure of spacetime is determined by the energy-momentum tensor of matter as well as the boundary data in general relativity; therefore, it might appear natural to suppose that the absolute translational and rotational accelerations of an observer occur *relative* to the mass-energy content of the universe.[12,17] The *relativity of translational acceleration* has been examined in a recent work[5] and found to be incompatible with the standard relativistic theory of gravitation; the problem of *relativity of rotation* is treated in the following section.

## 2. ORIGIN OF ROTATIONAL INERTIA
According to classical mechanics, accelerated motion occurs with respect to absolute space; moreover, inertial effects appear in an accelerated system due to fictitious forces that are all proportional to the inertial mass of the system. The equivalence of inertial and gravitational masses implies that all Newtonian inertial forces are proportional to the gravitational mass. This has led to the interpretation of these forces in terms of the relativistic gravitational influence of distant masses in the universe[17] in contrast to the Newtonian conception of absolute space.

In the standard geometric theory of gravitation, the inertial effects that an accelerated observer would measure have their origin in the absolute acceleration and rotation of the observer with respect to the spacetime whose metric is determined by the energy-momentum tensor of matter as well as the boundary data. This is directly related to, and a generalization of, the fact that in Newtonian mechanics an observer can determine by purely local measurements whether it is noninertial. Under these circumstances, the *relativity of arbitrary motion* clearly does not have a simple interpretation in general relativity.

What then is the physical origin of inertial effects in general relativity? Consider, for instance, a concrete situation namely the inertial effects measured by Earth-based observers that refer their measurements to coordinate axes that are fixed with respect to the Earth. The worldline of each such observer is accelerated and the observer's local frame rotates as demonstrated by the Foucault pendulum experiment. The origin of translational inertial effects has been the subject of a previous investigation[5]; therefore, in this paper attention is focused on the rotational inertial effects. Is there a way of interpreting the rotational inertial effects observed on the Earth (e.g., Coriolis and centrifugal forces) as being due to the gravitational influence of the rotation of the heavens about the Earth? This is not possible by a change of coordinates in our universe. However, let us imagine a *different* universe in which the Earth is completely at rest and the heavens actually rotate about the Earth so as to produce the same *relative* motion as in our universe. It is possible in principle to calculate the gravitational influence of rotating distant masses in such a universe once boundary data are properly specified. Following the approach of H. Thirring[18,19], let us assume that this other universe is asymptotically flat and the distant masses form a spherical shell of mass $M$ and radius $R$ that rotates with frequency $\omega_\oplus$. The Thirring-Lense papers have been discussed in detail in a previous investigation[20]; therefore, a comprehensive treatment will not be attempted here. Briefly, the Earth is assumed to be a test particle (with negligible mass) at rest at the center of the hollow sphere. The gravitomagnetic field generated by the rotating spherical shell can cause an ideal test gyroscope at the position of the Earth to precess in the same sense as the shell with a dragging frequency $\Omega_D = 2GJ/c^2 R^3$, where $J = \frac{2}{3}MR^2\omega_\oplus$ is the angular momentum of the shell. Let a hypothetical inertial test observer using a nonrotating tetrad system $\lambda^\mu_{(\alpha)}$ at the origin of the hollow sphere set up a Fermi coordinate system $\bar{x}^\alpha = (\tau, \bar{\mathbf{x}})$ in the neighborhood of the origin. The Earth-based spatial axes rotate with respect to the Fermi coordinate axes with frequency $\boldsymbol{\omega} = -\Omega_D = (4GM/3c^2R)\,\boldsymbol{\omega}_\oplus$. Let $x^i = R_{ij}\bar{x}^j$ be the natural Earth-based spatial coordinate system of the test observer, then the equation of motion of a free particle with respect to the test observer at the origin is

$$\frac{d^2 x^i}{d\tau^2} + 2\epsilon_{ijl}\omega_j\frac{dx^l}{d\tau} + (\omega_i\omega_j - \delta_{ij}\omega^2 + \epsilon_{ilj}\frac{d\omega_l}{d\tau} + k_{ij})x^j = 0 \quad , \tag{6}$$

where $(k_{ij})$ is the symmetric and traceless tidal matrix that represents the "electric" components of the Riemann curvature tensor as measured by the observer, i.e., $k_{ij} = R_{\mu\nu\rho\sigma}\lambda^\mu_{(0)}\lambda^\nu_{(i)}\lambda^\rho_{(0)}\lambda^\sigma_{(j)}$. The equation of motion (6) is the Jacobi equation expressed in terms of the Earth-based coordinate system used by the rotating geodesic observer that is stationary at the origin of the hollow sphere; therefore, this equation contains the tidal gravitational acceleration induced by the rotating shell in addition to the inertial accelerations. This result is to be compared with the idealized situation in our universe in which the spherical shell is static but the observer at the origin of the sphere rotates with frequency $\omega_\oplus$; the equation of motion of a free particle with respect to the rotating observer would only contain inertial accelerations, i.e.,

$$\frac{d^2 x^i}{d\tau^2} + 2\epsilon_{ijk}\,\omega_\oplus^j\,\frac{dx^k}{d\tau} + (\omega_\oplus^i\omega_\oplus^j - \delta_{ij}\omega_\oplus^2 + \epsilon_{ikj}\frac{d\omega_\oplus^k}{d\tau})x^j = 0 \quad, \tag{7}$$

since the Ricci-flat spherically symmetric spacetime inside the shell has zero curvature. Relativity of rotation would imply that equations (6) and (7) should have the same physical content, i.e., $\omega = \omega_\oplus$ and $(k_{ij}) = 0$. This is not possible for the rotational terms since in Thirring's work[18] $GM \ll c^2 R$; however, this restriction has been removed by Brill and Cohen[21], who have shown that it is possible to construct shells for which the rotational terms in equations (6) and (7) coincide. The radius of such a shell must be the same as its gravitational radius. It has been emphasized by these authors that such shells are rather special configurations of matter and that in general Mach's principle is only of heuristic value in general relativity.[22–24] An important consequence of the Brill-Cohen result should be noted: If the matter in the universe is such that a local gyroscope would precess in the gravitomagnetic field of the rotating distant masses with the same frequency as the rotation of the universe, then the dragging is complete and local measurements in connection with "fixed" stars would not reveal a rotation of the universe as a whole.[21] Observational results that place a stringent upper limit on the rotation of the universe have therefore been interpreted as supporting Mach's principle in general relativity.[25] The elucidation of the nature of the *dragging of the inertial frames* by rotating systems constitutes the basic physical significance of the Thirring-Lense[18–20] and Brill-Cohen[21,22] investigations.

It should be recognized that even if the rotational terms in equations (6) and (7) agree, the same physical results would be obtained only when the spacetime in the hollow interior of the *rotating* sphere is flat. It has been shown by Cohen and his coworkers that solutions exist that describe rotating *cylindrical* shells with flat interiors.[26,27] This problem has been further investigated by Pfister and Braun, who considered rotating quasi-spherical shells.[28,29] In a general investigation,[30] Pfister has shown that a *physically realistic* solution to the problem of rotating shells with flat interiors does not exist. This important result essentially brings to completion the solution of the problem first posed by Thirring[18] within the framework of general relativity.

The Earth has been treated thus far as a test particle of negligible influence on the spacetime structure; this restriction will be removed in the rest of this section.

Imagine therefore an ideal test gyroscope (e.g., a Foucault pendulum) at the north or south pole on the Earth, which is surrounded at a great distance by a shell of matter representing the rest of the matter in the universe. The post-Newtonian analysis contained in the Thirring-Lense papers[18-20] can be simply extended to take due account of the field of the Earth. If the shell is static and the Earth rotates with respect to the asymptotically flat region of spacetime, an Earth-based observer finds that the gyroscope precesses in the gravitomagnetic field of the Earth in addition to the Foucault precession; the existence of this extra precession may be regarded by the observer as a *relativistic inertial effect* that occurs in addition to the Newtonian rotational inertial effects. The frequency of this precession is given by[31,20]

$$\Omega_D^{\oplus}(\text{pole}) = \frac{2GJ_{\oplus}}{c^2 r_{\oplus}^3} \quad , \tag{8}$$

where the Earth's angular momentum ($J_{\oplus}$) is related to its frequency of rotation ($\omega_{\oplus}$) via its *moment of inertia*. On the other hand, if the Earth is static and the shell rotates with frequency $-\omega_{\oplus}$, the gyroscope precesses in the gravitomagnetic field of the shell. Relativity of rotation would then imply a connection between the moment of inertia of the Earth and the configuration of distant masses. The existence of such long-range correlations is physically impossible since only classical particle phenomena are involved here. It may be argued that this unreasonable consequence of the relativity of rotation might have been expected in asymptotically flat spacetimes. While the argument has been presented here in terms of Thirring's original analysis[18], it should be clear that the difficulty would remain regardless of boundary conditions. In fact, the argument may be applied to each rotating astronomical body: How could distant masses account for the moment of inertia of an isolated rotating mass?

It follows that the gravitomagnetic field is an absolute measure of rotation of a massive body; hence, a direct measurement of the gravitomagnetic field of the Earth by the GP-B would constitute observational proof of the absolute rotation of the Earth.

## 3. LARMOR'S THEOREM AND GRAVITATION

A *global* equivalence between inertia and gravitation is inadmissible since inertial effects are not due to the gravitational influence of distant masses. Instead, Einstein's principle of equivalence asserts the *local* indistinguishability of gravitational and inertial effects. This principle is of a heuristic nature and its precise mathematical formulation in general relativity amounts to the statement that each observer in a gravitational field is pointwise inertial ("local flatness").

In the post-Newtonian limit of general relativity, the gravitational field may be decomposed, in an observer—dependent way, into gravitoelectric ($E_g$) and gravitomagnetic ($B_g$) fields in close analogy with electromagnetism. It follows from Einstein's principle of equivalence that these fields must be locally equivalent to inertial effects; the standard account of this principle[12] corresponds to the local equivalence of the gravitoelectric field with a translational inertial acceleration. On the other hand, the gravitomagnetic field is locally equivalent to a rotational inertial effect. For instance, the gravitomagnetic field associated with a rotating mass is given by $\mathbf{B}_g = c\boldsymbol{\Omega}_D$, where the dragging frequency of the local inertial frames is

$$\boldsymbol{\Omega}_D = \frac{G}{c^2 r^5}[3(\mathbf{r}\cdot\mathbf{J})\mathbf{r} - r^2\mathbf{J}] \quad . \tag{9}$$

This amounts to a gravitational analog of Larmor's theorem[32] since the gravitomagnetic field is proportional to the rotational dragging of the local inertial frames.

Larmor's theorem follows from a comparison of Newton's law of motion for a free particle in an accelerated reference system with the Lorentz force law. *Local* equivalence of inertial and electromagnetic effects can be established to first order in the field strength for slowly varying fields and slowly moving charged particles with the same $q/m$. The motion of a charged particle in an electromagnetic field is then locally equivalent to the motion of a free particle with repsect to a reference frame that is accelerated with $\mathbf{g}_L = q\mathbf{E}/m$ and $\boldsymbol{\omega}_L = q\mathbf{B}/2mc$. The gravitational Larmor theorem has been described in a recent paper[32]; briefly, the analogy between the electromagnetic and gravitational fields leads to the adoption of a gravitational charge of magnitude $q_g(\text{electric}) = -m$ in the electric case and $q_g(\text{magnetic}) = -2m$ in the magnetic case. The latter convention is not surprising since a spin-2 field (gravity) has been approximated by a spin-1 field (electromagnetism). Moreover, in this convention the gravitational charges of the source of the field are assumed to be positive and the corresponding test particle charges are assumed to be negative in order to account for the attraction of gravity. The universality of the gravitational interaction is reflected in the fact that for any test particle $\mathbf{g}_L = -\mathbf{E}_g$ and $\boldsymbol{\omega}_L = -\mathbf{B}_g/c = -\boldsymbol{\Omega}_D$. The gravitational field can therefore be characterized in terms of the translational and rotational dragging of the local inertial frames.

The principle of equivalence is valid for classical particles and rays of radiation. On the basis of the gravitational Larmor theorem, deviations from the ray approximation have been studied in order to elucidate the differential deflection of polarized radiation in the gravitational field of a rotating mass.[32] The same phenomenon may be described for a particle, such as a thermal neutron[33], through the interaction Hamiltonian $\mathbf{S}\cdot\boldsymbol{\Omega}_D$, where $\mathbf{S}$ is the particle spin. For the Earth, this amounts to $\hbar\Omega_D^\oplus \sim 10^{-29}\text{eV}$. Experiments have searched for a coupling of intrinsic spin with the gravitational field of the Earth[34,35]; however, it appears that some eight orders of magnitude improvement in sensitivity is required in order to measure the coupling of intrinsic spin with the gravitomagnetic field of the Earth.

## 4. ABSOLUTE AND RELATIVE MOTION

In the Newtonian mechanics of classical point particles, there is no *a priori* relation between the intrinsic property of a particle, namely, its inertial mass $(m)$ and the state of the particle defined by its extrinsic space-time characteristics, namely, its position $\mathbf{Q}$ and velocity $\dot{\mathbf{Q}}$ at time $t$. This circumstance led Mach[6] to the idea that the absolute space and time of Newtonian mechanics must be superfluous in some way; instead, the relative motion of each classical particle with respect to the other classical particles should have ultimate physical significance since the particles are immediately connected via interactions. It is interesting to extend Mach's analysis to the propagation of electromagnetic waves as well since in the classical domain movement takes place via classical particles as well as electromagnetic waves. Furthermore, classical motion is either relative or absolute. Applying the same Machian criterion to electromagnetic wave motion, it would appear that intrinsic wave motion must be absolute since the wave characteristics are immediately connected with its existence in space and time. The situation may be stated as follows: Relativity of classical particle motion involves the possibility that an observer could be comoving with each particle in turn so that the same relative motion would be observed from different perspectives; however, an electromagnetic wave always moves with the fundamental speed $c$ with respect to an inertial observer. The motion of a wave would be absolute if no observer could ever be comoving with the wave. To stay at rest with a wave, an observer should not detect any temporal variation in the wave. For instance, a standing wave stands, but it does not stand still; thus a wave that stands still would have no variation in time.

According to quantum theory, the space-time description for the state of a particle is provided by its wave function. Generalizing the above discussion to the case of classical matter waves, one would expect the de Broglie-Schrödinger wave motion to be absolute. This is in fact consistent with Mach's fundamental conception since the intrinsic inertial properties of a particle (mass and spin) are related to its quantum state in an inertial frame in Minkowski spacetime.[36] Moreover, the absolute character of wave motion is in conformity with the nonrelativistic description of particle motion in an external potential $V(\mathbf{Q})$ according to the Heisenberg picture since the fundamental quantum condition, $[Q_j, P_k] = i\hbar\delta_{jk}$, may be written as

$$\sum_C [Q_j(A,C)\dot{Q}_k(C,B) - \dot{Q}_k(A,C)Q_j(C,B)] = \frac{i\hbar}{m}\delta_{jk}\langle A|B\rangle \quad , \qquad (10)$$

where the summation is over a complete set of orthonormal states and $O(A,B) = \langle A|O|B\rangle$ is a matrix element corresponding to an observable $O$. Equation (10) contains a direct connection between the inertial mass $m$ and the absolute space-time observables of the particle in the Heisenberg picture. As $m \to \infty$, the particle behaves classically and the connection disappears. From the standpoint of wave-particle duality, the wave aspect becomes negligible as $m \to \infty$. Consider, for instance, the passage of a free nonrelativistic particle (e.g., a thermal neutron) of mass $m$ and speed $v \ll c$ through a slit of width $D$; the angle of divergence of the corresponding beam due to diffraction is given by $\delta \sim \hbar/mvD$. As $m \to \infty$, $\delta \to 0$ and the classical (i.e., non-quantum) situation is approached so that the law of inertia is recovered in the classical limit. The importance of the ratio $\hbar/m \to 0$ in the derivation of the semiclassical limit of quantum mechanics should be noted.[37] Thus the application of Mach's basic insight to the quantum mechanics of a particle leads to a proper justification for the absolute space and time of Newtonian mechanics: To the extent that the classical mechanics of macroscopic systems can be thought of as a limiting form of quantum theory, Heisenberg's generalization of Hamiltonian mechanics to the quantum domain vindicates Newton's absolute space and time. This circumstance is a direct consequence of the nonlocality of a particle in the quantum theory: At any given instant of time, the Schrödinger wave function must be normalized over all space.

The relationship between the intrinsic properties of a system and its extrinsic space-time characteristics, in the sense of Mach's analysis, may be expressed through the existence of spatial and temporal scales $\hbar/mc$ and $\hbar/mc^2$, respectively, associated with the system. The intrinsic spatial scale must be compared with other significant length scales in the problem; for instance, it can be shown that if the system is a standard classical measuring device undergoing acceleration, the relevant acceleration length must exceed $\hbar/mc$. Therefore, the (translational or rotational) acceleration of the system cannot exceed a certain mass-dependent *maximal proper acceleration*. This notion is originally due to Caianiello.[38] On the other hand, the intrinsic temporal scale—when compared with the definition of time in terms of the period of electromagnetic radiation emitted in a certain standard atomic transition—implies an absolute standard of inertial mass. This concept as well as its metrological implications has been discussed by Wignall.[39] It should be pointed out here that the quantum of circulation, $h/m$, has been measured for helium-4 atoms in experiments involving superfluid helium,[40,41] and for superconducting electrons through a superconducting de Broglie wave interferometer.[42]

The classical particle picture leads to the relative motion of particles, while the classical wave picture leads to the absolute motion of the waves; the motion of a particle has classical manifestations in terms of absolute and relative motion in direct correspondence with wave-particle duality.[43] This viewpoint is in harmony with all available data as well as the standard extension of Lorentz invariance to accelerated systems and gravitational fields once such extension is confined to the eikonal regime in which the standard theory has been tested successfully thus far. Only a beginning has been made in attempting to incorporate these ideas in a consistent manner into the physical theory;[44] a complete synthesis remains a task for the future.

It appears from a preliminary analysis that such a complementarity of absolute and relative motion can be properly formulated only in Minkowski spacetime. It is expected that when wave phenomena are treated in the eikonal approximation within such a theory, the spacetime would appear to have an effective Riemannian curvature. In this way, contact could be established with the observational data that account for the success of general relativity. Thus the standard theory would be valid once phenomena involving classical particles and rays of radiation are only taken into account; beyond this approximation, new gravitational phenomena are expected which have not been explored as yet. Cosmology is based on the reception of null rays from cosmic sources; thus cosmological data—which can be interpreted according to general relativity—may indicate that our world is spatially open or closed. Either possibility should then have an interpretation in terms of the effective curvature of the underlying Minkowski spacetime.

## REFERENCES

1. W. de Sitter, Mon. Not. Roy. Astron. Soc. **77**, 155 (1916).
2. C.W.F. Everitt *et al.*, in *Near Zero: Festschrift for William M. Fairbank*, edited by C.W.F. Everitt (Freeman, San Francisco, 1986).
3. B. Mashhoon, H.J. Paik, and C.M. Will, Phys. Rev. D**39**, 2825 (1989).
4. V.B. Braginsky, A.G. Polnarev, and K.S. Thorne, Phys. Rev. Lett. **53**, 863 (1984).
5. B. Mashhoon, "On the Origin of Inertial Accelerations," UMC preprint (1991).
6. E. Mach, *The Science of Mechanics* (Open Court, La Salle, 1960), ch. II, sec. VI (n.b. part 11).
7. B. Mashhoon, "Quantum Theory and the Origin of Inertia," UMC preprint (1992).
8. N. Bohr and L. Rosenfeld, Mat.-fys. Medd. Dan. Vid. Selsk. **12**, no. 8 (1933).
9. B. Mashhoon, Phys. Lett. A**143**, 176 (1990).
10. B. Mashhoon, Found. Phys. (Bergmann Festschrift) **15**, 497 (1985).
11. B. Mashhoon, Phys. Lett. A**145**, 147 (1990).
12. A. Einstein, *The Meaning of Relativity* (Princeton University Press, Princeton, 1955).
13. J.A. Wheeler, in *Relativity, Groups and Topology*, edited by B. DeWitt and C. DeWitt (Gordon and Breach, New York, 1964).
14. J. Isenberg, Found. Phys. (Wheeler Festschrift) **16**, 651 (1986).
15. J.A. Wheeler, Int. J. Mod. Phys. A **3**, 2207 (1988).
16. B. Mashhoon, in *Quantum Gravity and Beyond: Essays in Honor of Louis Witten*, edited by F. Mansouri and J. Scanio (World Scientific, Singapore, 1993).
17. M. Born, *Einstein's Theory of Relativity* (Dover, New York, 1962).

18. H. Thirring, Phys. Z. **19**, 33 (1918).

19. J. Lense and H. Thirring, Phys. Z. **19**, 156 (1918).

20. B. Mashhoon, F.W. Hehl, and D.S. Theiss, Gen. Rel. Grav. **16**, 711 (1984).

21. D.R. Brill and J.M. Cohen, Phys. Rev. **143**, 1011 (1966).

22. J.M. Cohen and D.R. Brill, Nuovo Cimento **56B**, 209 (1968).

23. J.M. Cohen, Phys. Rev. **173**, 1258 (1968).

24. L. Lindblom and D.R. Brill, Phys. Rev. **D10**, 3151 (1974).

25. B. Mashhoon, Phys. Rev. **D11**, 2679 (1975).

26. J.M. Cohen and W.J. Sarill, Nature **228**, 849 (1970).

27. J.M. Cohen, W.J. Sarill, and C.V. Vishveshwara, Nature **298**, 829 (1982).

28. H. Pfister and K.H. Braun, Class. Quantum Grav. **2**, 909 (1985).

29. H. Pfister and K.H. Braun, Class. Quantum Grav. **3**, 335 (1986).

30. H. Pfister, Class. Quantum Grav. **6**, 487 (1989).

31. L.I. Schiff, Phys. Rev. Lett. **4**, 216 (1960).

32. B. Mashhoon, "On the Gravitational Analogue of Larmor's Theorem," Phys. Lett. A, in press (1993).

33. B. Mashhoon, Phys. Rev. Lett. **61**, 2639 (1988).

34. D.J. Wineland and N.F. Ramsey, Phys. Rev. **A5**, 821 (1972).

35. B.J. Venema *et al.*, Phys. Rev. Lett. **68**, 135 (1992).

36. E.P. Wigner, Annals of Math. **40**, 149 (1939).

37. P.S. Wesson, Space Sci. Rev. **59**, 365 (1992).

38. E.R. Caianiello, Riv. Nuovo Cimento **15**, no.4 (1992).

39. J.W.G. Wignall, Phys. Rev. Lett. **68**, 5 (1992).

40. W.F. Vinen, Proc. Roy. Soc. London A **260**, 218 (1961).

41. G.W. Rayfield and F. Reif, Phys. Rev. **136**, A 1194 (1964).

42. J.E. Zimmerman and J.E. Mercereau, Phys. Rev. Lett. **14**, 887 (1965).

43. B. Mashhoon, Phys. Lett. **A126**, 393 (1988).

44. B. Mashhoon, "A Nonlocal Theory of Accelerated Observers," Phys. Rev. A, in press (1993).

# Recent Progress and Open Problems in Linearization Stability

*V. Moncrief* *

## 1. INTRODUCTION

Of the many influential contributions made by Dieter Brill to the mathematical development of general relativity, one of particular significance was his discovery together with Stanley Deser, of the linearization stability problem for Einstein's equations [1]. Brill and Deser showed that the Einstein equations are not always linearization stable (in a sense we shall define more precisely below) and they initiated the long (and still continuing) technical program to deal with this problem when it arises.

Our aim in this article is not to review the extensive literature of positive results on linearization stability but rather simply to introduce the reader to this subject and then to discuss some recent research that has developed out of the study of linearization stability problems. These latter include the relationship of linearization stability questions to the problem of the Hamiltonian reduction of Einstein's equations and lead one directly to the study of a number of recent results in pure Riemannian geometry (e.g., the solution of the Yamabe problem by Schoen, Aubin, Trudinger and Yambe, the Gromov-Lawson results on the existence of metrics of positive scalar curvature and the still unfinished classification problem for compact 3-manifolds). They also include a study of the quantum analogue of the linearization stability problem which has been significantly advanced recently by the work of A. Higuchi on the quantization of the first order gravitational perturbations of de Sitter space [2].

Many of the questions we shall discuss remain open (or at least partially so). We hope this article may serve to stimulate further research in an interesting and subtle field of relativity research.

## 2. THE LINEARIZATION STABILITY PROBLEM FOR EINSTEIN'S EQUATIONS

The problem of linearization stability begins with the remarkable observation that, for certain given "background" Einstein spacetimes, there are some solutions of the

*Department of Physics, Yale University, 217 Prospect Street, New Haven, Connecticut 06511

associated linearized equations which cannot possibly be tangent to any differentiable curves of exact solutions. In other words some solutions of the linearized equations are "spurious" and should not be taken as first order approximations to nearby exact solutions. Furthermore one knows that, when such spurious solutions exist, the complementary non-spurious solutions do not form a linear space (even though they satisfy a linear equation) but must be constrained to obey a certain second order condition.

Fortunately one now knows to a large extent when to expect such linearization instabilities (by which we mean the existence of spurious solutions of the linearized equations) and what they signify about the space of exact solutions which one is attempting to approximate by linearization. For simplicity we shall discuss only the case of *vacuum* spacetimes having *compact* Cauchy surfaces and admitting at least one such surface of constant mean curvature. It is important to keep in mind the compactness assumption since results for non-compact (say asymptotically flat) hypersurfaces are in general quite different [3]. Whereas the compactness assumption is crucial, we have excluded material sources mostly for simplicity. Many of the results we shall describe extend naturally to the case of the Einstein-Maxwell or Einstein-Yang-Mills equations for example [4, 5].

Let M be a compact, orientable 3-manifold and let $\{g, \pi\}$ be the usual Hamiltonian Cauchy data defined on M for a vacuum solution of Einstein's equations. This data is subject to the Einstein constraint equations which we shall abbreviate as

$$\Phi(g, \pi) = \{\mathcal{H}(g, \pi), -2\,\delta \cdot \pi\} = 0 \qquad (1)$$

where $\mathcal{H}(g, \pi)$ is the Hamiltonian constraint and $-2\,\delta \cdot \pi$ the momentum constraint (a scalar density and one-form density respectively). Let $\mathcal{M}$ formally denote the set of smooth Riemannian metrics on M and let $T^*\mathcal{M}$ denote its "cotangent bundle" consisting of the smooth pairs $\{g, \pi\}$. For simplicity we shall not here enter into the details of natural choices of function spaces for these quantities. Equation (1) defines the constraint subset $\mathcal{C}$ of $T^*\mathcal{M}$

$$\mathcal{C} = \{(g, \pi) \in T^*\mathcal{M} \mid \Phi(g, \pi) = 0\}. \qquad (2)$$

The linearization stability problem hinges on the structure of the constraint set $\mathcal{C}$ and its failure to be globally a manifold. At regular points $(g_o, \pi_o)$ (which fortunately are generic) where $\mathcal{C}$ is a manifold, the tangent space to $\mathcal{C}$ is given by the kernel of the linearized constraint operator $D\Phi(g_o, \pi_o)$ and each solution $(h, p) \in \mathbf{T}_{(g_o, \pi_o)}T^*\mathcal{M}$ of

$$D\Phi(g_o, \pi_o) \cdot (h, p) = 0 \qquad (3)$$

is indeed tangent to a curve lying entirely in $\mathcal{C}$. The Einstein evolution equations produce no further complications and thus the solution of the full set of linearized equations generated by such Cauchy data (h, p) is tangent to a curve of exact solutions.

Fischer and Marsden [6, 7] gave a precise formulation and proof of the above remarks and showed that $(g_o, \pi_o)$ is a regular (manifold) point of $\mathcal{C}$ if $D\Phi(g_o, \pi_o)^*$, the $L^2$ - adjoint operator to $D\Phi(g_o, \pi_o)$, has trivial kernel. The author showed [8, 9] that the kernel of $D\Phi(g_o, \pi_o)^*$ corresponds precisely to (Cauchy data for) the global Killing fields admitted by the vacuum spacetime determined by the data $(g_o, \pi_o)$. Thus if this spacetime has no non-trivial global Killing fields then $(g_o, \pi_o)$ is definitely a regular (manifold) point of $\mathcal{C}$. The converse (that $(g_o, \pi_o)$ is singular if Killing fields exist) is known to be true at least for spacetimes admitting a Cauchy hypersurface of constant mean (extrinsic) curvature and is believed to be true in any case. To analyze the singular points having $\tau_o$ = mean curvature = constant in more detail one uses the following characterization (due to Fischer and Marsden) of the kernel of $D\Phi(g_o, \pi_o)^*$:

> For vacuum Cauchy data of constant mean curvature the kernel of $D\Phi(g_o, \pi_o)^*$ is non-trivial if and only if either

(i) $\pi_o = 0$ and $g_o$ is flat (in which case $(1,0)$ lies in kernel), or

(ii) $\exists$ a vector field $\mathbf{X}$ on M such that $\mathcal{L}_{\mathbf{X}} g_o = \mathcal{L}_{\mathbf{X}} \pi_o = 0$ where $\mathcal{L}_{\mathbf{X}}$ denotes the Lie derivative with respect to $\mathbf{X}$ (in which case $(0, \mathbf{X})$ lies in the kernel), or both.

In general the kernel, if non-trivial is spanned by a finite basis of elements of the above two types (of which both may be present or either one absent).

The second order conditions upon the first order perturbations (h, p) of a singular point $(g_o, \pi_o)$, which are now known to be both necessary and sufficient to exclude spurious perturbations [10, 11, 12], can be written in the form

$$\mathcal{E}_{(\mathbf{C},\mathbf{X})} := \int_\Sigma \ < (\mathbf{C},\mathbf{X}), D^2\Phi(g_o,\pi_o) \cdot ((h,p),(h,p)) > d^3\Sigma = 0$$

$$\forall (C,\mathbf{X}) \subset \ kernel \ D\Phi(g_o,\pi_o)^*. \tag{4}$$

Here the first order Cauchy data (h, p) are assumed to obey the usual linearized constraints, $D\Phi(g_o, \pi_o) \cdot (h, p) = 0$ (which are merely certain projections of the linearized Einstein equations) and it is possible to show that the second order expressions $\mathcal{E}_{(\mathbf{C},\mathbf{X})}$ are invariant with respect to the usual gauge transformations of the linearized theory. If the background data $(g_o, \pi_o)$, the perturbation data (h, p) and the Killing field data $(\mathbf{C}, \mathbf{X})$ are all propagated by the relevant evolution equations it is also possible to show that the above second order conditions are *conserved quantities* and thus independent of the Cauchy surface upon which they are first imposed. One can equivalently express conditions (4) as

$$\int_{\Sigma} <^{(4)} X \cdot (D^2 \ \mathbf{Einstein}(^{(4)}g) \cdot (^{(4)}h, \ ^{(4)}h)) \cdot \ ^{(4)}n > d^3\Sigma = 0 \qquad (5)$$

where $^{(4)}\mathbf{X}$, $^{(4)}g$ and $^{(4)}h$ are the Killing field, spacetime metric and spacetime perturbation respectively, $^{(4)}n$ is the unit future pointing normal field to the hypersurface $\Sigma$ and where $D^2 \ \mathbf{Einstein} \ (^{(4)}g) \cdot (,)$ signifies the second variation of the Einstein tensor of $^{(4)}g$.

In summary one can say that the *true linear approximation* to Einstein's vacuum field equation (about a background having compact Cauchy surfaces and non-trivial Killing symmetries) is not simply the usual linearized equation, D $\mathbf{Einstein}(^{(4)}g) \cdot (^{(4)}h) = 0$, but instead this equation supplemented by the second order conditions ((4) or equivalently (5)) described above. These conditions capture the fact that the space of solutions is not a manifold at $^{(4)}g$ but rather has conical singularities. Thus its associated "tangent space" is not the linear space that one expected but instead a linear space intersected with the second order conditions.

## 3.  LINEARIZATION STABILITY AND THE SPACE OF CONFORMAL STRUCTURES OF A COMPACT 3-MANIFOLD

The issue of linearization stability arises in a new and interesting way when one considers the global problem of *Hamiltonian reduction* of Einstein's equations for spacetimes having compact, orientable Cauchy hypersurfaces. Let M, as above, be a compact, orientable 3-manifold. In solving the constraints (one of the essential steps in reduction) through the use of the standard conformal method [13] one notes that only the conformal class of a metric on M needs to be specified (together with a transverse traceless (t-t) symmetric 2-tensor field) in order to construct appropriate Cauchy data. Lichnerowicz's equation determines the conformal factor which distinguishes the true "physical" metric from all of its conformal equivalents. Thus a natural candidate for the reduced configuration space for Einstein's equations should be the *space of conformal structures* on M [14].

One possible approach to realizing this space is to attempt to construct a submanifold of the space $\mathcal{M}(M)$ (of all smooth Riemannian metrics on M) which intersects each conformal class transversally in precisely one point and upon which the group $\mathcal{D}(M)$ (of smooth diffeomorphisms of M) acts naturally (through pull back of metrics). A realization of the space of conformal structures could then be defined, modulo technicalities which we shall not enter into here, as the quotient of the chosen submanifold by the given group action. In the model problem of Einstein's equations in $2 + 1$ dimensions this procedure leads (if $\mathcal{D}(M)$ is replaced by the subgroup $\mathcal{D}_0(M)$

of diffeomorphisms isotopic to the identity) to the Teichmüller space of the chosen compact 2-manifold as reduced configuration space ([15, 16]). In this case the cotangent bundle of Teichmüller space plays the role of the reduced phase space and one can identify, after making a suitable choice of time function, the Hamiltonian which generates the dynamical flow induced from Einstein's equations. This Hamiltonian turns out to be simply the area function of the hypersurfaces of the chosen foliation expressed in terms of the canonical Teichmüller parameters and the time through the solution of the constraints. In carrying out this program one finds that the sphere and the torus are exceptional whereas all the higher genus surfaces can be treated in a uniform fashion. This happens because the sphere problem is vacuous in $2 + 1$ dimensions (there are no solutions having constant mean curvature Cauchy surfaces) whereas the reduced phase space for the torus has conical singularities (associated precisely with linearization instabilities of exceptional, static solutions). By contrast, linearization instabilities are absent for the higher genus surfaces and there are no obstructions to completing the program outlined above.

To generalize the above model to $3 + 1$ dimensions one first needs to partition the compact 3-manifolds into various classes (analogous to the three classes mentioned above) corresponding to the different degrees of linearization instability that can occur. The simplest, and by far the largest, class of 3-manifolds to consider are those (like the higher genus surfaces) which are incompatible with continuous isometries altogether for solutions of Einstein's equations. These have been identified (modulo the still unproven Poincaré conjecture) by A. Fischer and the author [17]. Most of these manifolds (a subset we call *admissible* manifolds) have the additional property that they only admit Riemannian metrics of the negative Yamabe class (i.e., metrics g which are globally, uniquely conformal to metrics h having constant negative scalar curvature, $R(h) = -1$). It is easy to prove, for these admissible 3-manifolds, that the subspace $\mathcal{M}_{-1}(M)$ of the space of all Riemannian metrics $\mathcal{M}(M)$ on M satisfying the condition $R(g) = -1$ is in fact a smooth submanifold on which the group of diffeomorphism $\mathcal{D}(M)$ of M acts naturally (by pullback of metrics). Modulo "technicalities" one can define a space of conformal structures, $\mathcal{T}(M) = \mathcal{M}_{-1}(M)/\mathcal{D}_0(M)$, of M which plays a role completely analogous to the Teichmüller space for higher genus surfaces in Hamiltonian reduction. $\mathcal{T}(M)$ is the reduced configuration space and $T^*\mathcal{T}(M)$ is the reduced phase space on which an appropriate Hamiltonian (the volume functional for the chosen constant mean curvature foliation) generates the dynamical flow induced from Einstein's equations. There are a number of interesting open mathematical questions concerning this definition of $\mathcal{T}(M)$. For example is $\mathcal{T}(M)$ globally a manifold or does it admit exceptional "orbifold-type"singularities due to the occurence of non-trivial discrete isometries isotopic to the identity and does $\mathcal{T}(M)$, unlike the Teichmüller space for a surface, have non-trivial "topology" or is it merely diffeomorphic to an open ball in some Banach space? Orbifold singularities, even if they occur, will be complementary to an open dense set on which $\mathcal{T}(M)$ is a manifold but this complication, if it arises, suggests that one use $\mathcal{R}(M) = \mathcal{M}_{-1}(M)/\mathcal{D}(M)$, the analogue of Riemann moduli space, instead of $\mathcal{T}(M)$, the analogue of Teichmüller space, as reduced configuration space if the later choice has not successfully avoided (as it does in $2 + 1$ dimensions) the orbifold singular points. These questions are a challenge for future research.

The question we wish to address here however is how can one generalize this picture for reduction beyond the admissible manifolds to include all manifolds which topologically exclude linearization instabilities? According to the results in [17] (which assume the Poincaré conjecture and a conjecture that the only free finite group actions on $S^3$ are the standard orthogonal free actions of finite subgroups of $SO(4)$) one finds that any manifold of the form

$$M \approx S^3/\Gamma_1 \# \ldots \# S^3/\Gamma_q \#(S^2 \times S^1)_1 \# \ldots \#(S^2 \times S^1)_r \qquad (6)$$

where each $\Gamma_k$ is a finite subgroup of $SO(4)$ acting orthogonally on $S^3$ and where $q \geq 1$ and $q + r \geq 2$ (so that these manifolds are necessarily *composite* rather than *prime*) is *not admissible* (since it always admits metrics of the positive Yamabe class) but nevertheless (provided at least one of the $S^3/\Gamma_i$ is not a lens space) excludes solutions of Einstein's vacuum field equations with continuous isometry groups (and Cauchy surfaces of constant mean curvature). Furthermore it seems likely (though not yet proven) that every manifold with these properties is of the above type (it remains to be shown that a manifold which admits non-trivial $SO(2)$ actions always admits $SO(2)$-invariant Cauchy data). If the Poincaré conjecture or the algebraic conjecture mentioned above should prove to be false then it is possible that a larger set of manifolds could have the properties described though little seems yet to be known about the possible Yamabe classes admitted by such hypothetical counter-examples.

Any manifold of the type listed above admits metrics of all three Yamabe types. Thus one cannot hope to find a unique representative of every conformal class by imposing a condition such as $R(g) = -1$ on the scalar curvature. This worked for admissible manifolds (as for the higher genus surfaces) only because these manifolds excluded metrics of the positive and zero Yamabe classes altogether. One could perhaps try to replace this with the condition that $R(g) = $ constant together with the normalization (to freeze out constant conformal transformations) $vol_g(M) = 1$ where $vol_g(M)$ is the Riemannian volume of $(M, g)$. The problem however is that even though, thanks to the resolution of the Yamabe problem by Schoen, Aubin, Trudinger and Yamabe [18], every metric on M is globally conformal to one of constant scalar curvature and fixed volume, the conformal factor is not in general unique for metrics of the positive Yamabe class.

One source of non-uniqueness arises when g admits non-trivial, continuous, *conformal* isometries. Pulling back g by such a conformal isometry leaves the scalar curvature constant but sends g to an inequivalent element of the same conformal class. A theorem of Obata however shows that, except for the case of the standard sphere, the conformal isometry group of a compact, orientable Riemannian manifold $(M, g)$ is always a subgroup of the isometry group of some conformally related manifold $(M, \rho g)$ [19]. But the manifolds listed above do not admit any $SO(2)$ actions and so cannot admit metrics with continuous conformal isometries. Thus one source of non-uniqueness for the solution of the Yamabe problem has been excluded by our purely "topological" restriction on the choice of $M$. This however is not sufficient to render the choice of conformal representative unique.

To have any hope of making a unique choice of conformal representative one needs to introduce a further condition such as, for example, requiring the metric in question to minimize the value of the Yamabe functional. If this minimal choice (which exists by virtue of the solution of the Yamabe problem) is unique for the manifolds of interest we can choose it as our representative from each conformal class. Thus one would be led to consider the subspace of metrics of fixed volume, constant scalar curvature and minimal value (within each conformal class) of the Yamabe functional (which in fact already implies constancy of the scalar curvature). This space, if it proves to be a manifold, might be chosen to play a role analogous to that of $\mathcal{M}_{-1}(M)$ in the admissible cases described above.

The question however of whether a condition such as $R(g) = $ constant (together with a volume normalization) determines a smooth submanifold of the space $\mathcal{M}(M)$ is an interesting problem in pure differential geometry [20, 21]. One can show that g is a regular (manifold) point of this space provided the equation (which represents the $L^2$-adjoint of the derivative of the map which sends a metric to its scalar curvature)

$$- g\Delta C + Hess\ C - C\ Ric(g) = 0 \qquad (7)$$

(where Ric(g) is the Ricci tensor of g and $\Delta$ and Hess are its Laplacian and Hessian operators) admits only the trivial solution $C = 0$. Examples such as the standard sphere however are known for which this equation has non-trivial solutions. In fact all orientable *conformally flat* manifolds for which this equation has non-trivial solutions are known, thanks to work by O. Kobayashi [22] and J. Lafontaine [23]. Aside from the sphere however all compact, orientable, conformally flat 3-manifolds of this type are Riemannian products or warped products of the standard 2-sphere and the circle. Neither the 3-sphere nor $S^2 \times S^1$ is of interest at the moment however since these manifolds always admit solutions of Einstein's equations with continuous isometries.

It seems conceivable therefore that, for the manifolds listed above, the equation $R(g) = $ constant (supplemented by the volume normalization and the minimizing condition on the Yamabe functional which as we noted implies the constancy of $R(g)$) does lead to a smooth submanifold of $\mathcal{M}(M)$ consisting of a unique representative from each conformal class. Since $\mathcal{D}_0(M)$ acts on this space one could take its quotient with respect to this action and define a reduced configuration space by close analogy with the approach for admissible manifolds described above. Much analytical work would be needed to justify this proposal but fortunately the question of how the space of metrics having constant scalar curvature *branches* at points of bifurcation for the solutions of Yamabe's problem is an important topic of recent research in pure Riemannian geometry [24]. It seems likely that advances in this pure geometry problem will have immediate application to the reduction problem for Einstein's equations.

What has the above got to do with linearization instability given that the manifolds in question were chosen just so as to avoid the possibility of any such instabilities? An answer begins to emerge if we note that the equation $R(g) = \Lambda =$ constant, which appears in the approach to reduction described above, also arises in a completely different context in general relativity:

> Initial data $(g, 0)$ satisfying $R(g) = \Lambda =$ constant on an arbitrary compact, orientable 3-manifold M are solutions of the constraints for a time-symmetric initial Cauchy surface in the presence of a *cosmological constant* $\Lambda$. The associated criterion for the absence of linearization instabilities is that $\mathcal{L}_X g = 0 \rightarrow X = 0$ (to exclude space-like isometries) and that $-g\Delta C + Hess\ C - C\ Ric(g) = 0 \rightarrow C = 0$ (to exclude timelike or boost isometries).

In other words the question of whether Equation (7) admits non-trivial solutions is directly connected with the question of existence of linearization instabilities for the Einstein equations *in the presence of a cosmological constant*. Recognizing this one can, for example, appeal to the results of Kobayashi and Lafontaine to characterize all compact, *conformally flat*, time symmetric Cauchy data for a solution of Einstein's equations with cosmological constant $\Lambda$ and a *boost Killing field* [25]. These include the de Sitter solution of course but also a family of solutions defined over Riemannian products and warped products of the circle and standard 2-sphere. The branchings of the solution set for the pure geometry problem $R(g) = \Lambda =$ constant are directly related to the branchings of the space of solutions of the constraint equations for the special case of time symmetric data and a cosmological constant $\Lambda$. The existence of such branching signals the presence of boost type Killing fields in the corresponding spacetimes.

That solutions of Equation (7) imply *boost* rather than merely timelike Killing fields when $\Lambda$ is non zero follows from taking the trace of this equation which gives $2\Delta C = -\Lambda C$. Integrating this result over (M, g) one easily finds that any non zero solution must change sign somewhere on M. This corresponds to the Killing field with normal projection C changing from future directed timelike to past directed timelike in some region of the initial surface. For the special case of the standard sphere (which gives rise to de Sitter spacetime) the four independent solutions of Equation (7) are comprised of the four independent first order spherical harmonics and these correspond to the four boost Killing fields of de Sitter space.

# 4.   QUANTUM ASPECTS OF LINEARIZATION STABILITY

Lacking a well defined quantization of Einstein's equations we cannot expect to make rigorous statements about the quantum analogue (if there is one) of the linearization stability problem.  Since the source of the classical instabilities was the failure of the constraint set (and thus the reduced phase space) to be globally a manifold, it seems plausible that any successful approach to quantization would have to take this significant geometrical feature of the classical dynamics into account.

How should one quantize constrained systems whose constraint sets are not globally manifolds?  In the absense of a satisfactory general answer to this question one can ask the simpler question of how should one quantize the first order perturbations of some classical "background" solution, in the case of general relativity, a classical Einstein space time?  For simplicity assume the background is a vacuum metric $^{(4)}g$ defined on $M \times R$ with $M$ compact and that this spacetimes admits a foliation by Cauchy hypersurfaces of constant mean curvature. If $(M, ^{(4)}g)$ admits no global Killing symmetries the issue of linearization instabilities does not arise and the quantization of first order perturbations is, in principle, comparatively straightforward.

If Killing symmetries are present however then one must somehow impose a quantum analogue to the second order conditions described in section II. Otherwise one could construct "coherent states" representing essentially classical fluctuations about the background which violate the classical second order conditions (i.e., which correspond to spurious perturbations). The proposed quantization would then fail to have the proper classical limit and thus violate the elementary correspondence principle. Thus it seems clear that one needs to impose a quantum analogue of the second order conditions.

These conditions are, as we saw, nothing more than certain projections of the constraint equations expanded out to second order about the background. The particular projections taken are uniquely characterized by the fact that they are independent of the *second order* perturbations (which we are not yet attempting to quantize) and depend (quadratically) only upon the first order perturbations. These second order constraints supplement the usual linearized constraints which must of course also be imposed at the quantum level.

Formally one often thinks, following Dirac, of imposing constraints as linear restrictions upon the space of physical states.  One constructs operator analogues of the constraints (here both first and second order) and requires the physical states to be annihilated by these constraint operators.  This is a consistent procedure as long as there are no "anomalies" in the quantum commutator algebra of the constraints. There are no ordering ambiguities in the ordering of the linearized constraints and it is easy to verify that their commutators, one with another, vanish identically provided the classical background satisfies the exact constraints. This fact merely reflects the abelian character of linearized gauge transformations and the fact that the linearized constraints are the generators of such transformations. At the quantum level these constraints imply the invariance of the quantum states under the abelian group of linearized gauge transformations and thus they closely resemble the corresponding "Gauss law" conditions imposed upon physical states in quantized Maxwell theory for example.

At the classical level the second order constraints have a Poisson bracket algebra isomorphic to the Lie algebra defined by the Killing fields of the background spacetime:

$$\{\mathcal{E}_{(\mathbf{C},\mathbf{X})}, \mathcal{E}_{(\mathbf{C}',\mathbf{X}')}\} = \mathcal{E}_{[(\mathbf{C},\mathbf{X}),(\mathbf{C}',\mathbf{X}')]}$$

where $[(\mathbf{C}, \mathbf{X}), (\mathbf{C}', \mathbf{X}')]$ signifies the normal and tangential components of the Killing field defined by the Lie bracket of those having normal and tangential components $(\mathbf{C}, \mathbf{X})$ and $(\mathbf{C}', \mathbf{X}')$ respectively. This reflects the fact that the second order conditions are constants of the motion of the linearized equations by virtue of the invariance of these equations with respect to the continuous isometry transformations of the background spacetime (which are generated of course by the Killing fields from which the second order conditions are constructed). Furthermore the second order constraints are gauge invariant and thus Poisson commute with the linearized constraints (which, as we noted above, are just the generators of linearized gauge transformations).

Assuming, as seems to be true in practice, that one can find an anomaly-free ordering of the second order constraint operators, one is in a position to impose these as further restrictions upon the physical states which are consistent with the linearized constraints which we have already imposed. What do these new conditions imply about the physical states? Since the second order constraints are the generators of infinitesimal isometry transformations (as one can verify by computing the commutators of these constraints with the linearized canonical variables), *these constraints require the physical states to be invariant under isometries of the background spacetime.*

At first glance this sounds like an absurdly stringent condition upon the physical states. If the background were say time translationally invariant (i.e., stationary) then the physical states would be required to share this time translational invariance (i.e., to be annihilated by the infinitesimal generator of time translations). How can one live with a physical theory in which none of the physical states evolve in time? This question however leads to another—what does one mean by "time"? Are the physical states really stationary or must we only seek "time" amongst the dynamical variables upon which the physical states do depend? The first and second order conditions we are imposing are merely certain projections of the exact constraints expanded out to first and second order about the background and thus really represent certain approximations to the exact constraints which one wishes to impose in "quantum gravity," i.e., approximations to the formal Wheeler-DeWitt equations. It is always the case however, for quantization on spatially compact manifolds via the Wheeler-DeWitt approach, that one must seek an intrinsic time variable amongst the canonical variables in order to give dynamical meaning to the physical states. This is not easy to do in general and there is no universally accepted time function for this formal approach to quantum gravity. Nevertheless one should not be surprised to see the need for choosing an intrinsic time variable arising again in perturbation theory, at least in those cases for which the background is stationary.

As noted above the only stationary vacuum background spacetimes having compact Cauchy surfaces of constant mean curvature are in fact static and flat. The simplest are flat metrics on $T^3 \times R$ having one timelike and three spacelike commuting Killing fields. Quantization of the linearized perturbations of these models was considered by the author who showed that one could choose an intrinsic time function (the perturbed spatial volume of the universe) and thereby give a natural dynamical interpretation to the "time translationally invariant" Hilbert space of physical states [26]. The second order condition associated with the background timelike Killing field plays the role of the basic dynamical "Wheeler-Dewitt equation" for the linearized quantum theory and the physical states can be decomposed into "positive and negative frequency states" corresponding to expanding or contracting universes respectively. If we impose the super-selection rule of only admitting purely positive or purely negative frequency states (i.e., of not mixing states describing perpetually expanding and perpetually contracting universes) then we can easily construct a conserved, positive definite inner product for physical states. After the linearized constraints have also been taken into account, one can show that the resulting physical states consist of a "Fock-space" of states labeled by a complete set of transverse-traceless graviton quantum numbers together with a finite set of variables describing purely homogeneous perturbations ("infinite wavelength gravitons" [27]) and that these states evolve non-trivially with respect to intrinsic time (i.e., perturbed volume).

That volume should make an acceptable time function for this problem is suggested by the rigorous classical result that vacuum solutions on $T^3 \times R$ cannot admit a maximal hypersurface unless they are static and flat. Thus expanding solutions continue to expand and contracting solutions continue to contract. This classical result should have a perturbative quantum analogue and does by virtue of our imposition of the second order condition described above. No state except the "Fock vacuum", which represents a trivial perturbation of the static background, can fail to represent an expanding or collapsing universe. On the other hand, had we failed to impose the second order condition, imposing only the quantum analogues of the usual linearized equations, we could easily have constructed coherent states representing large amplitude, essentially classical gravitational waves which do not drive the universe into a state of either expansion or collapse and which thus violate the elementary correspondence principle. Thus the second order condition, far from being the liability it first seemed, provides the linearized analogue of the Wheeler-DeWitt dynamical equation and enforces consistency with the correspondence principle.

Actually, in this model, there are three more second order conditions (those associated with the spacelike Killing fields) which must also be imposed to complete the quantization of the linearized equations. They can be described equivalently as requiring the space translational invariance of the physical states or of demanding that the total spatial momentum of gravitons in the universe be zero. One should not be distressed at this point to be dealing with a Hilbert space which (because of space translation invariance) cannot localize graviton energy with respect to the background spacetime. It is perfectly feasible to describe, within this framework, relative concentrations of say graviton energy even though the absolute location of this energy, relative to the background, is not determined. An example from the ordinary non-relativistic quantum mechanics of many particle systems may help one to see this. Suppose one considers a wave function for a many particle system which is an eigenfunction of the total momentum with eigenvalue zero. Upon making the usual separation into relative and center of mass coordinates one easily sees that the wave function must be independent of the center of mass coordinates but can be an arbitrary function of the relative coordinates. Thus relative localization of particles is perfectly well describable in a Hilbert space of zero total momentum states but absolute localization (i.e., localization of the center of mass) is absent.

Of course there is no such zero momentum constraint to impose in ordinary quantum mechanics but the analogy shows that all we "lose" in imposing such a constraint in the quantization of linearized gravity is absolute localization of the center of mass relative to the background spacetime. On the other hand it is also clear that we cannot consistently think of "observers" within this framework who are localized relative to the background spacetime. Observers could only be incorporated into this framework by treating them as material sources coupled to gravity in the usual way and quantized by a natural extension of the pure vacuum problem discussed above. This should perhaps not be surprising given the universality of the gravitational interaction as captured by Einstein's strong equivalence principle. Nevertheless it seems surprising to be forced to something resembling an Everett- Wheeler [28] interpretation of quantum mechanics, in which observers are necessarily included as part of the quantum system under "observation" by considerations of linearization stability.

Perhaps the most fascinating application of the foregoing ideas is to the maximally symmetric, spatially compact solution of Einstein's equations with a cosmological constant—de Sitter spacetime. One can regard the cosmological constant as "put in by hand" or generated dynamically (as in inflationary models) but, in either case, the presence of ten Killing fields in the background leads to ten corresponding second order conditions to be imposed upon the physical states of the quantized linearized theory. These ten constraints *require the full de Sitter invariance of all physical states* which again seems like an absurdly restrictive condition to impose in linearized quantum gravity. Indeed, the usual Fock space only admits one fully de Sitter invariant state – the vacuum state. Our conclusion from this is not however that one must make do with only one physical state but rather that *Fock space is not the right Hilbert space for the quantized perturbations of de Sitter space.*

That it is possible to construct a Hilbert space consisting entirely of de Sitter invariant states has been demonstrated by A. Higuchi [29] who showed, roughly speaking, that one can define invariant states by "averaging non-invariant Fock space vectors over the de Sitter group." Higuchi showed that one can define a new inner product such that these states, which would have infinite "norms" as computed formally in the original Fock inner product, now all have finite inner products as computed using the new definition. Higuchi's Hilbert space is infinite dimensional and is presumably sufficiently rich in its structure to provide a "complete set of physical states" for the quantized perturbations of de Sitter space. Because "time translations," defined relative to the usual slicing of de Sitter space by standard spheres, are not isometries the Hamiltonian generating these translations is not constrained to annihilate the physical states as a consequence of the second order conditions. Thus, unlike in the previous example, there is no need to seek an intrinsic time function in order to give a dynamical interpretation to the quantum theory. The physical states do in general evolve relative to the background.

In a quantum theory containing only invariant states the only meaningful "observables" must themselves be de Sitter invariant. While one can define such observables and compute their matrix elements in the linearized theory much as in conventional quantum mechanics, there is the serious open problem of extending this formalism, in principle at least, to include higher order corrections. One must expect to face all the usual questions associated with the non-renormalizability of the "quantized Einstein equations" as well as confront new ones associated with the necessity of working with a non conventional Hilbert space and extending the treatment of constraints, observables and dynamics to the higher order problem.

## ACKNOWLEDGEMENT

This work was supported in part by NSF grants PHY-8903939, INT-9015153 and PHY-9201196 to Yale University.

## REFERENCES

1. D. Brill and S. Deser, "Instability of Closed Spaces in General Relativity," *Commun. Math. Phys.* **32**, 291-304 (1973).

2. A. Higuchi, "Quantum linearization instabilities of deSitter spacetime," *Class. Quantum Grav.* **8**, Part I: 1961-1981, Part II: 1983-2004, Part III: 2005-2021 (1991).

3. See for example the discussion in "Local linearization stability" by D. Brill, O. Reula and B. Schmidt in *J. Math. Phys.* **28**, 1844-1847 (1987) and references cited therein.

4. J. Arms, "Linearization Stability of the Einstein-Maxwell System," *J. Math. Phys.* **18**, 830-833 (1977).

5. J. Arms, J. Marsden and V. Moncrief, "The Structure of the Space of Solutions of Einstein's Equations II. Several Killing Fields and the Einstein-Yang-Mills Equations," *Ann. Phys.* **144**, 81-106 (1982).

6. A. Fischer and J. Marsden, *Bul. Am. Math. Soc.* **79**, 997-1003. (1973) and *Proc. Symp. Pure Math., Am. Math. Soc.* **27**, Part 2, 219-263 (1975).

7. For a recent review see also the article by A. Fischer and J. Marsden in *General Relativity* ed. S.W. Hawking and W. Israel (Cambridge, 1979).

8. V. Moncrief, "Space-time symmetries and linearization stability of the Einstein equations. I," *J. Math. Phys.* **16**, 493-498 (1975).

9. V. Moncrief, "Space-time symmetries and linearization stability of the Einstein equations. II," *J. Math. Phys.* **17**, 1893-1902 (1976).

10. This issue is treated by J. Arms, J. Marsden and V. Moncrief in *Ann. Phys.* **144**, 81-106 (1982) as well as in the following two references.

11. J. Arms, J. Marsden and V. Moncrief, *Commun. Math. Phys.* **78**, 445-478 (1981).

12. A. Fischer, J. Marsden and V. Moncrief, *Ann. Inst. Henri Poincaré* **33**, 147-194 (1980); see also the original references cited herein.

13. Y. Choquet-Bruhat and J. York, "The Cauchy Problem" in *General Relativity and Gravitation* ed. A. Held (Plenum, New York, 1980).

14. See, for example, "The manifold of conformally equivalent metrics" by A. Fischer and J. Marsden, *Can. J. Math.* **29**, 193-209 (1977).

15. V. Moncrief, "Reduction of the Einstein Equations in 2 + 1 Dimensions to a Hamiltonian System over Teichmüller Space," *J. Math. Phys.* **30**, 2907-2914 (1989).

16. V. Moncrief, "How solvable is (2 + 1)-dimensional Einstein gravity?", *J. Math. Phys.* **31**, 2978-2982 (1990).

17. See the article "Recent advances in ADM reduction" by V. Moncrief in the companion volume dedicated to C. W. Misner.

18. For a recent treatment, which includes references to the original articles, see J. Lee and T. Parker, "The Yamabe Problem," *Bull. Amer. Math. Soc.* **17**, 37-81 (1987).

19. M. Obata, *J. Math. Soc. Japan* **14**, 333-340 (1962).

20. See for example the discussion in chapter 4 of *Einstein Manifolds* by A. Besse (Springer-Verlag, 1987).

21. For a resolution of this problem in the special case of conformally flat Riemannian manifolds see Refs. [22] and [23] below.

22. O. Kobayashi, *J. Math. Soc. Japan* **34**, 665-675 (1982).

23. J. Lafontaine, *J. Math. Pure Appl.* **62**, 63-72 (1983).

24. For a discussion of recent developments on this problem see the article by R. Schoen in *Differential Geometry* ed. B. Lawson and K. Tenenblat (Longman Scientific & Technical, 1991).

25. V. Moncrief, *Class. Quantum Grav.* **9**, 2515-20 (1992).

26. V. Moncrief, *Phys. Rev. D* **18**, 983-989 (1978).

27. The natural splitting of the transverse-traceless perturbations into spatialy homogeneous modes and their $L^2$-orthogonal complement was not carried out in Ref. [26] but can easily be obtained by a straightforward refinement of the decomposition given therein; see Ref. [29] for details.

28. See, for example, *The Many-Worlds Interpretation of Quantum Mechanics*, B. S. DeWitt and N. Graham, eds. (Princeton Univ. Press, Princeton, 1973).

29. for a detailed treatment of the simpler problem of quantizing perturbations of the static, flat spacetime with toroidal spatial topology see A. Higuchi, *Class. Quantum Grav.* **8**, 2023-2034 (1991).

# Brill Waves

# NIALL Ó MURCHADHA

Physics Department, University College, Cork, Ireland

**Abstract**
Brill waves are the simplest (non-trivial) solutions to the vacuum constraints of general relativity. They are also rich enough in structure to allow us believe that they capture, at least in part, the generic properties of solutions of the Einstein equations. As such, they deserve the closest attention. This article illustrates this point by showing how Brill waves can be used to investigate the structure of conformal superspace.

## 1 INTRODUCTION

From time to time I amuse myself by mentally assembling a list of articles I would like to have written. The candidates for this list have to satisfy a number of criteria. Naturally, they have to be both important and interesting to me. Equally, they have to contain results that I can convince myself, however unreasonably, that I could have obtained. Every time I make my list I am struck again by the number of articles by Dieter Brill appearing on it. At first glance, this is explained by the large overlap between our interests. In reality, however, the explanation is to be found by considering the kind of article that Dieter has written over the years and the way in which he manages to convey major insights in a deceptively simple fashion. In this work I wish to return to an article that has a permanent place on my list, the famous Brill waves, and show how some of the the earliest work that Dieter did in general relativity continues to offer valuable insight a third of a century later.

## 2 BRILL

In his thesis (Brill, 1959) Dieter Brill considered axisymmetric, moment-of-time-symmetry, vacuum initial data for the Einstein equations.The starting point is an axially symmetric three-metric of the following form

$$g = e^{2Aq}(d\rho^2 + dz^2) + \rho^2 d\theta^2, \tag{1}$$

where $A$ is a constant and $q$ is an (almost) arbitrary function of $\rho$ and $z$. We only require that it satisfy $q = q_{,\rho} = 0$ along the z-axis, that it decay fairly rapidly at infinity (faster than $1/r$) and that it be reasonably differentiable.

This metric $g$ is to be conformally transformed to a metric $\bar{g} = \phi^4 g$ so that the metric $\bar{g}$ has vanishing scalar curvature, so as to satisfy the moment-of-time-symmetry initial

value constraint of the Einstein equations. This is equivalent to seeking a positive
solution $\phi$ to

$$8\nabla_g^2\phi - R\phi = 0, \quad \phi > 0, \quad \phi \to 1 \text{ at } \infty, \tag{2}$$

where $R$ is the scalar curvature of $g$. It is easy to calculate

$$R = -2Ae^{-2Aq}(q_{,\rho\rho} + q_{,zz}). \tag{3}$$

It is important to notice that $\sqrt{g} = \rho e^{2Aq}$ so that

$$\int R\,dv = \int \sqrt{g}R\,d^3x = -4A\pi \int \rho(q_{,\rho\rho} + q_{,zz})d\rho dz$$
$$= -4A\pi \int [(\rho q_{,\rho} - q)_{,\rho} + (\rho q_{,z})_{,z}]d\rho dz$$
$$= 0. \tag{4}$$

To obtain the last line we have to use both the regularity along the axis and the
asymptotic falloff of $q$. This is a remarkably powerful result, in that it gave Brill the
first positivity of energy result in General Relativity.

Let us assume that we obtain a regular solution to (2) (we will return to this issue
in Section 3). The total energy, the ADM mass, is contained in the $1/r$ part of the
physical metric $\bar{g}$. Since the base-metric $g$ falls off faster than $1/r$ the mass must be
contained in the $1/r$ part of the conformal factor $\phi$. More precisely we get

$$\phi \to 1 + \frac{M}{2r}. \tag{5}$$

Hence

$$2\pi M = -\oint_\infty \nabla^a\phi dS_a = -\oint_\infty \frac{\nabla^a\phi}{\phi}dS_a \tag{6}$$
$$= \int[-\frac{\nabla^2\phi}{\phi} + \frac{(\nabla\phi)^2}{\phi^2}]dv = \int[-\frac{R}{8} + \frac{(\nabla\phi)^2}{\phi^2}]dv \tag{7}$$
$$= \int \frac{(\nabla\phi)^2}{\phi^2}dv > 0. \tag{8}$$

Notice that we use (4) in going from (7) to (8). The key (as yet) unresolved question
is: When can we solve (2)?

## 3 CANTOR AND BRILL

The next issue to be considered is what choices of $q$ (or $Aq$) allow a regular solution
to (2). This question was first seriously discussed by Cantor and Brill (1981). They
derived the following

**Theorem I (Cantor and Brill).** *There is a $\bar{g}$ conformally equivalent to a given metric $g$ such that $R(\bar{g}) = 0$ if and only if*

$$\int [8(\nabla f)^2 + R(g)f^2]dv > 0, \tag{9}$$

*for every $f$ of compact support with $f$ not identically 0.*

Cantor and Brill further show that if $R(g)$ is small (in a precise sense) then the Sobolev inequality may be used to guarantee inequality (9). The Sobolev inequality states that for any asymptotically flat Riemannian three-manifold there exists a positive constant $S(g)$ such that

$$\int (\nabla f)^2 dv > S(g)[\int f^6 dv]^{\frac{1}{3}}, \tag{10}$$

for any $f$ of compact support. Let us now use the Hölder inequality on the second term in (9) to give

$$\int R(g)f^2 dv < [\int |R(g)|^{\frac{3}{2}} dv]^{\frac{2}{3}}[\int f^6 dv]^{\frac{1}{3}}. \tag{11}$$

Combining (10) and (11) shows that if

$$[\int |R(g)|^{\frac{3}{2}} dv]^{\frac{2}{3}} < 8S(g) \tag{12}$$

then expression (9) must be positive for any $f$ and a regular solution to (2) exists.

The Brill waves supply an obvious application of this result. Choose a metric of the form (1) with a fixed $q$ but allow $A$ to vary. It is clear from the expression (3) for the scalar curvature that we can always find a small enough $A$ (which can be either positive or negative) to guarantee that (12) holds.

The next interesting property of the Brill waves is that, for a fixed $q$, one can always find a large enough value of $A$ ( both positive and negative) such that inequality (9) cannot hold. We know that the scalar curvature integrates to zero. This means that it must have positive and negative regions. It is clear that we can always find an axisymmetric function $f_+$ which has support only on the positive regions of the scalar curvature and another function $f_-$ which has support only on the negative regions. This choice can be made independent of the value of $A$. We have that

$$\int (\nabla f_+)^2 dv = 2\pi \int [(f_{+,\rho})^2 + (f_{+,z})^2]\rho d\rho dz \tag{13}$$

entirely independent of $A$. We also have

$$\int R(g)f_+^2 dv = 4\pi A \int [-(q_{,\rho\rho} + q_{,zz})]f_+^2 \rho d\rho dz \tag{14}$$

where the integral on the right hand side of (14) is positive (from the choice of $f_+$) and independent of $A$. Therefore, with this choice of test-function there exists a number $|A_-|$ given by

$$|A_-| = \frac{\int[(f_{+,\rho})^2 + (f_{+,z})^2]\rho d\rho dz}{2\int[-(q_{,\rho\rho} + q_{,zz})]f_+^2\rho d\rho dz} \tag{15}$$

such that if $A$ is large and negative, i.e., $A < -|A_-|$, inequality (9) cannot hold. A similar bound, using $f_-$, can be derived showing that there exists an $|A_+|$ such that if $A > |A_+|$ (9) again breaks down.

This is not just a mathematical game, conformal transformations are the standard way of constructing solutions to the initial value constraints of General Relativity (Ó Murchadha and York, 1973) and for an interesting set of such data ('maximal' initial data) a necessary and sufficient condition is that the metric be conformally transformable to one with zero scalar curvature. What we are doing, therefore, is trying to map out conformal superspace $\tilde{S}$; more precisely, we are trying to delineate the axially symmetric subspace of conformal superspace (a so-called 'midisuperspace'). The functions $q(\rho, z)$ can be regarded as defining 'directions' in conformal superspace with $A$ acting as a 'distance' along each given ray. The 'point' $A = 0$ is flat space; there is an open interval around the origin on each ray which belongs to conformal superspace and each ray (after a finite 'distance') emerges from $\tilde{S}$.

We can even prove more; we can show that each ray passes only once out of $\tilde{S}$. Given $q$, let us assume that we have passed out of $\tilde{S}$, in other words we have an $A_0$ (assumed positive) such that inequality (9) does not hold. This means that there exists a function $f_m$ such that

$$\int[8(\nabla f_m)^2 + R(g)f_m^2]dv \leq 0. \tag{16}$$

In other words,

$$4\pi\int(4[(f_{m,\rho})^2 + (f_{m,z})^2] - A_0[(q_{,\rho\rho} + q_{,zz})]f_m^2)\rho d\rho dz \leq 0. \tag{17}$$

The second term must be negative to counter the positive first term. It is clear that if one increases $A$ while holding $f_m$ and $q$ fixed the inequality must worsen. Therefore, once a ray passes outside $\tilde{S}$, it stays outside. In other words, (the axially symmetric part of) $\tilde{S}$ seems to be simply connected with convex boundary.

## 4 BEIG AND Ó MURCHADHA

The next question that needs be asked is what happens as one moves along such a ray in $\tilde{S}$ and approaches the critical value of $A$. This question has been addressed is a number of recent articles (Ó Murchadha, 1987, 1989; Beig and Ó Murchadha, 1991).

There exists an object analogous to the Sobolev constant, $S(g)$, defined in equation (10), the conformal Sobolev or the Yamabe constant. The Yamabe constant is defined by

$$Y(g) = \inf \frac{\int [8(\nabla f)^2 + R(g)f^2]dv}{8(\int f^6 dv)^{\frac{1}{3}}}, \tag{18}$$

where the infimum is taken over all smooth functions of compact support. This object has a number of interesting properties. It is a conformal invariant; it achieves its maximum value [equalling $3(\frac{\pi^2}{4})^{2/3}$] at (conformally) flat space (Schoen, 1984). Flat space is its only extremum.

One reason for introducing this concept here is that it allows a different formulation of Theorem I.

**Theorem Ia.** *There is an asymptotically flat metric $\bar{g}$ conformally equivalent to a given metric $g$ such that $R(\bar{g}) = 0$ if and only if the Yamabe constant $Y(g)$ of $g$ is positive.*

One can immediately see this by comparing expressions (9) and (18).

The function that minimizes the Yamabe functional satisfies the non-linear equation

$$8\nabla_g^2 \theta - R\theta = -8\lambda\theta^5, \quad \theta \rightarrow 0 \ \ at \ \infty, \tag{19}$$

where $\lambda$ is a constant that is proportional to the Yamabe constant. It equals the Yamabe constant if $\theta$ is normalized via

$$\int \theta^6 dv = 1.$$

I will use this normalization from now on. It can be shown (Schoen, 1984) that $\theta$ exists, is everywhere non-zero, and falls off at infinity like $1/r$.

Let us now combine $\phi$, the solution of eqn.(2), and $\theta$, the solution of eqn.(18), using the Green identity

$$-8 \oint_\infty \nabla^a \theta dS_a = 8 \oint_\infty (\theta \nabla^a \phi - \phi \nabla^a \theta)dS_a \tag{20}$$

$$= 8 \int (\theta \nabla^2 \phi - \phi \nabla^2 \theta)dv \tag{21}$$

$$= \int [\theta R\phi - \phi(R\theta - Y\theta^5)]dv \tag{22}$$

$$= Y \int \theta^5 \phi dv. \tag{23}$$

As $Y$ approaches zero, both $\theta$ itself, and the surface integral of $\theta$ remain well-behaved. In particular, the $1/r$ part of $\theta$ does not vanish in the limit. Therefore the surface integral remains bounded away from zero. This means that the volume integral in (23) must blow up like $1/Y$. Since $\theta$ remains finite, $\phi$ must become unboundedly large. In other words, $\phi$ must blow up like $1/Y$.

It is not enough that $\phi$ blow up like a delta function at one point, it must become large on an extended region so that the integral $\int \theta^5 \phi dv$ blows up like $1/Y$. However, we know that $\phi$ is not a random object, it satisfies a differential equation which forces a broad blow-up on $\phi$. We can make this more precise by returning to eqn.(2). Using the fact that $\sqrt{g} = \rho e^{2Aq}$, we find that $\phi$ must satisfy

$$4\nabla_f^2 \phi + A(q_{,\rho\rho} + q_{,zz})\phi = 0, \quad \phi > 0, \quad \phi \to 1 \ at \ \infty, \tag{24}$$

where $\nabla_f^2$ is the flat-space laplacian.

Let us consider the situation where $q$ has compact support on some ball $B$. The scalar curvature, $R$, has the same compact support. Outside this region the manifold is flat and $\phi$ satisfies $\nabla_f^2 \phi = 0$. Hence the min-max principle tells us that the maximum of $\phi$ must be achieved inside the ball $B$. Further, since $\phi$ is positive and satisfies a linear elliptic equation on the compact set $B$ we can use the Harnack inequality (Gilbarg and Trudinger, 1983). This means that there exists a constant $C_B$ (which we can choose independent of $A$) such that

$$\max \ \phi = \max_B \phi \leq C_B \ \min_B \phi \leq C_B \ \min_{\delta B} \phi. \tag{25}$$

In other words, as the maximum of $\phi$ increases inside in $B$, $\phi$ everywhere in $B$ gets dragged up with it. In particular, $\phi$ on the boundary of $B$ becomes large.

As $A$ gets large, as $Y$ gets small, we reach a point where $\max \phi \geq C_B$ (the Harnack constant). From that point on we have that $min_{\delta B}\phi \geq 1$. Outside $B$ $\phi$ satisfies $\nabla^2 \phi = 0$. This means that the minimum and maximum of $\phi$ occur on the boundaries of the set. Since $\phi \geq 1$ on $\delta B$ and $\phi = 1$ at infinity, the minimum must be at infinity. Returning to (5), $\phi \to 1 + \frac{M}{2r}$, we can conclude that the constant $M$ must be positive. This is a positive energy proof which is not as strong as the original Brill proof, it only holds in the strong-field region, far away from flat space, as the Yamabe constant becomes small. However, it has the advantage of working for a much larger class of metrics.

By analogy with electrostatics we can regard $\phi$ (or rather $\xi = \phi - 1$) as the potential on and outside a (nonconducting) shell ($\delta B$). The potential on the shell is given and one solves for the potential outside. The ADM mass corresponds to the total charge that one has to distribute on the shell to give the specified potential distribution. If the potential is everywhere positive on the shell, the charge must be positive.

If the shell were conducting, the potential on it would be constant. Now we can talk about the capacitance $C$, the constant ratio between the 'charge' $M$ and the potential on the surface. The capacitance is a kind of Harnack constant because we have

$$C \min_{\delta B} \xi \le M. \tag{26}$$

$$C \max_{\delta B} \xi \ge M. \tag{27}$$

These are easy to derive. Replacing a non-constant potential on $\delta B$ by its minimum (maximum) value must decrease (increase) the charge while the decreased (increased) charge equals the capacitance multiplied by the potential. Since the capacitance depends only on the size and shape of the surface, not on the charge, we immediately see that as $\min_{\delta B} \xi$ (or $\min_{\delta B} \phi$) increases , so will the ADM mass. If we incorporate (25), (26) and (27) together we can see that the mass grows linearly with $\phi$. More precisely, we have

$$\frac{C}{C_B} \max \phi \le M \le C \max \phi, \tag{28}$$

$$C \min_B \phi \le M \le C.C_B \min_B \phi. \tag{29}$$

I can extract more information from (23). We have

$$Y \int \theta^5 \phi dv \ge Y \int_B \theta \phi dv \ge (Y \min_B \phi) \int \theta^5 dv. \tag{30}$$

The Harnack inequality (25) tells us that if $Y \max \phi$ diverges as $Y$ goes to zero, so also will $Y \min_B \phi$. Eqn. (28) now tells us that $Y \int \theta^5 \phi dv$ must become unboundedly large, which contradicts (23). Hence $Y \max \phi$ remains bounded away from both zero and infinity as $Y$ goes to zero. In turn, this means that both $Y \min_{\delta B} \phi$ and $Y \max_{\delta B} \phi$ remain finite and bounded away from zero as $Y$ goes to zero. Finally, this means that the ADM mass blows up like $1/Y$ as $Y$ goes to zero.

Not only does the monopole blow up like $1/Y$, all the other multipoles behave in a similar fashion. This allows us to show that a minimal two-surface must appear in the conformally transformed space if M becomes large enough. Let us consider a surface which has coordinate radius r in the background (flat) space. The area of this surface in the physical space is $A = 4\pi r^2 \phi^4$. $A_{,r}$ negative is equivalent to $\phi + 2r\phi_{,r} < 0$. The leading terms in $\phi$ are

$$\phi = 1 + \frac{M}{2r} + Q \frac{r^2 - 3z^2}{r^5} + \dots, \tag{31}$$

where Q is the quadrupole moment. Now we get

$$\phi + 2r\frac{d\phi}{dr} = 1 - \frac{M}{2r} - 5Q\frac{r^2 - 3z^2}{r^5} + \cdots \tag{32}$$

We need to find out if the right hand side of (32) can be negative. We know that $M$ will become large and positive but we have no control over $Q$ except that the ratio Q/M will remain bounded. A similar result holds for all the other multipole moments. We can bound the third term by $10|Q|/r^3$. Let us evaluate (32) at r = M/4. This gives

$$\phi + 2r\frac{d\phi}{dr} \le 1 - 2 + \frac{640|Q|}{M^3} + \cdots \tag{33}$$

It is clear that as M becomes large (as $Y$ becomes small) the third term (and all the higher order terms) becomes small. This implies that the area of the surface in question reduces when it is pushed outwards. (Technically we should evaluate the change in area along the outer normal rather than along the radial vector, however, the difference between the two becomes small as M and r become large.) This means that it is an outer trapped surface in the terminology of Penrose (1965). There is a minimal area surface outside all the trapped surfaces; this is the apparent horizon.

As M increases, the apparent horizon approximates more and more the spherical surface with radius $r = M/2$. The conformal factor on this surface equals 2, so the proper area of the apparent horizon approximates $16\pi M^2$. The location of the apparent horizon moves outward (relative to the flat background) and its area increases as M goes to infinity.

When M reaches infinity, when the horizon reaches infinity, when the Yamabe constant goes to zero, as A approaches the critical value, something catastrophic happens. The solution to (2) blows up and we can no longer construct an asymptotically flat manifold with zero scalar curvature. On the other hand we have a solution to (19) with $\lambda = 0$. This is a conformal factor that transforms the given base manifold into a regular compact manifold with zero scalar curvature. Thus one can construct a closed, vacuum cosmology at moment of time symmetry. If we increase A beyond the critical value, we recover a finite solution to (2), but at the price of allowing $\phi < 0$ in a region. This is John Wheeler's 'Bag of Gold'.

## 5 WHEELER

Let me move back in time. The earliest application I know of the Brill Wave solution was by John Wheeler in 1964. He considered eqn.(24) and brought his understanding of the Shrödinger equation to bear on it. He realised that it was just a scattering problem off a localized potential. The 'potential' $(R)$ averages to zero, but Wheeler realized that the negative parts of the potential were much more important than the positive parts. He threw away the positive parts and considered only a negative potential. He even further simplified the problem and considered the problem of scattering off a negative spherical square well potential. He wrote down an analytic solution to this problem.

Wheeler considered a spherically symmetric square well potential in flat space, i.e.,

$$u = -B, \quad r \le a$$

$$u = 0, \quad r > a. \tag{34}$$

It is easy to solve

$$\nabla^2 \psi - u\psi = 0 \quad \psi \to 1 \text{ at } \infty. \tag{35}$$

The solution is

$$\psi = \frac{1}{\cos B^{1/2}a} \frac{\sin B^{1/2}r}{B^{1/2}r} \quad r \le a$$

$$= 1 + \frac{a}{r}\left(\frac{\tan B^{1/2}a}{B^{1/2}a} - 1\right) \quad r > a. \tag{36}$$

So long as $B^{1/2}a < \pi/2$, $\psi$ in (36) is well behaved. As $B^{1/2}a$ approaches $\pi/2$, both $\psi$ itself, and the coefficient of the $1/r$ part (the ADM mass equivalent) become unboundedly large. Further, the surface on which $\psi = 2$ (the analogue of the apparent horizon) moves out towards infinity.

At the critical point ($B^{1/2}a = \pi/2$), $\psi$ ceases to exist. We now obtain a solution to a related equation

$$\nabla^2 \bar\psi - u\bar\psi = 0, \quad \bar\psi \to 0 \text{ at } \infty. \tag{37}$$

$$\bar\psi = \frac{\sin B^{1/2}r}{B^{1/2}r} \quad r \le a$$

$$= \frac{1}{B^{1/2}r} \quad r > a. \tag{38}$$

This is the equivalent of the Yamabe constant going to zero.

In retrospect, I am amazed how accurately this highly simplified model captures the behaviour of the actual situation. When the well is shallow, we have a regular scattering solution and the energy of the lowest eigenstate is positive. As the well is deepened the energy of the lowest state moves down towards zero. As this happens, we get the phenomenon of resonant scattering, the scattering solution grows bigger and bigger. When the well is just deep enough to give us a zero energy bound state, the scattering state blows up and ceases to exist. If the well is made even deeper, the scattering solution reappears but now it is no longer everywhere positive, it has a node. This pattern repeats itself as the energy of the next lowest energy state approaches zero, and so on.

The 1/r part of the scattering solution also blows up as one approaches the critical point. If one approaches the critical point from the other side, where the scattering solution has a node, the coefficient of the 1/r term approaches $-\infty$. This now allows one to give a fairly nonsensical answer to a fairly unreasonable question: What is the energy of a closed universe? My answer is zero, the average of $+\infty$ and $-\infty$!

We now have a fairly precise understanding of what happens as we pump up one of the Brill waves; the Yamabe constant approaches zero for some finite value of the parameter $A$, the mass approaches $+\infty$ and an apparent horizon appears which moves out towards infinity. The horizon, as it moves outwards, becomes more and more spherical. The breakdown of the system coincides with the horizon reaching spacelike infinity. At the critical point we get data which gives us a smooth compact manifold, without boundary, with zero scalar curvature.

I would like to point out here that we have less understanding of what happens as we further increase the factor $A$. Our only real guide is the Wheeler model. If this is valid, the mass should drop to $-\infty$ just on the other side (we would have initial data with a naked singularity), the mass would build up again until it became positive and then blow up to $+\infty$ as the eigenvalue of the first excited state approached zero, and so on. It would be nice to prove that this is (or is not) the correct picture.

The analysis given in Section 4 only deals with the situation when the Yamabe constant is close to zero. Does it hold in general? In particular, if one takes a Brill wave and increase the constant does the mass increase monotonically, does the Yamabe constant decrease monotonically? These are questions that are yet to be answered.

An even more fundamental question: What happens as one approaches the boundary of conformal superspace along a different direction, i.e., not by holding $q$ fixed and increasing $A$, but by changing $q$? I am unable to answer this question definitively, all I can say is that it is currently being investigated by numerical techniques.

## 6 ABRAHAMS, HEIDERICH, SHAPIRO AND TEUKOLSKY

Brill waves have been used by the numerical relativity community from its earliest days (e.g. Eppley, 1977) and this use has continued right up to the present. The standard approach (pioneered by Eppley) is to choose a metric of the form (1), with $q$ chosen to be some analytic function. Equation (2) is then solved numerically to find the conformal factor $\phi$. Finally, the 'physical' metric $\bar{g} = \phi^4 g$ is constructed. Eppley showed numerically that the overall structure described in the previous sections holds true. With a small scale factor $A$, equation (2) can be solved. As $A$ increases, the conformal factor increases in the middle. Eventually a minimal surface (an apparent horizon) appears in the physical space. Finally, at some finite value of $A$, the solution to (2) blows up.

Recently, Abrahams, Heiderich, Shapiro and Teukolsky (1992) showed that this picture of conformal superspace was grossly oversimplified. They chose as $q$ the function

$$q = \rho^2 \exp[-(\frac{\rho^2}{\lambda_\rho^2} + \frac{z^2}{\lambda_z^2})]. \tag{39}$$

Now, instead of holding $q$ fixed in the traditional fashion, they varied $q$ by reducing the 'characteristic wavelength' $\lambda_\rho$ slowly while holding $\lambda_z$ fixed. Further, they kept changing the scale factor $A$ so that the ADM mass of the physical metric remained at the constant value $M = 1$. They showed that the scale factor $A$ grew rapidly, $A \propto \lambda_\rho^{-2}$ as $\lambda_\rho \to 0$. It is clear that for all values of $\lambda_\rho \neq 0$, the base metric and the physical metric exist and are regular, asymptotically flat, axially symmetric Riemannian metrics with positive Yamabe constant. However, both the base metric and the physical metric become singular when $\lambda_\rho = 0$. Most importantly, nowhere along the sequence do apparent horizons appear in the physical metric.

This analysis raises major questions about the simple description of conformal superspace obtained in Section 4 by holding $q$ fixed and changing $A$. AHST are also probing the boundary of conformal superspace and seem to have discovered that the boundary, in addition to smooth metrics with $Y = 0$, also contains singular metrics.

In the spirit of AHST, let me consider the following sequence of metrics

$$g_\lambda = e^{2Cq_\lambda}(d\rho^2 + dz^2) + \rho^2 d\theta^2, \tag{40}$$

with

$$q_\lambda = \frac{\rho^2}{\lambda^2} \exp[-(\frac{\rho^2}{\lambda^2} + z^2)]. \tag{41}$$

It is clear that, for some fixed, small $C$, for each $g_\lambda$, as $\lambda \to 0$, one can get a regular solution to Eqn.(2). Further, the ADM mass along the sequence remains (more-or-less) constant and no apparent horizon appears.

Let me now make a (constant) coordinate transformation on each metric $g_\lambda$ replacing $\rho$ with $\lambda\rho$ and $z$ with $\lambda z$. This changes (40) to

$$g'_\lambda = \lambda^2 [e^{2Cq'_\lambda}(d\rho^2 + dz^2) + \rho^2 d\theta^2], \tag{42}$$

with

$$q'_\lambda = \rho^2 exp[-(\rho^2 + \lambda^2 z^2)].$$  (43)

Now I conformally transform $g'_\lambda$ by dividing it by $\lambda^2$ to finally get

$$g''_\lambda = e^{2Cq'_\lambda}(d\rho^2 + dz^2) + \rho^2 d\theta^2.$$  (44)

This combination of coordinate and conformal transformations cannot change the value of the Yamabe constant. This means that we can still solve (2) for each $g'_\lambda$. However, the value of the ADM mass is not a conformal invariant; it picks up a factor $\lambda^{-1}$. Thus the new mass, instead of remaining constant, blows up as $\lambda \to 0$.

When one looks at $q'_\lambda$, eqn.(43), it is clear that the limiting metric is no longer singular. However, it is still unpleasant, as it is no longer asymptotically flat. The 'z' dependence in $q'$ drops out and the metric becomes cylindrically symmetric rather than axially symmetric.

The ADM mass is measured with respect to some 'unit' (meters, lightyears, whatever). This combination of coordinate transformation together with a conformal transformation can be regarded as the equivalent of a change of units, and the value of the ADM mass will change appropriately. Thus any relationship between the mass and the Yamabe constant can only be valid in some very restrictive sense, one has to 'fix the units'.

Let me return to the other feature that AHST observed, the absence of apparent horizons. It does not matter whether one performs this combination of constant coordinate and conformal transformations on the base space or on the physical space. Apparent horizons at moment-of-time-symmetry are equivalent to minimal 2-surfaces. Minimal 2-surfaces are stable under constant scalings. Therefore, if AHST find no apparent horizons neither will I, even though the mass blows up for the sequence I consider. This does not contradict the analysis in Section 4. In that section I assumed the existence of an 'external' region where the physical geometry was dominated by the conformal factor. In the sequence of metrics considered here the support of the base scalar curvature spreads ever outwards, preventing us from taking advantage of the increase of M.

It is not even clear what happens to the Yamabe constant in the limit $\lambda \to 0$. All I can say is that it will not pass through zero at any finite value of $\lambda$. There is a hint in Wheeler's spherically symmetric toy model that the Yamabe constant may remain nonzero, even in the limit.

Let us return to (36), but instead of considering a sequence in which $B^{1/2}a$ approaches $\pi/2$, let us increase $a$ and simultaneously scale down $B$ so that $B^{1/2}a$ remains constant (equal to, say, $\pi/4$). Now, the coefficient of the $1/r$ term becomes unboundedly large (it scales with $a$). However, the value of $\psi$ itself never becomes large, we find $\psi \leq \sqrt{2}$ holds true no matter how large $a$ becomes. Hence we never obtain a surface on which $\psi = 2$. Further, the limit $a \to \infty$ does not correspond to the appearance of a zero energy bound state, a solution to (37). Thus, the limit point that AHST investigate almost certainly does not correspond to the Yamabe constant going to zero.

This trick, used by AHST, of scaling the sequence so that the mass remains constant, can also be implemented for the 'regular' sequences discussed in Section 4. Wheeler had already realised this in 1964. There are three equivalent choices of scalings:
(i) : Instead of letting the conformal factor $\phi$ go to 1 at infinity, solve (2) with the condition $\max \phi = 1$. This was the choice Wheeler made.
(ii) : Retain $\phi \to 1$ at infinity, but multiply all the coordinates with the Yamabe constant $Y$.
(iii): Retain $\phi \to 1$ at infinity, but divide all the coordinates by the ADM mass.

With each of these constant scalings, the appearance of minimal surfaces is uneffected. Now, however, as one moves along the sequence, as the minimal surface moves out to infinity, the area of the horizon shrinks to zero. When it pinches off at 'infinity', it really is at a finite proper distance from the middle. Further, it does so smoothly, we do not get a conical singularity, instead we get a regular closed manifold. As one moves further along the sequence, into the region of negative Yamabe constant, I expect that this pinched off point moves back from infinity and develops into a conical singularity. Thus we get again an asymptotically flat manifold, but now one with a naked singularity.

## 7 CONCLUSIONS
What I have been calling conformal superspace, the space of asymptotically flat, Riemannian, 3-manifolds that can be conformally transformed into (asymptotically flat) regular manifolds with positive scalar curvature, is an interesting object. It is the natural space one is left with if one, in a 3 + 1 analysis of the Einstein equations, factors out the Hamiltonian as well as the momentum constraints. If one does not wish to proceed this far, one can still think of it as an interesting subset of regular superspace, metrics on which we expect to find regular, classical solutions of the constraints.

Conformal superspace, $\tilde{S}$, consists of all asymptotically flat, smooth, Riemannian metrics with positive Yamabe constant. The boundary of this space, the limit points of (smooth) sequences of metrics in $\tilde{S}$ will, presumably, consist of metrics which violate one or other of the defining properties of $\tilde{S}$. In other words we expect to find metrics which are not asymptotically flat, metrics which are not smooth, metrics which are not uniformly elliptic and metrics for which the Yamabe constant equals zero.

What understanding we have of such issues as the 'size' and 'shape' of conformal superspace has basically been gained by looking closely at the Brill waves. I am convinced that yet more information can and will be extracted from them. In my introduction I stated that they "offer valuable insight a third of a century later". I have every expectation that this will continue to hold true for another third of a century.

## Acknowledgements

I would like to acknowledge my debt to Bobby Beig, significant parts of this article are just a recasting of our joint work. I would like to thank Edward Malec for critically reading the manuscript.

## REFERENCES
Abrahams, A., Heiderich, K., Shapiro, S. and Teukolsky, S. (1992), Phys. Rev. D**46**, 1452-1463.
Beig, R. and Ó Murchadha, N. (1991), Phys. Rev. Lett. **66**, 2421-2424.
Brill, D. (1959), Ann. Phys. (N.Y) **7**, 466-467.
Cantor, M. and Brill, D. (1981), Compositio Mathematica **43**, 317-330.
Eppley, K. (1977), Phys. Rev. D**16**, 1609-1614.
Gilbarg, D. and Trudinger, N. (1983), *Elliptic partial differential equations of second order* (Springer, Berlin).
Ó Murchadha, N. and York, J. W. (1974), Phys. Rev. D**10**, 428-436.
Ó Murchadha, N. (1987), Class. Quantum Grav. **4**, 1609-1622.
Ó Murchadha, N. (1989), in *Proceedings of the C.M.A.* Vol.**19**, edited by R. Bartnik (C.M.A., A.N.U., Canberra) 137-167.
Schoen, R. (1984), J. Diff. Geom. **20**, 479-495.
Wheeler, J. A. (1964), in *Relativity, Groups and Topology*, edited by B. DeWitt and C. DeWitt (Gordon and Breach, New York) 408-431.

# You Can't Get There from Here:
# Constraints on Topology Change

## KRISTIN SCHLEICH AND DONALD M. WITT

Department of Physics, University of British Columbia, Vancouver, BC V6T 1Z1

## Abstract

Recent studies of topology change and other topological effects have been typically initiated by considering semiclassical amplitudes for the transition of interest. Such amplitudes are constructed from riemannian or possibly complex solutions of the Einstein equations. This simple fact limits the possible transitions for a variety of possible matter sources. The case of riemannian solutions with strongly positive stress-energy is the most restrictive: no possible solution exists that mediates topology change between two or more boundary manifolds. Restrictions also exist for riemannian solutions with negative or indefinite stress-energy sources: all boundary manifolds must admit a metric with nonnegative curvature. This condition strongly restricts the possible topologies of the boundary manifolds given that most manifolds only admit metrics with negative curvature. Finally, the ability to construct explicit examples of topology changing instantons relies on the existence of a symmetry or symmetries that simplify the relevant equations. It follows that initial data with symmetry cannot give rise to a nonsymmetric solution of the Einstein equations. Moreover, analyticity properties of the Einstein equations strongly suggest that in general, complex solutions encounter the same topological restrictions. Thus the possibilities for topology change in the semiclassical limit are highly limited, indicating that detailed investigations of such effects should be carried out in terms of a more general construction of quantum amplitudes.

## 1 INTRODUCTION

One of the most striking features about the observed universe is that on distance scales ranging from fermis to megaparsecs its spatial topology is Euclidean. As there are a multitude of solutions to the Einstein equations with $\mathbb{R}^3$ or $S^3$ spatial topology, this observed feature is consistent with the classical dynamics of general relativity but is by no means required; one can show that all spatial topologies admit initial data suitable for finite evolution under the Einstein equations. In fact, even particularly nice spacetimes such as homogeneous inflationary solutions can have complicated spatial topologies [1]. Moreover, the topology of an initial data slice cannot change under such evolution. Thus, as topology change cannot occur by classical evolution from an initial data set, any explanation of the spatial topology of the observed universe must arise from its initial conditions, that is near its initial singularity. As the characteristic energy scales become large at this point, quantum fluctuations in both the matter and the metric become important to the dynamics of the universe. Such quantum fluctuations violate the classical energy and causality conditions and could lead to topology change. Therefore, it is natural to search for answers to issues concerning topology and topology change in terms of the quantum mechanics of gravity.

A favorite starting point for addressing such issues has been to study the low energy limit of quantum amplitudes for gravity. Such semiclassical amplitudes are formed from the action of classical solutions of the appropriate Einstein equations. Amplitudes between two boundarys of the same topology are formed from the action of classical Lorentzian solutions corresponding to the evolution from the specified initial geometry to the final geometry. However, amplitudes for two or more boundaries of different topologies cannot be formed from such solutions; instead they must be constructed from the action of classical Euclidean, or possibly complex, solutions [2]. Such amplitudes must satisfy certain boundary conditions whose generic form is that on each boundary manifold $\Sigma$, both the extrinsic curvature and the normal derivative of the matter fields vanish [3]. Thus the possibility for initial topologies and for topology change are dictated in the low energy limit by the existence or non-existence of such appropriately behaved Euclidean or complex solutions.

Only a few explicit topology changing Euclidean solutions are known [4, 5]; indeed one of Dieter Brill's recent contributions to relativity has been in finding such solutions [6]. However, these few examples have played an important role in the study of topology change and topological effects as they form the basis for other interesting arguments and calculations [7]. Understandably, such explicit examples of topology change are hard to construct and are special in nature; the existence of some type of symmetry such as an isometry or conformal isometry is used to simplify the Einstein equations into a tractable form. Therefore, as such solutions are not generic, it is interesting to know what kind of restrictions there are on the existence and properties of topology changing solutions that do not necessarily have any symmetry.

The first restriction on the existence of such solutions is purely topological; manifolds that lie in different cobordism classes cannot be boundaries of any mediating instanton. However, apart from this very weak restriction, additional constraints on the existence of such solutions arise from an interesting interplay of geometry with topology. This paper will discuss the possibilities for the existence of such instantons in several cases. It will become apparent that such general topology changing solutions are actually rather hard to come by, even if negative or indefinite stress-energy matter sources are allowed. In fact, the possibilities for topology change are highly restricted in the semiclassical limit. Finally, the assumption of symmetry further limits the possible riemannian or complex solutions.

## 2 EUCLIDEAN INSTANTONS

### 2.1 Strongly Positive Stress-Energy

It is useful to begin by discussing the case of Euclidean instantons with nonnegative Ricci tensor. Such instantons correspond to the Euclidean generalization of the familiar Lorentzian situation in which the matter satisfies the strong energy condition, i.e. the spacetime has nonnegative Ricci tensor. Thus such solutions can be aptly described as having strongly positive Euclidean stress-energy. It can be proven that such Euclidean solutions cannot produce topology changing instantons [8]. This result is most readily illustrated for the case of compact instantons with vanishing extrinsic curvature on all boundaries. For this special class of instantons, the proof follows from the following generalization of Bochner's theorem:

**Theorem 1.** *Let $M^n$ be a closed n-manifold with a riemannian metric such that $R_{ab} \geq 0$. Then the universal cover of $M^n$, $\widetilde{M}^n = \mathbb{R}^k \times \Sigma^{n-k}$ where $k = \dim H^1(M^n; \mathbb{R})$ and $\Sigma^{n-k}$ is a simply connected manifold. Furthermore, the metric is a product $g = f + h$ where $f$ is a flat metric on $\mathbb{R}^k$ and $h$ is a metric with $R_{ab} \geq 0$ on $\Sigma^{n-k}$.*

First, note that the integer $k$ is just the first betti number $b_1(M^n)$. Next each nontrivial generator $\omega$ of $H^1$ corresponds uniquely to a harmonic 1-form $\sigma$ on $M^n$; that is $\sigma = \omega + d\gamma$ where $\gamma$ is a unique function on $M^n$. To show this recall that $\sigma$ is harmonic if $\Delta\sigma = (\delta d + d\delta)\sigma = 0$, where $d$ is the exterior derivative and $\delta$ is its dual in the $L^2$ inner product of forms. Equivalently, a form is harmonic if and only if $\delta\sigma = 0$ and $d\sigma = 0$. Therefore, in order for there to be a harmonic $\sigma$ corresponding to $\omega$, both $d\omega + dd\gamma = 0$ and $\delta\omega + \delta d\gamma = 0$. The first equation is satisfied trivially, the second requires that there exist a unique function $\gamma$ such that $\delta d\gamma = -\delta\omega$. If the exterior derivative is represented by the anti-symmetrized covariant derivative of a form, then $\delta$ is the negative divergence with respect to the first index of the form. Hence, the equation for $\gamma$ is equivalently written as $-\nabla^2\gamma = \nabla^a\omega_a$. Since Gauss' law implies $\int_{M^n}\delta\omega = 0$, it follows that $\delta\omega$ is orthogonal to the kernel of $-\nabla^2$ in the $L^2(M^n)$ inner product. Hence, there exists a unique weak solution $\gamma$ on $M^n$. Finally, elliptic regularity theorems imply that this solution is a strong smooth solution. Therefore, given any nontrivial element of $H^1(M^n)$ there is unique harmonic 1-form in $\mathcal{H}^1(M^n)$ such that $\sigma = \omega + d\gamma$. Hence $\mathcal{H}^1(M^n) \cong H^1(M^n)$.

Next, observe that for any harmonic $\sigma$,

$$0 = \int_{M^n} \sigma(\delta d + d\delta)\sigma = \int_{M^n}\left(-\sigma^c\nabla^b(\nabla_b\sigma_c - \nabla_c\sigma_b) - \sigma^c\nabla_c\nabla^b\sigma_b\right)d\mu \tag{1}$$

where $d\mu$ is the measure on the manifold $M^n$. Using the relation $\nabla^a\nabla_b v_a - \nabla_b\nabla^a v_a = R_b^d v_d$, (1) reduces to

$$0 = \int_{M^n} \left( (\nabla_b \sigma_c)^2 + \sigma^a \sigma^b R_{ab} \right) d\mu \tag{2}$$

after an integration by parts. If $R_{ab}$ is positive, the right hand side of (2) is positive. Therefore, by contradiction there must not exist any such harmonic form $\sigma$ and thus $H^1(M^n; R) = 0$. If $R_{ab}$ is nonnegative then the right hand side of (2) can vanish if

$$\nabla_a \sigma_b = 0. \qquad \sigma^a \sigma^b R_{ab} = 0 \tag{3}$$

It follows directly that $\sigma_b$ must be a Killing vector of constant length. Therefore for each such harmonic form, $M^n$ must locally be a product manifold or to be precise, the universal cover $\widetilde{M}^n$ must contain one copy of $\mathbb{R}$. Moreover, the metric on the cover can be split using this Killing vector. Consequently if there are $k$ such nontrivial 1-forms, $\widetilde{M}^n = \mathbb{R}^k \times \Sigma^{n-k}$ where $\Sigma^{n-k}$ is a simply connected manifold and its metric is a product $g = f + h$ where $f$ is a flat metric on $\mathbb{R}^k$ and $h$ is the metric on $\Sigma^{n-k}$. Q.E.D.

If $b_1(M^n) \neq 0$ then the above theorem implies that there is at least one zero eigenvalue of the Ricci curvature because the metric locally splits on $M^n$. Therefore, if $R_{ab} > 0$ for some riemannian metric on $M^n$, it follows that $b_1(M^n) = 0$. Thus the geometry and topology are closely linked for closed manifolds with nonnegative Ricci. Given the above theorem, the following result for compact Euclidean instantons can be easily proven:

**Theorem 2.** *Given any compact $n$-manifold $W^n$ with two or more boundaries, riemannian metric $g$ with $R_{ab} \geq 0$ and vanishing extrinsic curvature on the boundaries, then $W^n = I \times \Sigma^{n-1}$ and $g = d\lambda^2 + h$ where $I$ is a closed interval and $h$ is the metric on $\Sigma^{n-1}$ and is independent of $\lambda$.*

As the extrinsic curvature vanishes on all boundaries, observe that the compact manifold $W^n$ can be doubled over at these boundaries to form the closed manifold $N^n$. The continuity of the induced metric and the matching of the extrinsic curvature on each boundary imply that the metric on the doubled manifold is continuous and has at least two generalized derivatives. Therefore, metric is continuous and in the Sobolev space $L_2^2(N^n)$ and $R_b^a$ is well defined. Thus as $N^n$ satisfies the conditions of theorem 1, it can be applied directly: Take any curve which joins any two of the boundaries in $W^n$. In $N^n$ this curve, which will be called $c_0$, is a closed curve which is not contractible. It follows that the 1-form $\alpha_0$ which is the dual to the tangent vector of $c_0$, is a nonzero element of $H^1(N^n)$. Consequently theorem 1 implies that the metric on $W^n$ must locally be a product metric. Furthermore, there is a harmonic 1-form $\alpha = \alpha_0 + d\gamma$ on $N^n$ such that $\alpha$ is parallel. This means $\alpha$ has a nonzero constant length. Now, $\alpha$ restricted to $W^n$ is trivial, that is $\alpha = d(\gamma_0 + \gamma)$; however $\alpha$ restricted to $W^n$ still has constant nonzero length. Hence, $\nabla^a f \nabla_a f = $ constant $\neq 0$, where $f = \gamma_0 + \gamma$. Therefore, $f : W^n \to \mathbf{R}$ is a smooth function with no critical points. Hence $W^n$ is not only locally a product manifold but globally a product manifold. Q.E.D.

Again, as in the closed case, it is seen that the geometry and topology are closely linked for compact Euclidean instantons with strongly positive stress-energy; there are no such instantons with two or more boundaries of different topology. Moreover, the only such instanton with two boundaries of the same topology is a product manifold. Finally, although discussed and proven above for the compact case with vanishing extrinsic curvature, versions of these theorems hold for more general boundary conditions, for manifolds with asymptotic regions and for manifolds where the Ricci curvature is negative on small scales [8]. Consequently, strongly positive Euclidean stress-energy implies that Euclidean instantons are in general product manifolds.

## 2.2 Negative or Indefinite Stress-Energy

The results of the previous subsection lead to the conclusion that negative or indefinite Euclidean stress-energy for the matter source is needed for there to be any possibility of topology change. Indeed, this conclusion was used by Giddings and Strominger to motivate the use of the massless axion field in their explicit construction of a topology changing instanton [4]. However, it turns out that negative or indefinite stress-energy matter sources need not lead to amplitudes for arbitrary sets of boundary manifolds. Indeed, the set of possible boundary manifolds is restricted to be a small set [3]:

**Theorem 3.** *Let $W^n$ be a n-manifold with riemannian metric $g$ with geodesically complete boundaries that have vanishing extrinsic curvature and nonpositive normal component of stress-energy. Furthermore let $g$ satisfy the Einstein equations with negative or indefinite Euclidean stress-energy tensor. Then each boundary manifold $\Sigma^{n-1}$ must admit a geodesically complete metric with nonnegative scalar curvature.*

In a sufficiently small neighborhood of each boundary $\Sigma^{n-1}$, the topology of $W^n$ is the product topology $I \times \Sigma^{n-1}$. Correspondingly, in this neighborhood of the boundary the equations of motion can be decomposed in terms of the in terms of the extrinsic curvature $k_{ab}$ and the induced (n-1)-metric $h_{ab}$. Let $n^a$ be the normal vector field in this neighborhood. Then the $n^a n^b G_{ab}$ component of the riemannian Einstein equations is the Hamiltonian constraint;

$$k^2 - k_{ab}k^{ab} - \mathcal{R}(h) = 8\pi G n^a n^b T_{ab} \tag{4}$$

where $\mathcal{R}(h)$ is the scalar curvature of $h_{ab}$. Imposing the boundary conditions that the normal component of stress-energy is nonpositive, that is $n^a n^b T_{ab} \leq 0$, and the extrinsic curvature vanishes, it follows immediately that

$$-\mathcal{R}(h) \leq 0. \tag{5}$$

Thus the scalar curvature $\mathcal{R}$ of $\Sigma^{n-1}$ is locally nonnegative everywhere. Therefore $\Sigma^{n-1}$ must admit a geodesically complete metric with nonnegative scalar curvature. Q.E.D.

This theorem can be expected to hold quite generally for many matter sources with negative or indefinite stress-energy tensor. The specific element entering into it is that the normal component of stress-energy in the Hamiltonian constraint was manifestly nonpositive on each boundary $\Sigma^{n-1}$. Indeed, the Euclidean stress-energy tensors for a minimally coupled scalar field $T_{ab} = \frac{1}{2}(\nabla_a\phi\nabla_b\phi - \frac{1}{2}g_{ab}\nabla_c\phi\nabla^c\phi)$, electromagnetism $T_{ab} = \frac{1}{2}(F_{ac}F_b^c - \frac{1}{4}g_{ab}F_{cd}F^{cd})$, and the axion field $T_{ab} = (3H_{acd}H^{bcd} - \frac{1}{2}g_{ab}H_{cde}H^{cde})$ all have nonpositive contributions when the standard instanton boundary conditions are imposed. This is immediately obvious for the case of the scalar field as its stress-energy tensor is manifestly nonpositive when $n^a\nabla_a\phi = 0$. Similarly the generic form of the boundary condition on matter fields reduces in the case of the electromagnetic and axion fields to $n^aF_{ab} = 0$ and $n^aH_{ade} = 0$. This stems from the fact that the electromagnetic field strength and the axion field strength are locally derived from 1-form and 2-form potentials respectively. Thus the boundary conditions in theorem 2 on the normal component of the Euclidean stress-energy tensor correspond to the generic boundary conditions on the matter fields for topology changing solutions for these sources. Consequently, one expects similar results for many other sources of negative or indefinite stress-energy as well as the above examples. Finally, note that theorem 3 applies to open manifolds as well as compact manifolds; the condition on the scalar curvature is completely local.

The main consequence of theorem 3 is that it places a strong restriction on the topology of the boundary (n-1)-manifolds; they must either admit a metric with positive scalar curvature or all metrics with $R \geq 0$ must have vanishing Ricci curvature [9]. Thus if $n^an^bT_{ab}$ is nonzero somewhere on the (n-1)-manifold, then it follows that the (n-1)-manifold must admit a metric with positive scalar curvature. If $n^an^bT_{ab}$ vanishes identically everywhere on the boundary, then the scalar curvature must also vanish identically. In this case the (n-1)-manifold must admit either a metric with positive scalar curvature or all metrics with $R \geq 0$ must have vanishing Ricci curvature. (Note that any manifold admitting a metric with positive scalar curvature also admits one with zero scalar curvature. Thus the net effect of identically vanishing curvature is to enlarge the set of possible boundary manifolds.) Thus this theorem places especially strong restrictions on the possible topologies for boundary (n-1)-manifolds. Although these restrictions hold for geodesically complete n-manifolds, it is useful to first discuss their meaning in the case of closed manifolds in three dimensions where the consequences can be concretely illustrated.

In three dimensions, up to the Poincare conjecture the only closed 3-manifolds that admit positive curvature are the non-orientable manifolds whose double covers are $S^3$, $S^2 \times S^1$, $S^3/\Gamma$ (where $\Gamma$ is a finite group), these orientable manifolds themselves or a connected sum of these manifolds. This result also follows from the work of Gromov and Lawson [9]. Secondly, if the scalar curvature vanishes everywhere, the set of 3-manifolds includes not only those that admit positive curvature but also the ten flat 3-manifolds of which six are orientable and four are nonorientable [10]. Thus any closed boundary 3-manifold of a Euclidean instanton with nonnegative scalar curvature must be one of these 3-manifolds.

Now, most closed 3-manifolds do not admit metrics with positive scalar curvature; closed 3-manifolds with negative scalar curvature constitute a much larger class. In order to understand what is meant by this statement, one uses the following definitions and theorems from differential geometry. A *link $L$* is the disjoint union of a finite number of circles that are continuously embedded in a 3-manifold. For example, a knot is a link that consists of one continuously embedded circle. A *hyperbolic link $L$* is one for which the manifold $S^3 - L$ has a complete hyperbolic metric. A complete hyperbolic metric is a complete metric with constant negative sectional curvature. These definitions are needed for the following theorem that directly follows from the work of Thurston:

**Theorem (Thurston).** *All compact 3-manifolds are the result of Dehn surgeries on hyperbolic links. Moreover, given a fixed hyperbolic link $L$, all but a finite number of compact 3-manifolds generated by Dehn surgeries are hyperbolic 3-manifolds [11].*

Therefore, if one considers all closed 3-manifolds generated from a given hyperbolic link $L$ one has the quantitative result that a finite number of 3-manifolds constructed from this link will admit metrics with non-negative curvature and a countably infinite number of 3-manifolds will admit metrics with negative curvature. Thus only a very small set of 3-manifolds admit metrics with positive scalar curvature.

Similarly, but in a less precisely quantifiable sense, only a very small set of closed manifolds admit metrics with nonnegative scalar curvature in four or more dimensions. For example, it is known that given any n-manifold $M^n$ that admits positive scalar curvature, one can add any of a certain infinite class of $K(\pi,1)$ prime factors to it to form a new manifold that will admit only negative curvature [9]. As there are a countably infinite number of such $K(\pi,1)$ manifolds, a countably infinite family of manifolds can be constructed from $M^n$ by adding all $K(\pi,1)$ factors to it. Then $M^n$ is the only member of this family that admits positive scalar curvature. Therefore, the case of manifolds admitting positive curvature in higher dimensions parallels the three dimensional case. Therefore, in any dimension, theorem 3 places a strong restriction on the allowed boundary topologies for Euclidean instantons.

## 3 RESTRICTIONS ON SYMMETRIC INSTANTONS

It is clear that general solutions to the Einstein equations cannot be explicitly constructed; however, solutions with symmetries often can be found. Thus it is natural to ask the question, what restrictions are there on instantons with a local symmetry or a conformal symmetry? Obviously, one must begin by assuming that the boundaries $\Sigma^{n-1}$ are members of a restricted set of topologies as discussed in section 2.2. Even so, further restrictions arise. For example, one can prove a generalized form of Bochner's theorem for closed riemannian manifolds with nonpositive Ricci tensor:

**Theorem 4.** *A closed manifold $M^n$ with $n \geq 2$ and riemannian metric $g$ and $R_{ab} \leq 0$ can have no Killing vectors or conformal Killing vectors unless its universal cover $\widetilde{M}^n = \mathbf{R}^k \times \Sigma^{n-k}$ with product metric $g = f + h$.*

Assume that there is a conformal Killing vector field on this manifold,

$$\nabla_{(a} k_{b)} = \frac{1}{n} \nabla_c k^c g_{ab}. \tag{6}$$

Next, consider the quantity

$$\int_{M^n} (\nabla_{(a} k_{b)})^2 d\mu = \frac{1}{n} \int_{M^n} (\nabla_a k^a)^2 d\mu \tag{7}$$

that follows from squaring both sides of (6) and integrating over the manifold. Integrating by parts and rearranging terms, one finds that

$$0 = \int_{M^n} \left( \frac{1}{2} (\nabla_a k_b)^2 + (\frac{1}{2} - \frac{1}{n})(\nabla^a k_a)^2 - \frac{1}{2} R_{ab} k^a k^b \right) d\mu. \tag{8}$$

The first two terms are manifestly positive semidefinite everywhere on the manifold $M^n$ for $n \geq 2$. The sign of the third term depends on the sign of $R_{ab}$ and by the conditions of the theorem is manifestly positive semidefinite as well. Consequently, if $R_{ab} < 0$ there can be no Killing vector or conformal Killing vector. If $R_{ab}$ has a zero eigenvalue, then the equation can be satisfied if

$$\nabla_a k_b = 0 \qquad k^a k^b R_{ab} = 0 \tag{9}$$

This implies that $k_a$ is a Killing vector of constant length and it follows by the same line of argument as in theorem 1 that the universal cover $\widetilde{M}^n$ is a product space with product metric. Q.E.D.

Again the topology and geometry are strongly coupled; if $R_{ab} < 0$ then the above theorem implies that $M^n$ can have no Killing or conformal Killing vectors. Additionally, this theorem can be applied directly to compact instantons $W^n$ with nonpositive Ricci tensor by doubling over the instanton at each of its boundaries to obtain a closed manifold $M^n$. As the Einstein equations are satisfied on $W^n$, the boundary conditions are sufficient to ensure that they also hold on $M^n$. Thus it follows by a simple argument similar to that in theorem 4 that compact instantons $W^n$ with nonpositive Ricci and two or more boundaries of different topology can have no local symmetries or conformal symmetries [8].

For example, this result applies to the case of orientable compact axionic instantons in four dimensions: For this case, the sign of $R_{ab}$ is determined by the equations of motion. As $M^4$ is orientable, $H$ can be defined everywhere as $H_{abc} = \epsilon_{abcd} a^d$ where $a$ is a 1-form and $\epsilon$ is the volume 4-form on $M^4$. Substituting for $H$ into the Einstein equations, one finds that the Ricci curvature is manifestly negative indefinite, $R_{ab} = -96\pi G a_a a_b$. Therefore the conditions of theorem 4 are satisfied and consequently $\widetilde{M}^4$ must be a product manifold. Note that by (9) the axion field must be perpendicular to the vector $k^a$ everywhere and therefore as $a_a$ is normal to the boundary by construction, $\mathbb{R}^k$ must be a component of the cover of the boundary manifold. Finally, similar results hold for nonorientable compact axionic instantons in four dimensions as can be shown using their double cover in the above argument. Therefore, theorem 4 provides a restriction on instantons with symmetry that is physically relevant.

Although the above results are a useful guide to the possible topologies of symmetric Euclidean instantons, more generally one would like to consider the existence of a solution, either Euclidean or complex, on a general manifold with boundary $W^n$ that has symmetric initial data. Additionally such solutions need not satisfy the instanton boundary condition of vanishing extrinsic curvature. First, consider the case of riemannian Einstein metrics; for such solutions, the Einstein equations imply that the metric must be analytic, that is $g_{ab}$ can be expanded in a Taylor series in a neighborhood of every point. The induced metric $h_{ab}$ on $\Sigma^{n-1} = \partial W^n$ and the extrinsic curvature $k_{ab}$ will satisfy the constraints and the slicing of $W^n$ near the boundary will yield an evolution of these tensors. This slicing is given by the level surfaces of a "time function" which is generated by a vector field $t^a = Nn^a + s^a$, where $N$ and $s^a$ are the lapse and shift. The evolution is just the effect of pushing the boundary along $t^a$. The evolution equations are of the form

$$\mathcal{L}_t h_{ab} = 2mNk_{ab}$$
$$\mathcal{L}_t k_{ab} = Q_{ab}(h, k, \rho, J) \tag{10}$$

supplemented by the equations of evolution for the matter if applicable. Note that the tensor $Q_{ab}$ is analytic in its arguments and $m$ is a constant determined by the signature of the metric; $m = 1$ for riemannian metrics and $m = -1$ for Lorentzian metrics. The above two equations and the constraints are equivalent to the Einstein equations.

Since $W^n$ is not necessarily a product manifold, this evolution will eventually break down.

Conversely, given $h_{ab}$ and $k_{ab}$ on the boundary $\Sigma^{n-1}$ which satisfy the constraints and certain regularity conditions, one can evolve the initial data inward to obtain an Einstein metric on a small open set of $W^n$:

**Theorem 5.** *Let $(h_{ab}, k_{ab}, \rho, J^a)$ be analytic initial data on the hypersurface $\Sigma^{n-1}$ satisfying the constraints along with analytic equations describing the matter sources and any additional analytic initial data necessary for this matter. Then the initial data evolves for some finite time into a unique analytic metric on $I \times \Sigma^{n-1}$.*

In order prove this, choose synchronous gauge, that is $N = 1$ and $s^a = 0$. The metric in this gauge is $\mathbf{g} = dt^2 + h_{ij} dx^i dx^j$. Now combine the evolution equations for the geometry (10) by replacing $k_{ab}$ with a time derivative of $h_{ab}$ in order to obtain a single second order equation for $h_{ab}$, namely,

$$\pounds_t \pounds_t h_{ab} = Q_{ab}(h, (2m)^{-1}\pounds_t h, \rho, J). \tag{11}$$

If analytic matter sources are present corresponding to nonzero $\rho$ and $J$, one would additionally reexpress these equations for the matter in terms of first or second order equations. Next, in order to prove that there is a solution to this new set of equations, one can apply the Cauchy-Kowalewski theorem because all fields have been assumed to be analytic. Equivalently, one can prove this directly by expanding (11) and the corresponding matter equations in power series expansions and showing that there is a common radius of convergence. Q.E.D.

Now the goal is to consider initial data with local symmetries:

**Definition.** *A tensor field $T_{ab\cdots}{}^{cd\cdots}$ on a manifold $M^n$ is locally symmetric with respect to $G$ if and only if every point in $M^n$ has an open neighborhood $U$ such that the following conditions are satisfied:*

(i) *There is a finite set of vector fields $\{\xi_i{}^a\}$ on $U$ which generate a faithful representation of the Lie algebra of $G$.*

(ii) $\pounds_{\xi_i} T_{ab\cdots}{}^{cd\cdots}\big|_U = 0$ *for these vectors.*

Similarly, one defines a local discrete symmetry to be a local discrete invariance in an open set. If the tensor with the local symmetry is the metric on the manifold, then the vectors are local Killing vectors. An initial data set has a local $G$ symmetry if the tensors $h_{ab}$ and $k_{ab}$ on $\Sigma^{n-1}$ have a local symmetry with respect to $G$. If sources are present other than a cosmological constant, then the initial data for the other fields most also be invariant under the local $G$ symmetry.

The locally symmetric initial data on $\Sigma^{n-1}$ evolves into a locally symmetric metric on an open subset of $W^n$ near the boundary when the initial data is pushed along some time flow. This is stated in a precise form in the following theorem:

**Theorem 6.** *Given an analytic initial data set which is locally symmetric under the Lie group $G$, then the initial data evolves into a unique analytic metric on $I \times \Sigma^{n-1}$ which is locally invariant under $G$ if the matter equations preserve the local symmetry.*

To preserve the local invariance explicitly, choose a gauge which is also invariant under $G$ locally, for example synchronous gauge. Let $\{\tilde{\xi}_l{}^a\}$ be a set of vector fields on $\Sigma^{n-1}$ which generate the local $G$ symmetry. Next, define the evolution of the vector fields to be $\xi_l{}^\mu = (0, \tilde{\xi}_l^k)$. This well defined and consistent with the equations of motion (11). These new vector fields are extensions of the initial fields to $I \times \Sigma^{n-1}$ and by construction still represent the local symmetry $G$.

Since the initial data has an analytic evolution, it can be written as a convergent Taylor series in a small neighborhood. In particular,

$$h_{ab}(t) = h_{ab}(0) + t\mathcal{L}_t h_{ab}(0) + \frac{t^2}{2}\left(\mathcal{L}_t\right)^2 h_{ab}(0) + \cdots . \tag{12}$$

Using the Taylor series and synchronous gauge, one can verify that $\mathcal{L}_{\xi_l}\left(\mathcal{L}_t\right)^p h_{ab}(0) = 0$ for all $p$ is equivalent to the metric on $I \times \Sigma^{n-1}$ being locally $G$ invariant. In order to prove that each term is invariant, observe the following: The symmetry of the initial data implies that $\mathcal{L}_{\xi_l} h_{ab}(0) = 0$ and $\mathcal{L}_{\xi_l}\left(\mathcal{L}_t\right) h_{ab}(0) = 0$. Directly substituting this into the evolution equations implies that $\mathcal{L}_{\xi_l}\left(\mathcal{L}_t\right)^2 h_{ab}(0) = 0$. By applying operator $(\mathcal{L}_t)$ to the evolution equations $p$ times and substituting the $p - 1$ results into the equations, one can verify that $\mathcal{L}_{\xi_l}\left(\mathcal{L}_t\right)^p h_{ab}(0) = 0$. Therefore, $\mathcal{L}_{\xi_l} h_{ab}(t) = 0$. Since $\mathbf{g} = dt^2 + h_{ij}dx^i dx^j$, it follows that $\mathcal{L}_{\xi_l} g_{ab} = 0$. Q.E.D.

Now, all of these facts can be combined to prove

**Theorem 7.** *An analytic riemannian solution $W^n$ with boundary data with a local $G$ symmetry must also have a local $G$ symmetry everywhere.*

Pick a neighborhood of one of the boundaries of $W^n$ and evolve the boundary data inward a finite amount. This can always be done even though $W^n$ is not assumed to a product manifold. The fact that an open subset of $W^n$ has the local symmetry and that this evolution is unique implies that the evolved metric is the same as the riemannian solution on an open subset of $W^n$. Therefore, by analyticity $W^n$ must have the local symmetry everywhere. Q.E.D.

Now, two important observations can be made. First, if the solution is a riemannian Einstein metric assumed to be twice continuously differentiable, for example, then the solution must be analytic without any further assumptions because of the regularity theorems for the elliptic operators. This yields the following uniqueness theorem for riemannian solutions with local $SO(4)$ symmetry:

**Theorem 8.** *Given any smooth riemannian solution on $W^4$ such that $R_{ab} = \Lambda g_{ab}$ with $\Lambda > 0$ and a local $SO(4)$ symmetry for $h_{ab}$ and $k_{ab}$, then its metric $g$ is locally the round 4-sphere metric.*

By the previous discussion, the metric $g$ is analytic. Next, the local symmetry theorems imply that $W^4$ has the local symmetry $SO(4)$ and in the neighborhood of the boundary, the metric can be can be written locally as $\mathbf{g} = dt^2 + a(t)^2 h_{ij} dx^i dx^j$. Finally, one can calculate $a(t)$ using the Einstein equations to obtain a unique answer locally corresponding to the round 4-sphere metric. Q.E.D.

Secondly, all of the above theorems apply to complex metrics if they are assumed to be analytic; the only change will be that constant $m$ in the evolution equations (10) and (11) will now become a complex number. An interesting consequence is that theorem 8 when generalized to the complex case implies that $a(t)$ can have one other solution for real $h$; this solution is simply the Lorentzian deSitter metric. Therefore, there are no solutions with locally $SO(4)$ symmetric real initial $h$ other than the Euclidean and Lorentzian deSitter solutions.

## 4 CONCLUSION

It is clearly not an easy matter to change topology semiclassically, even allowing for matter fields with negative or indefinite stress-energy tensor. However, although this result is unfortunate for those who might wish to compute effects from general topology change semiclassically, it is not unexpected. After all, topology change is a quantum effect and thus it would be surprising to find it a common occurrence in the low energy limit. Note that these results by no means imply that more general topology change is not possible, but rather indicate that such effects must be studied without the assumption of the semiclassical limit. Indeed possible methods for doing so exist; for example one can model general topology changing amplitudes by using Regge calculus in a sums over histories construction of quantum amplitudes [12]. Thus although semiclassical amplitudes for topology change have proven useful, questions about the consequences of general topology change require a more general formulation of the corresponding quantum amplitudes.

## REFERENCES

[1] J. Morrow-Jones and D. Witt, "Inflationary Initial Data for Generic Spatial Topology", UBC preprint (1992).

[2] There may exist Lorentzian solutions that interpolate between boundary manifolds

of different topology. However, such solutions necessarily have closed timelike curves (R. Geroch, *J. Math. Phys.* **8**, 782, (1967)) are thus not evolutions of initial data. Furthermore, generically ones which obey energy conditions are still product manifolds (F. Tipler, *Ann. Phys.* **108**, 1, (1977)). One can view such solutions as special cases of complex solutions.

[3] K. Schleich and A. Anderson, *Class. Quant. Grav.* **9**, 89, (1992).

[4] S. Giddings and A. Strominger, *Nucl. Phys.* B **306**, 890 (1988).

[5] R. Myers, *Phys. Rev.* D **38**, 1327, (1988), S. Coleman and K. Lee, *Nuc. Phys.* B **329**, 387, (1990), J. J. Halliwell and R. Laflamme, *Class. Quant. Grav.* **6**, 1839, (1989), R. Laflamme and B. Keay, U. British Columbia preprint, (1989).

[6] D. Brill, *Phys. Rev.* D **46**, 1560, (1992).

[7] S. Coleman, *Nucl. Phys.* B **310**, 643 (1988), W. Fischler and L. Susskind, *Phys. Lett.* B **212**, 407, (1988), J. Preskill, *Nucl. Phys.* B **323**, 141 (1989).

[8] K. Schleich and D. Witt, in preparation.

[9] M. Gromov and H. B. Lawson, Jr., *Inst. Hautes Etudes Sci. Publ. Math.* **58**, 83 (1983). See also the earlier work of R. Schoen and S. T. Yau, *Ann. Math.* **110**, 127 (1979) and *Manuscripta Math.* **28**, 159 (1979).

[10] J. A. Wolf, *Spaces of Constant Curvature*, (McGraw-Hill, New York, 1967).

[11] W. Thurston, *Bull. Amer. Math. Soc*, **6**, 357, (1982).

[12] K. Schleich and D. Witt "Generalized Sums over Histories for Quantum Gravity II. Simplicial Conifolds ", *Nucl. Phys.* B, to appear.

# Time, measurement and information loss in quantum cosmology

*Lee Smolin*
Department of Physics, Syracuse University,
Syracuse NY USA 13244

## Abstract

A framework for a physical interpretation of quantum cosmology appropriate to a nonperturbative hamiltonian formulation is proposed. It is based on the use of matter fields to define a physical reference frame. In the case of the loop representation it is convenient to use a spatial reference frame that picks out the faces of a fixed simplicial complex and a clock built with a free scalar field. Using these fields a procedure is proposed for constructing physical states and operators in which the problem of constructing physical operators reduces to that of integrating ordinary differential equations within the algebra of spatially diffeomorphism invariant operators. One consequence is that we may conclude that the spectra of operators that measure the areas of physical surfaces are discrete independently of the matter couplings or dynamics of the gravitational field.

Using the physical observables and the physical inner product, it becomes possible to describe singularities, black holes and loss of information in a nonperturbative formulation of quantum gravity, without making reference to a background metric. While only a dynamical calculation can answer the question of whether quantum effects eliminate singularities, it is conjectured that, if they do not, loss of information is a likely result because the physical operator algebra that corresponds to measurements made at late times must be incomplete.

Finally, I show that it is possible to apply Bohr's original operational interpretation of quantum mechanics to quantum cosmology, so that one is free to use either a Copenhagen interpretation or a corresponding relative state interpretation in a canonical formulation of quantum cosmology.

# 1.    Introduction

What happens to the information contained in a star that collapses to a black hole, after that black hole has evaporated? This question, perhaps more than any other, holds the key to the problem of quantum gravity. Certainly, no theory could be called a successful unification of quantum theory and general relativity that does not confront it. Nor does it seem likely that this can be done without the introduction of new ideas. Furthermore, in spite of the progress that has been made on quantum gravity on several fronts over the last years, and in spite of some recent attention focused directly on it[1] this problem remains at this moment open.

In this paper I would like to ask how this problem may be addressed from the point of view of one approach to quantum gravity, which is the nonperturbative approach based on canonical quantization[AA91, CR91, LS91]. This approach has been under rapid development for the last several years in the hopes of developing a theory that could address such questions from first principles. What I hope to show here is that this approach has recently come closer to being able to address problems of physics and cosmology. To illustrate this, I hope to show here that the canonical approach may lead to new perspectives about the problem of what happens when a black hole evaporates that come from thinking carefully about how such questions can be asked from a purely diffeomorphic and nonperturbative point of view.

One reason why the canonical approach has not, so far, had much to say about this and other problems is that there is a kind of discipline that comes from working completely within a nonperturbative framework that, unfortunately, tends to damp certain kinds of intuitive or speculative thinking about physical problems. This is that, as there is no background geometry to make reference to, one cannot say anything about physics unless it is said using physical operators, states and inner products. Unfortunately, while we have gained some nontrivial information about the physical states of the theory, there has been, until recently, rather little progress about the problem of constructing physical operators.

From a conceptual point of view, the problem of the physical observables is difficult because it is closely connected to the problem of time[2]. It is difficult to construct physical observables because in a diffeomorphism invariant theory one cannot be naive about where and when an observation takes place. Coordinates have no meaning so that, to be physically meaningful, an operator must locate the information it is to measure by reference to the physical configuration of the system. Of course, this is not necessary if we are interested only in global, or topological, information about the fields, but as general relativity is a local field theory, with local degrees of freedom,

---

[1]An unsystematic sampling of the interesting papers that have recently appeared are [CGHS, EW91, B92, GH92, DB92, HS92] [JP92, SWH92, ST92, 'tH85].

[2]Two good recent reviews of the problem of time with many original references are [KK92, I92]. The point of view pursued here follows closely that of Rovelli in [CRT91].

and as we are local observers, we must have a practical way to construct operators that describe local measurements if we are to have a useful quantum theory of gravity.

Thus, to return to the opening question, if we are, within a nonperturbative framework, to ask what happens *after a black hole evaporates,* we must be able to construct spacetime diffeomorphism invariant operators that can give physical meaning to the notion of *after the evaporation.* Perhaps I can put it in the following way: the questions about loss of information or breakdown of unitary evolution rely, implicitly, on a notion of time. Without reference to time it is impossible to say that something is being lost. In a quantum theory of gravity, time is a problematic concept which makes it difficult to even ask such questions at the nonperturbative level, without reference to a fixed spacetime manifold. The main idea, which it is the purpose of this paper to develop, is that the problem of time in the nonperturbative framework is more than an obstacle that blocks any easy approach to the problem of loss of information in black hole evaporation. It may be the key to its solution.

As many people have argued, the problem of time is indeed the conceptual core of the problem of quantum gravity. Time, as it is conceived in quantum mechanics is a rather different thing than it is from the point of view of general relativity. The problem of quantum gravity, especially when put in the cosmological context, requires for its solution that some single concept of time be invented that is compatible with both diffeomorphism invariance and the principle of superposition. However, looking beyond this, what is at stake in quantum gravity is indeed no less and no more than the entire and ancient mystery: *What is time?* For the theory that will emerge from the search for quantum gravity is likely to be the background for future discussions about the nature of time, as Newtonian physics has loomed over any discussion about time from the seventeenth century to the present.

I certainly do not know the solution to the problem of time. Elsewhere I have speculated about the direction in which we might search for its ultimate resolution[LS92c]. In this paper I will take a rather different point of view, which is based on a retreat to what both Einstein and Bohr taught us to do when the meaning of a physical concept becomes confused: reach for an operational definition. Thus, in this paper I will adopt the point of view that time is precisely no more and no less than that which is measured by physical clocks. From this point of view, if we want to understand what time is in quantum gravity then we must construct a description of a physical clock living inside a relativistic quantum mechanical universe.

This is, of course, an old idea. The idea that physically meaningful observables in general relativity may be constructed by introducing a physical reference system was introduced by Einstein[AE]. To my knowledge it was introduced to the literature on quantum gravity in a classic paper of DeWitt[BD62] and has recently been advocated by Rovelli[CRM91], Kuchar and Torre[KT91], Carlip[SC92] and other authors. However, what I hope becomes clear from the following sections is that this is not just a nice idea which can be illustrated in simple model systems with a few degrees of freedom. There is, I believe, a good chance that this proposal can become the heart of a viable strategy to construct physical observables, states and inner products in the real animal-the quantum theory of general relativity coupled to an arbitrary set of matter fields. Whether any of those theories really exist as good diffeomorphism invariant quantum field theories is, of course, not settled by the construction of an approach to their interpretation. However, what I think emerges from the following is a workable strategy to construct the theory in a way that, if the construction works, what we will have in our hands is a physical theory with a clear interpretation.

The interpretational framework that I will be proposing is based on both technical and conceptual developments. On the technical side, I will be making use of recent developments that allow us to construct finite operators that represent diffeomorphism invariant quantitites[CR93, LS93, VH93]. These include spatially diffeomorphism invariant operators that measure geometrical quantities such as the areas of surfaces picked out by the configurations of certain matter fields. By putting this together with a simple physical model of a field of synchronized clocks, we will see that we are able to implement in full quantum general relativity the program of constructing physical observables based on quantum reference systems .

It may be objected that real clocks and rulers are much more complicated things than those that are modeled here; in reality they consist of multitudes of atoms held together by electromagnetic interactions. However, my goal here is precisely to show that useful results can be achieved by taking a shortcut in which the clocks and rulers are idealized and their dynamics simplified to the point that their inclusion into the nonperturbative dynamics is almost trivial. At the same time, no simplifications or approximations of any kind are made concerning the dynamics of the gravitational degrees of freedom. What we will then study is a system in which toy clocks and rulers interact with the fully nonlinear gravitational field within a nonperturbative framework.

However, while I will be using toy clocks and rulers, the main results will apply equally to any system in which certain degrees of freedom can be used to locate events relative to a physical reference system. The chief of these results is that the construction of physical observables need not be the very difficult problem that it has sometimes been made out to be. In particular, it is not necessary to exactly integrate Einstein's equations to find the observables of the coupled gravity-reference matter system. Instead, I propose here an alternative approach which consists of the following steps: i) Construct a large enough set of spatially diffeomorphism invariant operators to represent any observations made with the help of a spatial physical reference system; ii) Find the reality conditions among these spatially diffeomorphism invariant operators, and find the diffeomorphism invariant inner product that implements them; iii) Construct the projection of the Hamiltonian constraint as a finite and diffeomorphism invariant operator on this space. iv) Add degrees of freedom to correspond to a clock, or to a field of synchronized clocks. The Hamiltonian constraint for both states and operators now become ordinary differential equations for one parameter families of states and operators in the diffeomorphism invariant Hilbert space parametrized by the physical time measured by this clock. v) Define the physical inner product from the diffeomorphism invariant inner product by identifying the physical inner products of states with the diffeomorphism invariant inner products of their data at an initial physical time.

The steps necessary to implement this program are challenging. But the recent progress, concerning both spatially diffeomorphism invariant operators [CR93, LS93, VH93] and the form of the Hamiltonian constraint operator[RS88, BP92, RG90, Bl] suggests to me that each step can be accomplished. If so, then we will have a systematic way to construct the physical theory, together with a physical interpretation, from the spatially diffeomorphism invariant states and operators.

In the next section I review recent results about spatially diffeomorphism invariant observables which allow us to implement the idea of a spatial frame of reference. In section 3 I show how a simple model of a field of clocks can be used to promote these to physical observables[3].

Let me then turn from technical developments to conceptual developments. As is well known, there are two kinds of spatial boundary conditions that may be imposed in a canonical approach to quantum gravity: the open and the closed, or cosmological. The use of open boundary conditions, such as asymptotic flatness, avoids some of the main conceptual issues of quantum gravity because there is a real Hamiltonian which is tied to the clock of an observer outside the system, at spatial infinity. However, the asymptotically flat case also introduces additional difficulties into the canonical quantization program, so that it has not, so far, really helped with the construction of the full theory[4]. Furthermore, it can be argued that the asymptotically flat case represents an idealization that, by breaking the diffeomorphism invariance and postulating a classical observer at infinity, avoids exactly those problems which are the keys to quantum gravity. Thus, for both practical and philosophical reasons, it is of interest to see if it is possible to give a physical interpretation to quantum gravity in the cosmological context.

There has been a great deal of discussion recently about the interpretational problems of quantum cosmology[5]. However, most of it is not directly applicable to the project of this paper, either because it is tied to the path integral approach to quantum gravity, because it is applicable only in the semiclassical limit or because it breaks, either explicitly or implicitly, with the postulate that only operators that commute with the

---

[3]In an earlier draft of this paper there was an error in the treatment of the gauge fixed quantization in this section. The present treatment corrects the error and is, in addition, considerably simplified with respect to the original version.

[4]However, there are some interesting developments along this line, see [JB92].

[5]See, for example, [GMH, HPMZ, AS91].

Hamiltonian constraint can correspond to observable quantities. What is required to turn canonical quantum cosmology into a physical theory is an interpretation in terms of expectation values, states and operators that describes what observers inside the universe can measure.

At the time I began thinking about this problem it seemed to me likely that what was required was some modification of the relative state idea of Everett[HE57], perhaps along the lines sketched in [LS84], which avoided commitment to the metaphysical idea of "many worlds" and incorporated some of the recent advances in understanding of the phenomena of "decoherence"[HPMZ]. The reason for this was that it seemed that the original interpretation of quantum mechanics, as developed by Bohr, Heisenberg, von Neumann and others could not be applied in the cosmological context. However, I have come to believe that this is too hasty a conclusion, and that, at least in the context in which physical observables are constructed by explicit reference to a physical reference frame and physical clocks, it is possible to apply directly to quantum cosmology the point of view of the original founders of quantum mechanics. The key idea is, as Bohr always stressed [B], to keep throughout the discussion an entirely operational point of view, so that the quantum state is never taken as a description of physical reality but is, instead, part of a description of a process of preparation and measurement involving a whole, entangled system including both the quantum system and the measuring devices.

I want to make it clear from the start that I do not intend here to take up the argument about different interpretations of quantum mechanics. In either ordinary quantum mechanics or in quantum cosmology there may be good reasons to prefer another interpretation over the original interpretation of Bohr. What I want to argue here is only that the claim that it is necessary to give up Bohr and von Neumann's interpretation in order to do quantum cosmology is wrong. As in ordinary quantum mechanics, once a strictly operational interpretation such as that of Bohr and von Neumann has been established, one can replace it with any other interpretation that makes more substantive claims about physical reality, whether it be a relative state interpretation, a statistical interpretation, or anything else. For this reason, I will give, in section 4, a sketch of an interpretation of quantum cosmology following the original language of Bohr and von Neumann. The reader who wants to augment this with the more substantive language of Everett, or of decoherence, will find that they can do so, in quantum cosmology no more and no less than in ordinary quantum mechanics[6].

Of course, one test that any proposed interpretation of quantum cosmology must

---

[6]The question of modeling the measurement process in parametrized systems is discussed in a paper in this volume by Anderson[AAn93]. Although Anderson warns against a too naive application of the projection postulate that does not take into account the fact that measurements take a finite amount of time, I do not think there is any inconsistency between his results and those of the present paper.

satisfy is that it give rise to conventional quantum mechanics and quantum field theory in the appropriate limits. In section 5 I show how ordinary quantum field theory can be recovered by taking limits in which the gravitational degrees of freedom are treated semiclassically.

Having thus set out both the technical foundations and the conceptual bases of a physical interpretation of quantum cosmology, we will then be in a position to see what a fully nonperturbative approach may be able to contribute to the problems raised by the existence of singularities and the evaporation of black holes[SWH75]. While I will certainly not be able to resolve these problems here, it is possible to make a few preliminary steps that may clarify how these problems may be treated within a nonperturbative quantization. In particular, it is useful to see whether there are ways in which the existence of singularities and loss of information or breakdown of quantum coherence could manifest themselves in a fully nonperturbative treatment that does not make reference to any classical metric.

What I will show in section 6 is that there are useful notions of singularity and loss of information that make sense at the nonperturbative level. As there is no background metric, these must be described completely in terms of certain properties of the physical operator algebra. The main result of this section is that this can be done within the context of the physical reference systems developed in earlier sections. Furthermore, one can see at this level a relationship between the two phenomena, so that it seems likely the existence of certain kinds of singularities in the physical operator algebra can lead to effects that are naturally described as "loss of information." These results indicate that the occurrence of singularities and of loss of information are not necessarily inconsistent with the principles of quantum mechanics and general relativity. Whether they actually occur is then a dynamical question; it is possible that some consistent quantum theories of gravity allow the existence of singularities and the resulting loss of information, while others do not.

In order to focus the discussion, the results of section 6 are organized by the statements of two conjectures, which I call the *quantum singularity conjecture* and the *quantum cosmic censorship conjecture*. They embody the conditions under which we would want to say that the full quantum theory of gravity has singularities and the consequent loss of information.

The concluding section of this paper then focuses on two questions. First, are there approaches to an interpretation of quantum cosmology which, not being based on an operational notion of time, may avoid some of the limitations of the interpretation proposed here? Second, are there models and reductions of quantum cosmology in which the ideas presented here may be tested in detail?

It is an honor to contribute this paper to a volume in honor of Dieter Brill, who I have known for 16 years, first as a teacher of friends, then as a colleague as I became a frequent visitor to the Maryland relativity group. I am grateful for the warm hospitality I have felt from Dieter and the Maryland group on my many visits there.

## 2.   A quantum reference system

In this section I will describe one example of a quantum reference system in which relative spatial positions are fixed using the configurations of certain matter fields. While I mean for this example to serve as a general paradigm for how reference systems might be described in quantum cosmology, I will use a coupling to matter and a set of observables that we have recently learned can be implemented in nonperturbative quantum gravity. Although I do not give the details here, every operator described in this section may be constructed by means of a regularization procedure, and in each case the result is a finite and diffeomorphism invariant operator[CR93, LS93, VH93].

In this section I will speak informally about preparations and measurements, however the precise statements of the measurement theory are postponed to section 4; as this depends on the operational notion of time introduced in the next section.

In this and the following sections, I am describing a canonical quantization of general relativity coupled to a set of matter fields. The spatial manifold, $\Sigma$, has fixed topological and differential structure, and will be assumed to be compact. For definiteness I will make use of the loop representation formulation of canonical quantum gravity[AA86, RS88]. Introductions to that formalism are found in [AA91, CR91, LS91]; summaries of results through the fall of 1992 are found in [AA92, LS92b].

## 2.1.   Some operators invariant under spatial diffeomorphisms

In the last year we have found that while it seems impossible to construct operators that measure the gravitational field at a point, there exist well defined operators that measure nonlocal observables such as areas, volumes and parallel transports. In order to make these invariant under spatial diffeomorphisms we can introduce a set of matter fields which will label sets of open surfaces in the three manifold $\Sigma$. I will not here give details of how this is done, but ask the reader to assume the existence of matter fields whose configurations can be used to label a set of $N$ open surfaces, which I will call $S_I$, where $I = 1, ..., N$. The boundaries to these surfaces will also play a role, these are denoted $\partial S_I$.

There are actually three ways in which such surfaces can be labeled by matter fields. One can use scalar fields, as described by Rovelli in [CR93] and Husain in [VH93], one can use antisymmetric tensor gauge fields, as is described in [LS93] or one can use abelian gauge fields in the electric field representation, as discussed by Ashtekar and Isham [AI92]. In each case we can construct finite diffeomorphism invariant operators which measure either the areas of these surfaces or the parallel transport of the spacetime connection around their boundaries. I refer the reader to the original papers for the technical details.

The key technical point is that, in each of these cases, the matter field can be quantized in a *surface representation*, in which the states are functionals of a set of $N$ open surfaces in the three dimensional spatial manifold $\Sigma$[RG86, LS93]. For each of the $N$ matter fields, a general bra will then be labeled by an unordered open surface, which may be disconnected, and will be denoted $< S_I|$ We assume that the states in the surface representation satisfy an identity which is analogous to the Abelian loop identities [GT81, AR91, AI92]. This is that whenever two, possibly disconnected, open surfaces $S^1$ and $S^2$ satisfy, for every two form $F_{ab}$, $\int_{S^1} F = \int_{S^2} F$, we require that $< S_I^1| =< S_I^2|$.

A general bra for all $N$ matter fields is then labeled by $N$ such surfaces, and will be denoted $< S| =< S_1, ..., S_N|$ so that the general state may be written

$$\Psi[S] =< S|\Psi > \qquad (1)$$

It is easy to couple this system to general relativity using the loop representation[RS88]. The gravitational degrees of freedom are incorporated by labeling the states by both surfaces and loops, so that a general state is of the form

$$\Psi[\gamma, S] =< \gamma, S|\Psi > \qquad (2)$$

where $\gamma$ is a loop in the spatial manifold $\Sigma$. We assume that all of the usual identities of the loop representation [RS88, AA91, CR91, LS91] are satisfied by these states.

The next step is to impose the constraints for spatial diffeomorphism invariance. By following the same steps as in the pure gravity case, it is easy to see that the exact solution to the diffeomorphism constraints for the coupled matter-gravity system is that the states must be functions of the diffeomorphism equivalence classes of loops and $N$ labeled (and possibly disconnected) surfaces. As in the case of pure gravity the set of these equivalence classes is countable. If we denote by $\{\gamma, S\}$ these diffeomorphism equivalence classes, every diffeomorphism invariant state may be written,

$$\Psi[\{\gamma, S\}] =< \{\gamma, S\}|\Psi > \qquad (3)$$

Later we will include other matter fields, which will be denoted generically by $\phi$. In that case states will be labeled by diffeomorphism equivalence classes, which I will denote by $\{S, \gamma, \phi\}$. The space of such states will be denoted $\mathcal{H}_{diffeo}$.

A word of caution must be said about the notation: the expression $< \{\gamma, S\}|\Psi >$ is not to be taken as an expression of the inner product. Instead it is just an expression for the action of the bras states on the kets. The space of kets is taken to be the space of functions $\Psi[\{S, \gamma\}]$ of diffeomorphism invariant classes of loops and surfaces. The space of bras are defined to be linear maps from this space to the complex numbers, and given some bra $< \chi|$, the map is defined by the pairing $< \chi|\Psi >$. The space of bras has thus a natural basis which is given by the $< \{\gamma, S\}|$, whose action is defined by (3). For the moment, the inner product remains unspecified, so there is no isomorphism between the space of bras and kets. At the end of this section I will give a partial specification of the inner product.

On this space of states it is possible to construct two sets of diffeomorphism invariant observables to measure the gravitational field. The first of these are the areas of the $I$ surfaces, which I will denote $\hat{A}^I$. Operators which measure these areas can be constructed in the loop representation. The details are given in [CR93, LS93] where it is shown that after an appropriate regularization procedure the resulting quantum operators are diffeomorphism invariant and finite.

The bras $< \{\gamma, S\}|$ are in fact eigenstates of the area operators, as long as the loops do not have intersections exactly at the surfaces. In references [CR93, LS93] it is shown that in this case,

$$< \{\gamma, S\}|\hat{A}^I = \frac{l^2_{Planck}}{2} \mathcal{I}^+[S^I, \gamma] < \{\gamma, S\}| \tag{4}$$

where $\mathcal{I}^+[S^I, \gamma]$ is the unoriented, positive, intersection number between the surface and the loop that simply counts the intersections between them[7].

A second diffeomorphism invariant observable that can be constructed is the Wilson loop around the boundary of the $I$'th surface, which I will denote $\hat{T}^I$. As shown in [LS93] it has the action

$$< \{\gamma, S\}|\hat{T}^I = < \{\gamma \cup \partial S^I, S\}| \tag{5}$$

In addition, diffeomorphism invariant analogues of the higher loop operators have recently been constructed by Husain [VH93].

It is easy to see that the algebra of these operators has the form,

$$[\hat{T}^J, \hat{A}^I] = \frac{l^2_{Planck}}{2} \mathcal{I}^+[S^I, \partial S^J] \left( \hat{T}^J + \text{ intersection terms} \right) \tag{6}$$

where intersection terms stands for additional terms that arise if it happens that the loop $\gamma$ in the quantum state acted on intersects the boundary $\partial S^J$ exactly at the surface $S^I$ or if the boundary itself self-intersects at that surface. We will not need the detailed form of these terms for the considerations of this paper.

---

[7]In the case that there is an intersection at the surface, the eigenbra's and eigenvalues can still be found, following the method described in [LS91].

## 2.2.   Construction of the quantum reference system

With these results in hand, we may now construct a quantum reference system. The problem that we must face to construct a measurement theory for quantum gravity is how to give a diffeomorphism invariant description of the reference frame and measuring instruments because, as the geometry of spacetime is the dynamical variable we wish to measure, there is no background metric available to use in their description. The key idea is then that a reference frame must be specified by a particular topological arrangement of the matter fields that go into its construction. In the simple model we are considering here it is very easy to do this. As our reference frame is to consist of surfaces, what we need to give to specify the reference frame is a particular topological arrangement of these surfaces.

One way to do this is the following[LS93]. Choose a simplicial decomposition of the three manifold $\Sigma$ which has $N$ faces, which we may label $\mathcal{F}_I$. For reasons that will be clear in a moment, it is simplest to restrict this choice to simplicial decompositions in which the number of edges is also equal to $N$. Let me call such a choice $\mathcal{T}$.

Now, for each $\mathcal{T}$ with $N$ faces there is a subspace of the state space $\mathcal{H}_{diffeo}$ which is spanned by basis elements $< \{\gamma, \mathcal{S}\}|$ in which the surfaces $\mathcal{S}_I$ can be put into a one to one correspondence with the faces $\mathcal{F}_I$ of $\mathcal{T}$ so that they have the same topology of the faces of the simplex. We may call this subspace $\mathcal{H}_{\mathcal{T},diffeo}$.

As I will describe in section 4, the preparation of the system is described by putting the system into such a subspace of the Hilbert space associated with an arrangement of the surfaces. Once we know the state is in the subspace $\mathcal{H}_{\mathcal{T},diffeo}$ any measurements of the quantities $\hat{A}^I$ or $\hat{T}^I$ can be interpreted in terms of areas of the faces $\mathcal{F}^I$ or parallel transports around their edges.

## 2.3.   How do we describe the results of the measurements?

Given a choice of the simplicial manifold $\mathcal{T}$ and the corresponding subspace $\mathcal{H}_{\mathcal{T},diffeo}$ we may now make measurements of the gravitational field. As I will establish in section 4, there will be circumstances in which it is meaningful to say that we have, at some particular time, made a measurement of some commuting subset of the operators $\hat{A}^I$ and $\hat{T}^I$ which we described above. We may first note that these operators are block diagonal in $\mathcal{H}_{diffeo}$ in that their action preserves the subspaces $\mathcal{H}_{\mathcal{T},diffeo}$. From the commutation relations (6) we may deduce that if we restrict attention to one of these subspaces and these observables there are two maximal sets of commuting operators; we may measure either the $N$ $\hat{A}^I$ or the $N$ $\hat{T}^I$. This corresponds directly to the fact that the canonical pair of fields in the Ashtekar formalism are the spatial frame field and the self-dual connection.

How are we to describe the results of these measurements? Each gives us $N$ numbers, which comprise partial information that we can obtain about the geometry of spacetime as a result of a measurement based on the quantum reference frame built on $\mathcal{T}$. Now, as in ordinary quantum mechanics, we would like to construct a classical description of the result of such a partial measurement of the system. I will now show that in each case it is possible to do this. What we must do, in each case, is associate to the results of the measurements, a set of classical gravitational fields that are described in an appropriate way by $N$ parameters.

It is simplest to start with the measurements of the areas, which give us a partial measurement of the spatial geometry.

## Classical description of the output of the measurements of the areas

The output of a measurement of the $N$ areas will be $N$ rational numbers, $a_I$, (times the Planck area), each from the discrete series of possible eigenvalues in the spectra of the $\hat{A}^I$. Let us associate to each such set of areas a piecewise flat three geometry $\mathcal{Q}(a)$ that can be constructed as follows. $\mathcal{Q}(a)$ is the Regge manifold constructed by putting together flat tetrahedra according to the topology given by $\mathcal{T}$ such that the areas of the $N$ faces $\mathcal{F}_I$ are given by $a_I l^2_{Planck}$. Since such a Regge manifold is defined by its edge lengths and since we have fixed $\mathcal{T}$ so that the number of its edges is equal to the number of its faces, the $N$ areas $a_I l^2_{Planck}$ will generically determine the $N$ edge lengths.

Note that we are beginning with the assumption that all of the areas are positive real numbers, so the triangle inequalities must always be satisfied by the edge lengths. At the same time, the tetrahedral identities may not be satisfied, in that there is no inequality which restricts the areas of the faces of an individual tetrahedron in $\mathcal{T}$. For example, there exist configurations $\{\gamma, \mathcal{S}\}$ in which the loop only intersects one of the faces, giving that one face a finite area while the remaining faces have vanishing area. Thus, we must include the possibility that $\mathcal{Q}(a)$ contains tetrahedra with flat metrics with indefinite signatures. The emergence of a geometry that, at least when measured on large scales, may be approximated by a positive definite metric must be a property of the classical limit of the theory.

There may also be special cases in which more than one set of edge lengths are consistent with the areas. In which case we may say that the measurement of the quantum geometry leaves us with a finite set of possible classical geometries. There is nothing particularly troubling about this, especially as this will not be the generic case.

Thus, in general, the outcome of each measurement of the $N$ area operators may be describe by a particular piecewise flat Regge manifold, which represents the partial measurement that has been made of the spatial geometry.

## Classical description of the output of the measurements of the self-dual parallel transports

What if we measure instead the $N$ Wilson loops, $\hat{T}^I$? The output of such a measurement will be $N$ complex numbers, $t^I$. Can we associate these with a classical construction? I want to show here that the answer is yes.

Any such classical construction must not involve a spatial metric, as we have made the spatial geometry uncertain by measuring the quantities conjugate to it. So it must be a construction which is determined by $N$ pieces of information about the self-dual part of the spacetime curvature.

Such a construction can be given, as follows. We may construct a dual graph to $\mathcal{T}$ in the natural way by associating to each of its tetrahedra a vertex and to each of its faces, $\mathcal{F}_I$ an edge, called $\alpha_I$, such that the 4 $\alpha_I$ associated with the faces of a given tetrahedra have one of their end points at the vertex that corresponds to it. As each face is part of two tetrahedra, we know where to put the two end points of each edge, so the construction is completely determined. We may call this dual graph $\Gamma_{\mathcal{T}}$.

Now, to each such graph we may associate a distributional self-dual curvature which is written as follows,

$$F^i_{ab}(x) = \sum_I \int d\alpha^c_I(s) \, \epsilon_{abc} \, \delta^3(x, \alpha_I(s)) \, b^i_I \tag{7}$$

which is determined by giving $N$ $SL(2, C)$ Lie algebra elements $b^i_I$. If we use the non-abelian Stokes theorem [YaA80], we may show that the moduli of the $N$ complex $b^i_I$ are determined by the $N$ complex numbers $t_I$ that were the output of the measurement by

$$\frac{1}{2} Tr e^{\imath b^i_I \tau^i} = cos|b_I| = t_I \tag{8}$$

where $\tau^i$ are, of course, the three Pauli matrices. The remaining information about the orientation of the $b^i_I$ is gauge dependent and is thus not fixed by the measurement.

The reader may wonder whether a connection field can be associated with a distributional curvature of the form of (7). The answer is yes, what is required is a Chern-Simon connection with source given by (7). For any such source there are solutions to the Chern-Simon equations, which, however, require additional structure to be fully specified. One particularly simple way to do it, which does not depend on the imposition of a background metric, is the following[LS91, MS93]. Let us give an arbitrary specification of the faces of the dual graph $\mathcal{G}_{\mathcal{T}}$, which I will call $\mathcal{K}_I$. We may note that there is one face of the dual graph for each edge of $\mathcal{T}$, so that their number is also equal to $N$. Then we may specify a distributional connection of the form,

$$A^i_a(x) = \sum_I \int d^2 \mathcal{K}^{bc}_I(\sigma)\epsilon_{abc}\delta^3(x, \mathcal{K}_I(\sigma))a^i_I \tag{9}$$

where the $a^i_I$ are, again, $N$ Lie algebra elements. It is then not difficult to show that the usual relationship between the connection and the curvature holds, in spite of their distributional form and that the $b^i_I$'s may be expressed in terms of the $a^i_I$'s. For details of this the reader is referred to [MS93].

Thus, we have shown that to each measurement of the $N$ Wilson loops of the self-dual connection, we can also associate a classical geometry, whose construction is determined by $N$ pieces of information (in this case complex numbers.) The result may be thought of as a partial determination of the geometry of a spacetime Regge manifold. If we add a dimension, and allow time to be discrete than the construction we have just given can be thought of as a spatial slice through a four dimensional Regge manifold. In that case, each edge in our construction becomes a face in the four dimensional construction and, as in the Regge case, the curvature is seen to be distributional with support on the faces. However, in this case it is only the self-dual part of the curvature that is given, because its measurement makes impossible the measurement of any conjugate information. In fact, a complete construction of a Regge-like four geometry can be given along these lines, for details, see [MS93].

## 2.4.   The spatially diffeomorphism invariant inner product

In all of the constructions so far given, the inner product has played no role because we have been expressing everything in terms of the eigenbras of the operators in a particular basis, which are the $< \{\gamma, \mathcal{S}\}|$. However, as in ordinary quantum mechanics, a complete description of the measurement theory will require an inner product. The complete specification of the inner product must be done at the level of the physical states, which requires that we take into account that the states are solutions to the Hamiltonian constraint. This problem, which I would like to claim is essentially equivalent to the problem of time, is the subject of the next section. But it is interesting and, as we shall see, useful to see how much can be determined about the inner product at the level of spatially diffeomorphism invariant states.

Now, in order to determine the inner product at the diffeomorphism invariant level we should be able to write the reality conditions that our classical diffeomorphism invariant observables satisfy. For the $\hat{T}^I$ this is an unsolved problem, these operators are complex, but must satisfy reality conditions which are determined by the reality conditions on the Ashtekar connections. To solve this problem it will be necessary to adjoin additional diffeomorphism invariant operators to the $\hat{T}^I$ and $\hat{\mathcal{A}}^I$ in order to enable us to write down a complete star algebra of diffeomorphism invariant observables.

However, the reality conditions the area operators satisfy are very simple: they must be real. As a result we may ask what restrictions we may put on the inner product such that

$$\hat{A}^{I\dagger} = \hat{A}^I?$$ (10)

To express this we must introduce characteristic kets, which will be denoted $|\{\gamma, S\} >$. They are defined so that

$$\Psi_{\{\gamma, S\}}[\{\gamma, S\}'] \equiv < \{\gamma, S\}'|\{\gamma, S\} >= \delta_{\{\gamma, S\}\{\gamma, S\}'}$$ (11)

Here the meaning of the delta is follows. Fix a particular, but arbitrary set of surfaces and loops $(\gamma, S)$ within the diffeomorphism equivalence class $\{\gamma, S\}$. Then the $\delta_{\{\gamma, S\}\{\gamma, S\}'}$ is equal to one if and only if there is an element $S', \gamma'$ of the equivalence class $\{\gamma, S\}'$ such that a) for every two form $F_{ab}$ on $\Sigma$, $\int_{S'} F = \int_S F$ and b) for every connection $A_a^i$ on $\Sigma$ and every component $\gamma_I$ of $\gamma$ (and similarly for $\gamma'$), $T[\gamma_I'] = T[\gamma_I]$. If the condition is not satisfied then $\delta_{\{\gamma, S\}\{\gamma, S\}'}$ is equal to zero.

Let me denote the diffeomorphism invariant inner product by specifying the adjoint map from kets to bras,

$$< \{\gamma, S\}^\dagger| \equiv |\{\gamma, S\} >^\dagger .$$ (12)

It is then straightforward to show that the condition (10) that the $N$ operators must be hermitian restricts the inner product so that, in the case that the loops $\gamma$ have no intersections with each other,

$$< \{\gamma, S\}^\dagger| =< \{\gamma, S\}|.$$ (13)

## 3.    Physical observables and the problem of time

In this section I would like to describe an operational approach to the problem of time in quantum cosmology, which is based on the point of view that time is no more and no less than that which is measured by physical clocks. The general idea we will pursue is to couple general relativity to a matter field whose behavior makes it suitable for use as a clock. One then turns the Hamiltonian constraint equations into evolution equations that proscribe how spatially diffeomorphism invariant quantities evolve according to the time measured by this physical clock.

We will then try to build up the physical theory with the clock from the spatially diffeomorphism invariant theory for the gravitational and matter degrees of freedom, in which the clock has been left out. To do this we will need to assume several things about the diffeomorphism invariant theory, which are motivated by the results of the last section.

a) In the loop representation we have the complete set of solutions to the spatial diffeomorphism constraints coupled to more or less arbitrary matter fields. Given some choice of matter fields, which I will denote generically by $\phi$, I will write the general spatially diffeomorphism invariant state by $\Psi[\{\gamma, \phi\}]_{diffeo}$, where the brackets $\{...\}$ mean spatial diffeomorphism equivalence class. The reference frame fields discussed in the previous section are, for the purpose of simplifying the formulas of this section, included in the $\phi$. However, the fields that represent the clock are not to be included in these diffeomophism invariant states. As in the previous section, the Hilbert space of diffeomorphism invariant states will be denoted $\mathcal{H}_{diffeo}$.

b) I will assume that an algebra of finite and diffeomorphism invariant operators, called $\mathcal{A}_{diffeo}$, is known on $\mathcal{H}_{diffeo}$. The idea that spatially diffeomorphism invariant operators are finite has become common in the loop representation, where there is some evidence for the conjecture that diffeomorphism invariance requires finiteness [ARS92, CR93, LS93, LS91]. The area and parallel transport operators discussed in the previous section are examples of such operators. I will use the notation $\hat{\mathcal{O}}^I_{diffeo}$ to refer to elements of $\mathcal{A}_{diffeo}$, where $I$ is an arbitrary index that labels the operators.

c) I will assume also that the classical diffeomorphism invariant observables $\mathcal{O}^I_{diffeo}$ that correspond to the quantum $\hat{\mathcal{O}}^I_{diffeo}$ are known. This lets us impose the reality conditions on the algebra, as described in [AA91, AI92, CR91].

d) Finally, I assume that the inner product $< \mid >_{diffeo}$ on $\mathcal{H}_{diffeo}$ has been determined from the reality conditions satisfied by an appropriate subset of the $\hat{\mathcal{O}}^I_{diffeo}$.

## 3.1.  A classical model of a clock

I will now introduce a new scalar field, whose value will be defined to be time. It will be called the clock field and written $T(x)$, and it will be assumed to have the unconventional dimensions of time. Conjugate to it we must have a density, which has dimensions of energy density, which will be denoted $\tilde{\mathcal{E}}(x)$ such that[8],

$$\{\tilde{\mathcal{E}}(x), T(y)\} = -\delta^3(x,y). \tag{14}$$

To couple these fields to gravity we must add appropriate terms to the diffeomorphism and Hamiltonian constraints. I will assume that $T(x)$ is a free massless scalar field, so that

$$\mathcal{C}(x) = \frac{1}{2\mu}\tilde{\mathcal{E}}^2 + \frac{\mu}{2}\tilde{q}^{ab}\partial_a T \partial_b T + \mathcal{C}_{grav}(x) + \mathcal{C}_{matter}(x), \tag{15}$$

where $\mathcal{C}_{grav}(x)$ and $\mathcal{C}_{matter}(x)$ are, respectively, the contributions to the Hamiltonian constraint for the gravitational field and the other matter fields and $\mu$ is a constant

---

[8]We adopt the convention that the delta function is a density of weight one on its first entry. Densities will usually, but not always, be denoted by tildes.

with dimensions of *energy density*. Note that the form of (16) is dictated by the fact that in the Ashtekar formalism the Hamiltonian constraint is a density of weight two.

Similarly, the diffeomorphism constraint becomes

$$\mathcal{D}(v) = \int_\Sigma v^a(\partial_a T)\tilde{\mathcal{E}} + \mathcal{D}_{grav}(v) + \mathcal{D}_{matter}(v) \tag{16}$$

The reader may check that these constraints close in the proper way.

We may note that because there is no potential term for the clock field we have a constants of motion,

$$\mathcal{E} \equiv \int_\Sigma d^3x \tilde{\mathcal{E}}(x). \tag{17}$$

This generates the symmetry $T(x) \to T(x) + $ constant. It is easy to verify explicitly that

$$\{\mathcal{C}(N), \mathcal{E}\} = \{\mathcal{D}(v), \mathcal{E}\} = 0, \tag{18}$$

(where $\mathcal{C}(N) \equiv \int N\mathcal{C}$) .

We would now like to chose a gauge in which the time slicing of the spacetime is made according to surfaces of constant $T$. We thus choose as a gauge condition.

$$\partial_a T(x) = 0. \tag{19}$$

This may be solved by setting $T(x) = \tau$, where $\tau$ will be taken to be the time parameter. We may note that the condition that the evolution follows surfaces of constant $T(x)$ fixes the lapse because,

$$(\dot{\partial_a T})(x) = \{\mathcal{C}(N), \partial_a T(x)\} = \partial_a(N(x)\tilde{\mathcal{E}}(x)) = 0 \tag{20}$$

Thus, our gauge condition can only be maintained if

$$N(x) = \frac{c}{\tilde{\mathcal{E}}(x)}. \tag{21}$$

where $c$ is an arbitrary constant.

One way to say what this means is that all but one of the infinite number of Hamiltonian constraints have been broken by imposing the gauge condition (19). The one remaining component of the Hamiltonian constraint is the one that satisfies (21). However, since we must have eliminated the nonconstant piece of $T(x)$ by the gauge fixing (19) we must solve the constraints which have been so broken to eliminate the fields which are conjugate to them. Thus, all but one of the degrees of freedom in $\tilde{\mathcal{E}}(x)$ must be eliminated by solving the Hamiltonian constraint. The one which is kept independent can be taken to be the global constant of motion $\mathcal{E}$ defined by (17).

Up to this one overall degree of freedom, the local variations in $\mathcal{E}(x)$ must be fixed by solving the Hamiltonian constraint locally, which gives us[9],

$$\tilde{\mathcal{E}}(x) = \sqrt{-2\mu[\mathcal{C}_{grav}(x) + \mathcal{C}_{matter}(x)]} \tag{22}$$

Note that, to keep the global quantity $\mathcal{E}$ independent, we should solve this equation at all but one, arbitrary, point of $\Sigma$.

The idea is then to reduce the phase space and constraints so that the local variations in $T(x)$ and $\tilde{\mathcal{E}}(x)$ are eliminated, leaving only the global variables $\tau$ and $\mathcal{E}$. We may note that

$$\{T(x), \mathcal{E}\} = 1. \tag{23}$$

so that the reduced Poisson bracket structure must give

$$\{\tau, \mathcal{E}\} = 1 \tag{24}$$

while the Poisson brackets of the gravitational and matter fields remain as before. After the reduction, we then have one remaining Hamiltonian constraint, which is

$$\begin{aligned}
\mathcal{C}_{g.f.} = \mathcal{C}(c/\tilde{E}) &= \frac{c}{2\mu}\mathcal{E} + c\int \frac{d^3x}{\mathcal{E}(x)}\left(\mathcal{C}_{grav}(x) + \mathcal{C}_{matter}(x)\right) \\
&= \frac{c}{2\mu}\mathcal{E} + \frac{c}{\sqrt{2\mu}}\int d^3x\sqrt{-[\mathcal{C}_{grav}(x) + \mathcal{C}_{matter}(x)]} 
\end{aligned} \tag{25}$$

By (24) $\mathcal{C}_{g.f.}$ generates reparametrization of the global time variable $\tau$. The effect of the reduction on the diffeomorphism constraint is simply to eliminate the time field so that

$$\mathcal{D}(v)_{g.f.} = \mathcal{D}_{grav}(v) + \mathcal{D}_{matter}(v) \tag{26}$$

We may note that the reduced Hamiltonian constraint is invariant under diffeomorphisms and hence commutes with the reduced diffeomorphism constraint.

To summarize, the gauge fixed theory is based on a phase space which consists of the original gravitational and matter phase space, to which the two conjugate degrees of freedom $\tau$ and $\mathcal{E}$ have been added. The diffeomorphism constraint remains the original one while there remains one Hamiltonian constraint given by (25).

It is now straightforward to construct physical operators. They must be of the form

$$\mathcal{O} = \mathcal{O}[A, E, \phi, \tau, \mathcal{E}] \tag{27}$$

---

[9]We make here a choice in taking the positive square root, which is to restrict attention to a subspace of the original phase space. This choice, which we carry through as well in the quantum theory, is the equivalent of a positive frequency condition.

where $A, E$ are the canonical variables that describe the gravitational field and $\phi$ stands for any other matter fields. The diffeomorphism constraints imply that

$$\mathcal{O}[A, E, \phi, \tau, \mathcal{E}] = \mathcal{O}[\{A, E, \phi\}, \mathcal{E}, \tau] \qquad (28)$$

where the brackets $\{...\}$ indicate that the observable can depend only on combinations of the gravitational and matter fields that are spatially diffeomorphism invariant. The requirement that the observable commute with the reduced Hamiltonian constraint gives us

$$\frac{d\mathcal{O}[A, E, \phi, \tau, \mathcal{E}]}{d\tau} = \sqrt{2\mu} \left\{ \int d^3x \sqrt{-[\mathcal{C}_{grav}(x) + \mathcal{C}_{matter}(x)]}, \mathcal{O}[A, E, \phi, \tau, \mathcal{E}] \right\}. \qquad (29)$$

Thus, we have achieved the following result concerning physical observables:

*For every spatially diffeomorphism invariant observable $\mathcal{O}[\{A, E, \phi\}]_{diffeo}$ which is a function of the gravitational and matter fields (but not the clock degrees of freedom) there is a physical observable whose expression in the gauge given by (19) and (21) is the two parameter family of diffeomorphism invariant observables[10], of the form $\mathcal{O}[\{A, E, \phi\}, \tau, \mathcal{E}]_{g.f.}$ which solves (29) subject to the initial condition that*

$$\mathcal{O}[\{A, E, \phi\}, \tau = 0, \mathcal{E}]_{g.f.} = \mathcal{O}[\{A, E, \phi\}]_{diffeo} \qquad (30)$$

By construction, we may conclude that the observable $\mathcal{O}[\{A, E, \phi\}, \tau]_{g.f.}$ is the value of the diffeomorphism invariant function $\mathcal{O}[\{A, E, \phi\}]_{diffeo}$ evaluated on the surface $T(x) = \tau$ of the spacetime gotten by evolving the constrained initial data $\{A, E, \phi, T = 0\}$.

Now, $\mathcal{O}[\{A, E, \phi\}, \tau, \mathcal{E}]_{g.f.}$ is the value of a physical observable only in the gauge picked by (19) and (21)). However, once we know the value of any observable in a fixed gauge we may extend it to a fully gauge invariant observable. To do so we look for a gauge invariant function[11] $\mathcal{O}[A, E, \phi, T(x), \mathcal{E}(x)]_{Dirac}$ that commutes with the full diffeomorphism and Hamiltonian constraints, with arbitrary lapses $N$, that has the same physical interpretation as our gauge fixed observable $\mathcal{O}[\{A, E, \phi\}, \tau, \mathcal{E}]_{g.f.}$. This means that

$$\mathcal{O}[A, E, \phi, T(x), \tilde{\mathcal{E}}(x)]_{Dirac}\big|_{T(x)=\tau, \tilde{\mathcal{E}}(x)=\sqrt{-2\mu[\mathcal{C}_{grav}(x)+\mathcal{C}_{matter}(x)]}} = \mathcal{O}[A, E, \phi, \tau, \mathcal{E}]_{g.f.} \qquad (31)$$

Once we have one such physical observable, we may follow Rovelli [CRT91, CRM91] and construct a one parameter family of physical observables called "evolving constants of motion". These are fully gauge invariant functions on the phase space

---

[10]The subscript $g.f.$ will be used throughout this paper to refer to observables that commute with the full spatial diffeomorphism constraints but only the gauge fixed Hamiltonian constraint.

[11]The subscript *Dirac* will always refer to an observable that commutes with the full set of constraints without gauge fixing.

that, for each $\tau$ tell us the value of the spatially diffeomorphism invariant observable $\mathcal{O}[\{A, E, \phi\}]_{diffeo}$ on the surface $T(x) = \tau$ as a function of the data of the initial surface. That, is for each $\tau$ we seek a function $\mathcal{O}'[A, E, \phi, T(x), \mathcal{E}(x)](\tau)$ that commutes with the Hamiltonian constraint for all $N$ (with $\tau$ taken as a parameter and not a function on the phase space) and which satisfies

$$\mathcal{O}'[A, E, \phi, T(x), \tilde{\mathcal{E}}(x)](\tau)\big|_{T(x)=0, \tilde{\mathcal{E}}(x)=\sqrt{-2\mu[\mathcal{C}_{grav}(x)+\mathcal{C}_{matter}(x)]}} = \mathcal{O}[A, E, \phi, \tau, \mathcal{E}]_{g.f.}. \tag{32}$$

## 3.2.    Quantization of the theory with the time field

We would now like to extend the result of the previous section to the quantum theory. To do this we must introduce an appropriate representation for the clock fields and construct and impose the diffeomorphism and Hamiltonian constraint equations.

We will first construct the quantum theory corresponding to the reduced classical dynamics that follows from the gauge fixing (19) and (21). After this we will discuss the alternative possibility, which is to construct the physical theory through Dirac quantization in which no gauge fixing is done.

In the gauge fixed quantization the states will be taken to be functions $\Psi[\gamma, \phi, \tau]$ so that

$$\hat{\tau}\Psi[\gamma, \phi, \tau] = \tau\Psi[\gamma, \phi, \tau], \tag{33}$$

$$\hat{\mathcal{E}}\Psi[\gamma, \phi, \tau] = -\imath\hbar\frac{\partial}{\partial\tau}\Psi[\gamma, \phi, \tau] \tag{34}$$

and all the other defining relations are kept. The space of these states, prior to the imposition of the remaining constraints, will be called $\mathcal{H}_{reduced}$.

We now apply the reduced diffeomorphism constraints (26). The result is that the states be functions of diffeomorphism equivalence classes of their arguments, so that

$$\mathcal{D}(v)_{g.f.}\Psi[\gamma, \phi, \tau] = 0 \Rightarrow \Psi[\gamma, \phi, \tau] = \Psi[\{\gamma, \phi\}, \tau] \tag{35}$$

where, again, the brackets indicate diffeomorphism equivalence classes.

We may then apply the reduced Hamiltonian constraint (25). By (34) this implies that, formally

$$\frac{\imath\hbar\partial\Psi[\{\gamma, \phi\}, \tau]}{\partial\tau} = \int d^3x\sqrt{-2\mu[\hat{\mathcal{C}}_{grav}(x) + \hat{\mathcal{C}}_{matter}(x)]}\Psi[\{\gamma, \phi\}, \tau] \tag{36}$$

As this is the fundamental equation of the quantum theory, we must make some comments on its form. First, and most importantly, as the Hamiltonian constraint involves operator products, this equation must be defined through a suitable regularization procedure. Secondly, to make sense of this equation requires that we define an operator square root. Both steps must be done in such a way that the result is a finite and diffeomorphism invariant operator. That is, for this approach to quantization to work, it must be possible to regulate the gravitational and matter parts of the Hamiltonian constraint in such a way that the limit

$$\lim_{\epsilon \to 0} \int d^3x \sqrt{-2\mu[\hat{C}^\epsilon_{grav}(x) + \hat{C}^\epsilon_{matter}(x)]} = \hat{W} \qquad (37)$$

(where the $\epsilon$'s denote the regulated operators) exists and gives a well defined (and hence finite and diffeomorphism invariant) operator $\hat{W}$ on the space of spatially diffeomorphism invariant states of the gravitational and matter fields.

It may seem that to ask that it be possible to both define a good regularization procedure and define the operator square root is to be in danger of being ruled out by the "two miracle" rule: it is acceptable practice in theoretical physics to look forward to the occurrence of one miracle, but to ask for two is unreasonable. However, recent experience with constructing diffeomorphism invariant operators in the loop representation of quantum gravity suggests the opposite: these two problems may be, in fact, each others solution. Perhaps surprisingly, what has been found is that the only operators which have been so far constructed as finite, diffeomorphism invariant operators involve operator square roots.

The reason for this is straightforward. In quantum field theory operators are distributions. In the context of diffeomorphism invariant theories, distributions are densities. As a result, there is an intrinsic difficulty with defining operator products in a diffeomorphism invariant theory through a renormalization procedure. Any such procedure must give a way to define the product of two distributions, with the result being another distribution. But this means that the procedure must take a geometrical object which is formally a density of weight two, and return a density of weight one. What happens as a result is that there is a grave risk of the regularization procedure breaking the invariance under spatial diffeomorphism invariance, because the missing density weight ends up being represented by functions of the unphysical background used in the definition of the renormalization procedure. Many examples are known in which exactly this happens[LS91].

However, it turns out that in many cases the square root of the product of two distributions can be defined as another distribution without ambiguity due to this problem of matching density weights. It is, indeed, exactly this fact that makes it possible to define the area operators I described in the previous section, as well as other operators associated with volumes of regions and norms of one form fields[ARS92, LS91].

I do not know whether the same procedures that work in the other cases work to make the limit (37) exist. We may note, parenthetically, that if $\hat{W}$ can be defined as an operator on diffeomorphism invariant states in the context of a separable Hilbert structure, the problem of the finiteness of quantum gravity will have been solved. I will assume here that the problem of constructing a regularization procedure such that this is the case can be solved, and go on.

Assuming then the existence of $\hat{\mathcal{W}}$, we may call the space of solutions to (35) and (36) $\mathcal{H}_{g.f.}$, where the subscript denotes again that we are working with the gauge fixed quantization. We will shortly be discussing the inner product on this space. For the present, the reader may note that once $\hat{\mathcal{W}}$ exists, the problem of finding states that solve the reduced Hamiltonian constraint is essentially a problem of ordinary quantum mechanics. For example, there will be solutions of the form

$$\Psi[\{\gamma, \phi\}, \tau] = \Phi[\{\gamma, \phi\}]e^{-\imath \omega \tau} \qquad (38)$$

This will solve (36) if $\Phi[\{\gamma, \phi\}]$ is an eigenstate of $\hat{\mathcal{W}}$, so that

$$\hat{\mathcal{W}}\Phi = \hbar\omega\Phi \qquad (39)$$

More generally, given any diffeomorphism invariant state $\Psi[\{\gamma, \phi\}] \in \mathcal{H}_{diffeo}$, which is a function of the gravitational and matter fields alone, there is a physical state, $\Psi[\{\gamma, \phi\}, \tau] \in \mathcal{H}_{g.f}$ in our gauge fixed quantization which is the solution to (36) with the initial conditions

$$\Psi[\{\gamma, \phi\}, \tau]_{\tau=0} = \Psi[\{\gamma, \phi\}] \qquad (40)$$

Thus what we have established is that there is a map

$$\Lambda : \mathcal{H}_{diffeo} \to \mathcal{H}_{g.f.} \qquad (41)$$

in which every diffeomorphism invariant state of the gravitational and matter fields is taken into its evolution in terms of the clock fields. Furthermore, there is an inverse map,

$$\Theta : \mathcal{H}_{g.f.} \to \mathcal{H}_{diffeo} \qquad (42)$$

which is defined by evaluating the physical state at $\tau = 0$.

## 3.3.    The operators of the gauge fixed theory

The physical operators in the gauge fixed quantum theory may be found analogously to the classical observables of the gauge fixed theory. A physical operator is an operator on $\mathcal{H}_{g.f.}$, which may be written $\hat{\mathcal{O}}[\hat{A}, \hat{E}, \hat{\phi}, \hat{\tau}, \hat{\mathcal{E}}]$. The requirement that it commute with the reduced diffeomorphism constraints restricts it to be of the form $\hat{\mathcal{O}}[\{\hat{A}, \hat{E}, \hat{\phi}\}, \hat{\tau}, \hat{\mathcal{E}}]$. The requirement that it commute with the reduced Hamiltonian constraint becomes the evolution equation,

$$\imath\hbar\frac{d\hat{\mathcal{O}}[\hat{A}, \hat{E}, \hat{\phi}, \tau, \hat{\mathcal{E}}]}{d\tau} = \left[\hat{\mathcal{W}}, \hat{\mathcal{O}}[\hat{A}, \hat{E}, \hat{\phi}, \tau, \hat{\mathcal{E}}]\right]. \qquad (43)$$

Using this equation we may find a physical operator that corresponds to every spatially diffeomorphism invariant operator on $\mathcal{H}_{diffeo}$, which depends only on the non-clock fields. Given any such operator, $\hat{\mathcal{O}}[\{\hat{A}, \hat{E}, \hat{\phi}\}]_{diffeo}$ we may construct an operator on $\mathcal{H}_{g.f.}$ which we denote $\hat{\mathcal{O}}[\{\hat{A}, \hat{E}, \hat{\phi}\}, \tau]_{g.f.}$ which solves (43) subject to the initial condition

$$\hat{\mathcal{O}}[\{\hat{A}, \hat{E}, \hat{\phi}\}, \tau = 0]_{g.f.} = \hat{\mathcal{O}}[\{\hat{A}, \hat{E}, \hat{\phi}\}]_{diffeo} \tag{44}$$

We may, indeed, solve (43) to find that

$$\hat{\mathcal{O}}[\{\hat{A}, \hat{E}, \hat{\phi}\}, \tau]_{g.f.} = e^{-\imath \hat{W} \tau / \hbar} \hat{\mathcal{O}}[\{\hat{A}, \hat{E}, \hat{\phi}\}]_{diffeo} e^{+\imath \hat{W} \tau / \hbar}. \tag{45}$$

## 3.4. The physical interpretation and inner product of the gauge fixed theory

In the classical theory, the physical observables $\mathcal{O}[\{A, E, \phi\}, \tau]_{g.f.}$ were found to correspond to the values of diffeomorphism invariant functions of the non-clock fields on the surface defined by the gauge condition (19) and the value of the time parameter $\tau$. As they satisfy the analogous quantum equations we would like to interpret the corresponding quantum operators that solve (43) and (44) in the same way. That is, we will take $\hat{\mathcal{O}}[\{\hat{A}, \hat{E}, \hat{\phi}\}, \tau]_{g.f.}$ to be the operator that measures the diffeomorphism invariant quantity $\hat{\mathcal{O}}[\hat{E}, \hat{A}, \hat{\phi}]_{diff}$ after a physical time $\tau$.

Once this interpretation is fixed, there is a natural choice for the physical inner product. The idea, advocated by Ashtekar [AA91], is that the physical inner product is to be picked to satisfy the reality conditions for a large enough set of physical observables. The difficult part of the definition is the meaning of "large enough", but study of a number of examples shows that large enough means a complete set of commuting operators, and an equal number of operators conjugate to them [AA91, AA92, RT92]. Now, as in ordinary quantum field theory, it is very unlikely that any two operators defined at different physical times commute. Thus, we may postulate that the largest set of operators for which reality conditions can be imposed for the physical theory are two conjugate complete sets defined at a single moment of physical time.

Thus, we could define an inner product by imposing the reality conditions for a complete set of operators $\hat{\mathcal{O}}[\{\hat{A}, \hat{E}, \hat{\phi}\}, \tau]_{g.f.}$ at any physical time $\tau$. Now, the nicest situation would be if the resulting inner products were actually independent of $\tau$. However, there are reasons, some of which are discussed below, to believe that this may not be realized in full quantum gravity. If this is the case then the most natural assumption to make is that the physical inner product must be determined by a complete set of operators at the initial time, $\tau = 0$, as that will correspond to the time of preparation of the physical system.

However, by (44) we see that to impose the reality conditions on the operators $\hat{O}[\{\hat{A}, \hat{E}, \hat{\phi}\}, \tau]_{g.f.}$ at $\tau = 0$ is to simply impose the diffeomorphism invariant reality conditions for the non-clock fields. Thus, we may propose that for any two physical states $\Psi$ and $\Psi'$ in $\mathcal{H}_{g.f.}$

$$< \Psi | \Psi' >_{g.f.} = < \Theta \circ \Psi | \Theta \circ \Psi' >_{diffeo} \qquad (46)$$

As a result, we may conclude that the physical expectation value of the operator that measures the diffeomorphism invariant quantity $\mathcal{O}[\{A, E, \phi\}]_{diffeo}$ at the time $\tau$ in the state $\Psi[\{\gamma, \phi\}, \tau]$ in $\mathcal{H}_{g.f.}$ is given by

$$< \Psi | \hat{O}[\{\hat{A}, \hat{E}, \hat{\phi}\}, \tau]_{g.f.} | \Psi >_{g.f.} = < \Theta \circ \Psi | \Theta \circ \left( \hat{O}[\{\hat{A}, \hat{E}, \hat{\phi}\}, \tau]_{g.f.} | \Psi > \right)_{diffeo} \qquad (47)$$

## 3.5.   A word about unitarily

The reader may notice that in fixing the physical inner product there was one condition I might have imposed, but did not. This was that the operator $\hat{W}$ that generates evolution for the non-clock fields be hermitian. The reason for this is one aspect of the conflict between the notions of time in quantum theory and general relativity. From the point of view of quantum theory, it is natural to assume that the time evolution operator is unitary. However, this means that all physical states of the form $\Psi[\{\gamma, \phi\}, \tau]$ exist for all physical clock times $\tau$. This directly contradicts the situation in classical general relativity, in which for every set of initial data which solves the constraints for compact $\Sigma$ (and which satisfies the positive energy conditions) there is a time $\tau$ after which the spacetime has collapsed to a final singularity so that no physical observable could be well defined.

Furthermore, we may note that we may not be free to choose the physical inner product such that $\hat{W}$ is hermitian. The physical inner product has already been restricted by the reality conditions applied to a certain set of observables of the diffeomorphism invariant theory. As $\hat{W}$ is to be defined through a limit of a regularization procedure, it is probably best to fix the inner product first and then define the limit inside this inner product space[12]. However, then it is not obvious that the condition that $\hat{W}$ be hermitian will be consistent with the conditions that determine the inner product.

How a particular formulation of quantum gravity resolves these conflicts is a dynamical problem. This is, indeed, proper, as the evolution operator for quantum gravity could be unitary only if the quantum dynamics avoided complete gravitational collapse in every circumstance, and whether this is the case or not is a dynamical problem. The implications of this situation will be the subject of section 6.

---

[12]Note that this is different from the problem of finding the kernel of the constraints, for which it is sufficient to define the limit in a pointwise topology, as was done in [JS88, RS88]. There we were content to let the limit be undefined on the part of the state space not in the kernel. As we now want to construct the whole operator we probably need an inner product to control the limit.

## 3.6.   The physical quantum theory without gauge fixing

As in the classical theory, once we have the gauge fixed theory it is easier to see how to construct the theory without gauge fixing. Here I will give no technical details, but only sketch the steps of the construction.

In the Dirac approach, we first construct the kinematical state space, which consists of all states with a general dependence on the variables, of the form $\Psi[\gamma, \phi, T(x)]_{Dirac}$. Instead of (33) and (34) we have the defining relations

$$\hat{T}(x)\Psi[\gamma, \phi, T(x)]_{Dirac} = T(x)\Psi[\gamma, \phi, T(x)]_{Dirac} \tag{48}$$

and

$$\hat{\mathcal{E}}(x)\Psi[\gamma, \phi, T(x)]_{Dirac} = -\imath\hbar\frac{\delta}{\delta T(x)}\Psi[\gamma, \phi, T(x)]_{Dirac}. \tag{49}$$

The physical state space, $\mathcal{H}_{Dirac}$ then consists of the subspace of states that satisfy the full set of constraints,

$$\hat{\mathcal{D}}(v)\Psi_{Dirac} = 0 \tag{50}$$

and

$$\lim_{\epsilon \to 0} \hat{\mathcal{C}}^\epsilon(N)\Psi_{Dirac} = 0 \tag{51}$$

for all $v^a$ and $N$. As in the gauge fixed formalism, we will be interested only in those solutions that arise from initial data of the non-clock fields, so that they can represent states prepared at an initial clock time. Thus, we will be interested in the subspace of Dirac states such that

$$\Psi[\{\gamma, \phi, T(x) = 0\}]_{Dirac} = \Psi[\{\gamma, \phi\}]_{diffeo}. \tag{52}$$

is normalizable in an appropriate inner product that gives a probability measure to the possible preparations of the system we can make at the initial time.

There is a complication that arises in the case of the full constraints, because (51) is a second order equation in $\delta/\delta T(x)$. This is the familiar problem of the doubling of solutions arising from the Klein-Gordon like form of the full Hamiltonian constraint. In the gauge fixed quantization we studied in section 3.2-3.4 this problem did not arise because the reduced Hamiltonian constraint was first order in the derivatives of the reduced time variable $\tau$.

However, there is a way to deal with this problem in the full, Dirac, quantization, because we have the constant of motion $\mathcal{E}$ defined by (17). We can use this to impose a positive frequency condition on the physical states. Thus, using $\hat{\mathcal{E}}$ we can split the Hilbert space $\mathcal{H}_{Dirac}$, to be the direct product of two subspaces, $\mathcal{H}_{Dirac}^\pm$, where $\mathcal{H}_{Dirac}^+$ is spanned by the eigenstates of $\hat{\mathcal{E}}$ whose eigenvalues have positive real part, and $\mathcal{H}_{Dirac}^-$ is spanned by the eigenstates of $\hat{\mathcal{E}}$ with eigenvalues with negative real part. Associated to this splitting we have projection operators $P^\pm$ that project onto each of these subspaces[13]. From now on, we will restrict attention to states and operators in the positive frequency part of the physical Hilbert space.

As in the gauge fixed case we thus define a map $\Theta : \mathcal{H}^+_{Dirac} \to \mathcal{H}_{diffeo}$ by

$$(\Theta \circ \Psi_{Dirac})[\{\gamma, \phi\}] = \Psi[\{\gamma, \phi, T(x) = 0\}]_{Dirac} \tag{53}$$

Further, if there is a unique solution to the constraints that satisfies also the positive frequency condition,

$$P^+ \Psi_{Dirac} = \Psi_{Dirac} \tag{54}$$

there is a corresponding inverse map $\Lambda : \mathcal{H}_{diffeo} \to \mathcal{H}^+_{Dirac}$ which takes each state in $\mathcal{H}_{diffeo}$ into its positive frequency evolution under the full set of constraints.

The operators on this space, which we can call the Dirac operators are as well solutions to the full set of constraints,

$$\left[\hat{\mathcal{D}}(v), \hat{\mathcal{O}}_{Dirac}\right] = 0 \tag{55}$$

$$\lim_{\epsilon \to 0} \left[\hat{\mathcal{C}}^\epsilon(N), \hat{\mathcal{O}}_{Dirac}\right] = 0 \tag{56}$$

We will impose as well the positive frequency condition,

$$\left[P^+, \hat{\mathcal{O}}_{Dirac}\right] = 0 \tag{57}$$

which converts (56) from a second order to a first order functional differential equation.

We may then seek to use these equations to extend diffeomorphism invariant operators on $\mathcal{H}_{diffeo}$, which act only on the non-clock degrees of freedom, to positive frequency Dirac operators. That is, given an operator $\hat{\mathcal{O}}[\{\hat{E}, \hat{A}, \hat{\phi}\}]_{diffeo}$ we seek operators of the form $\hat{\mathcal{O}}[\{\hat{E}, \hat{A}, \hat{\phi}, T(x), \hat{\mathcal{E}}\}]_{Dirac}$ which solve (55), (56) and (57) which have the property that

$$\hat{\mathcal{O}}[\{\hat{E}, \hat{A}, \hat{\phi}, T(x) = 0, \hat{\mathcal{E}}\}]_{Dirac} = \hat{\mathcal{O}}[\{\hat{E}, \hat{A}, \hat{\phi}\}]_{diffeo} \tag{58}$$

Furthermore, we can construct quantum analogues of the evolving constants of motion (32). These are corresponding one parameter families of Dirac observables $\hat{\mathcal{O}}'[\{\hat{E}, \hat{A}, \hat{\phi}, T(x) = 0, \hat{\mathcal{E}}\}](\tau)_{Dirac}$ that satisfy (55), (56) and (57) (with $\tau$, again, treated just as a parameter) and the condition

$$\hat{\mathcal{O}}'[\{\hat{E}, \hat{A}, \hat{\phi}, T(x) = 0, \hat{\mathcal{E}}\}](\tau)_{Dirac} = \hat{\mathcal{O}}[\{\hat{E}, \hat{A}, \hat{\phi}, T(x) = \tau, \hat{\mathcal{E}}\}]_{Dirac} \tag{59}$$

---

[13]Note that I have not assumed that $\hat{\mathcal{E}}$ is hermitian, for the reasons discussed in the previous section.

As in the classical case, one can relate these operators also to the operators of the gauge fixed theory. However, as there are potential operator ordering problems that come from the operator versions of the substitutions in (32), and as the gauge fixed and Dirac operators act on different state spaces, it is more convenient to make the definition in this way.

We may note that these may not be all of the physical operators of the Dirac theory, as there may be operators that satisfy the constraints and positive frequency condition for which there is no diffeomorphism invariant operator of only the non-clock fields such that (58) holds. But this is a large enough set to give the theory a physical interpretation based on the use of the clock fields.

To finish the construction of the Dirac formulation, we must give the physical inner product. The same argument that we gave in the gauge fixed case leads to the conclusion that we may impose a physical inner product $< \ | \ >_{Dirac}$ such that, if $\Psi$ and $\Phi$ are two elements of $\mathcal{H}^+_{Dirac}$

$$< \Psi | \Phi >_{Dirac} = < \Theta \circ \Psi | \Theta \circ \Phi >_{diffeo} \tag{60}$$

For the reason just stated, this may not determine the whole inner product on $\mathcal{H}^+_{Dirac}$, but it is enough to do some physics because we may conclude that in the Dirac formalism the expectation value of the operator that corresponds to measuring the diffeomorphism invariant quantity $\mathcal{O}_{diffeo}$ of the nonclock field a physical time $\tau$ after the preparation of the system in the state $|\Psi >$ (which, by definition is in $\mathcal{H}^+_{Dirac}$) is

$$< \Psi | \hat{O}'(\tau)_{Dirac} | \Psi >_{Dirac} = < \Theta \circ \Psi | \Theta \circ \left( \hat{O}'(\tau)_{Dirac} | \Psi > \right)_{diffeo} . \tag{61}$$

## 4.   Outline of a measurement theory for quantum cosmology

I will now move away from technical problems, and consider the question of how a theory constructed according to the lines of the last two sections could be interpreted physically. In order to give an interpretation of a quantum theory it is necessary to describe what mathematical operations in the theory correspond to preparation of the system and what mathematical operations correspond to measurement. This is the main task that I hope to fulfil here. I should note that I will phrase my discussion entirely in the traditional language introduced by Bohr and Heisenberg concerning the interpretation of quantum mechanics. As we will see, with the appropriate modifications, there is no barrier to using this language in the context of quantum cosmology. However, if the reader prefers a different language to discuss the interpretation of quantum mechanics, whether it be the many worlds interpretation or a statistical interpretation, she will, as in the case of ordinary quantum mechanics, be able to rephrase the language appropriately.

The interpretation of quantum cosmology that I would like to describe is based on the following four principles:

**A) The measurement theory must be completely spacetime diffeomorphism invariant.** The interpretation must respect the spacetime diffeomorphism invariance of the quantum theory of gravity. Thus, we must build the interpretation entirely on physical states and physical operators.

**B) The reference system, by means of which we locate where and when in the universe measurements take place, must be a dynamical component of the quantum matter plus gravity system on which our quantum cosmology is based.** This is a consequence of the first principle, because the diffeomorphism invariance precludes the meaningful use of any coordinate system that does not come from the configuration of a dynamical variable.

**C) As we are are studying a quantum field theory, any measurement we can make on the system must be a partial measurement.** This is an important point whose implications will play a key role in what follows. The argument for it is simple: a quantum field theory has an infinite number of degrees of freedom. Any measurement that we make returns a finite list of numbers. The result is that any measurement made on a quantum field theory can only result in a partial determination of the state of the system.

**D) The inner product is to be determined by requiring that a complete set of physical observables for the gravity and matter degrees of freedom satisfy the reality conditions at the initial physical time corresponding to preparation of the state.**

For concreteness I will phrase the measurement theory in terms of the particular type of reference frames and clock fields described in the last two sections. However, I will use a language that can refer to either the gauge fixed formalism described in subsections 3.2-3.4 or the Dirac formalism described in subsection 3.6. I will use a general subscript *phys* to refer to the physical states, operators and inner products of either formalism. If one wants to specify the gauge fixed formalism then read *phys* to mean *g.f.* so that operators $\hat{\mathcal{O}}(\tau)_{phys.}$ will mean the gauge fixed operators $\hat{\mathcal{O}}[\hat{E}, \hat{A}, \hat{\phi}, \hat{\mathcal{E}}, \tau]_{g.f.}$ defined in subsection 3.3. Alternatively, if one wants to think in terms of the Dirac formalism then read *phys* to mean *Dirac* everywhere, so that the operators $\hat{\mathcal{O}}(\tau)_{phys.}$ refer to the $\tau$ dependent "evolving constants of motion" $\hat{\mathcal{O}}'(\tau)_{Dirac}$. Furthermore, I will always assume that reference is being made to states and operators in the positive frequency subspace of the Dirac subspace. I will use this notation as well in section 6.

Thus, putting together the results of the last two sections, I shall assume, for purposes of illustration, that we have available at least two sets of $\tau$ dependent physical observables $\hat{A}^I(\tau)_{phys}$ and $\hat{T}^I(\tau)_{phys}$ which measure, respectively, the areas of the simplices of the reference frame, and parallel transport around them, at a physical time $\tau$.

However, while I refer to a particular form of clock dynamics and a particular set of observables, I expect that the interpretation given here can be applied to any theory in which the physical states, observables and inner product are related to their diffeomorphism invariant counterparts in the way described in the last section.

Let us now begin with the process of preparing a system for an observation.

## 4.1. Preparation in quantum cosmology

Let me assume that at time $\tau = 0$ we make a preparation prior to performing some series of measurements on the quantum gravitational field. This means that we put the quantum fields which describe the temporal and spatial reference system into appropriate configurations so that the results of the measurements will be meaningful. There are two parts to the preparation: arrangement of the spatial reference system and synchronization of the physical clocks.

As in ordinary quantum mechanics, we can assume that we, as observers, can move matter around as we choose in order to do this. This certainly does not contradict the assumption that the whole universe including ourselves could be described by the quantum state $\Psi$ for, if it did, we would be simply unable to do quantum cosmology because, *ipso facto* we are in the universe and we do move things such as clocks and measuring instruments around more or less as we please.

In ordinary quantum mechanics the act of preparation may be described by projecting the quantum states of the reference system and measuring instruments into appropriate states, after which the direct product with the system state is taken. In the case of a diffeomorphism invariant theory we cannot do this because there is no basis of the diffeomorphism invariant space $\mathcal{H}_{diffeo}$ whose elements can be written as direct products of matter states and gravity states. Thus, the requirement of diffeomorphism invariance has entangled the various components of the whole system even before any interactions occur.

However, this entanglement does not prevent us from describing in quantum mechanical terms the preparation of the reference system and clock fields. What we must do is describe the preparation by projecting the physical states into appropriate subspaces, every state of which describes a physical situation in which the matter and clock fields have been prepared appropriately.

Let us begin with synchronizing the clocks. We may assume that we are able to synchronize a field of clocks over as large a volume of the universe as we please, or even over the whole universe (if it is compact). The difficulty of doing this is, *a priori* a practical problem, not a problem of principle. So we assume that we may synchronize our clock field so that there is a spacelike surface everywhere on which $T(x) = 0$.

In terms of the formalism, this act of preparing the clocks corresponds to assuming that the state is normalizable in the inner product defined by either (46) or (60). That is, there may be states in either the solution space to the gauge fixed constraints or the Dirac constraints that are not normalizable in the respective inner products defined by the maps to the diffeomorphism invariant states of the non-clock fields. Such states cannot correspond to preparations of the matter and gravitational fields made at some initial time of the clock fields.

Once the clocks are synchronized we can prepare the spatial reference frame. As described in subsections 2.2 and 2.3 we do this by specifying that the $N$ surfaces are arranged as the faces of a simplicial complex $\mathcal{T}$. In the formalism this is described by the statement that the state of the system is to be further restricted to be in a subspace $\mathcal{H}_{phys,\mathcal{T}}$ of $\mathcal{H}_{phys}$.

It is interesting to note that, at least in principle, the preparation of the spatial and temporal reference frames can be described without making any assumption about the quantum state of the gravitational field. Of course, this represents an ideal case, and in practice preparation for a measurement in quantum cosmology will usually involve fixing some degrees of freedom of the gravitational field. But, it is important to note that this is not required in principle. In particular, no assumption need be made restricting the gravitational field to be initially in anything like a classical or semiclassical configuration to make the measurement process meaningful.

This completes the preparation of the spatial and temporal reference frames[14].

## 4.2.    Measurement in quantum cosmology

After preparing the system we may want to wait a certain physical time $\tau$ before making a measurement. Let us suppose, for example, that we want to measure the area of one of the surfaces picked out by the spatial reference system at the time $\tau$ after the preparation. How are we to describe this? To answer this question we need to make a postulate, analogous to the usual postulates that connect measurements to

---

[14]Note that, in a von Neumann type description of the measuring process, we must also include the measuring instruments in the description of the system. These are contained in the dependence of the physical states on additional matter fields in the set $\phi$ that represent the actual measuring instruments. I will not here go through the details of adopting the von Neumann description of measurement to the present case, but there is certainly no obstacle to doing so. However, as pointed out by Anderson [AAn93] it is necessary to take into account the fact that a real measuring interaction takes a finite amount of physical time. There is also no obstacle to including in the preparation as much information about our own existence as may be desired, for example, there is a subspace of $\mathcal{H}_{phys,\mathcal{T}}$ in which we are alive, awake, all our measurement instruments are prepared and we are in a mood to do an experiment. There is no problem with assuming this and requiring that the system be initially in this subspace. However, as in ordinary quantum mechanics, as long as we do not explicitly make any measurements on ourselves, there is no reason to do this.

the actions of operators in quantum mechanics. It seems most natural to postulate the following:

**Measurement postulate of quantum cosmology**: *The operator that corresponds to the making of a measurement of the spatially diffeomorphism invariant quantity represented by $\hat{O}_{diffeo} \in \mathcal{H}_{diffeo}$ at the time that the clock field reads $\tau$, is the physical (meaning either gauge fixed or Dirac) operator $\hat{O}_{phys}(\tau)$. Thus, we postulate that the expected value of making a measurement of the quantity $\mathcal{O}$ at a time $\tau$ after the preparation in the physical state $|\Psi >$ is given by $< \Psi|\hat{O}_{phys}(\tau)|\Psi >_{phys}$.*

It is consistent with this to postulate also that: *the only possible values which may result from a measurement of the physical quantity $\hat{O}_{phys}(t)$ are its eigenvalues, which may be found by solving the physical states equation*

$$\hat{O}_{phys}(\tau)|\lambda(\tau) >_{phys} = \lambda(\tau)|\lambda(\tau) >_{phys} \tag{62}$$

inside the physical state space $\mathcal{H}_{phys}$.

Having described how observed quantities correspond to the mathematical expressions of quantum cosmology, we have one more task to fulfil to complete the description of the measurement theory. This is to confront the most controversial part of measurement theory, which is the question of what happens to the quantum state after we make the measurement. In ordinary quantum mechanics there are two points of view about this, depending on whether one wants to employ the projection postulate or some version of the relative state idea of Everett. This choice is usually, but perhaps not necessarily, tied to the philosophical point of view that one holds about the quantum state. If one believes, with Bohr and von Neumann, that the quantum state is nothing physically real, but only represents our information about the system, then there is no problem with speaking in terms of the projection postulate. There is, in this way of speaking, only an abrupt change in the information that we have about the system. Nothing physical changes, i.e. collapse of the wave function is not a physical event or process.

On the other hand, if one wants to take a different point of view and postulate that the quantum state is directly associated with something real in nature, the projection postulate brings with it the well known difficulties such as the question of in whose reference frame the collapse takes place. There are then two possible points of view that may be taken. Some authors, such as Penrose [RP], take this as a physical problem, to be solved by a theory that is to replace, and explain, quantum mechanics. Therefore, these authors want to accept the collapse as being something that physically happens. The other point of view is to keep the postulate that the quantum state is physically real but to give an interpretation of the theory that does not involve the projection postulate. In this case one has to describe measurement in terms of the correlations that are set up during the measurement process between the quantum state of the measuring instrument and the quantum state of the system as a result of their interaction during the measurement.

Of course, these two hypotheses lead, in principle, to different theories as in the second it is possible to imagine doing experiments that involve superpositions of states of the observer, while in the first case this is not possible. Nevertheless, there is a large set of cases in which the predictions of the two coincide. Roughly, these are the cases in which the quantum state, treated from the second point of view, would decohere. Indeed, if one takes the second point of view then some version of decoherence is necessary to recover what is postulated from the first point of view, which is that the observer sees a definite outcome to each experiment.

I do not intend to settle here the problem of which of these points of view corresponds most closely to nature. However, I would like to make two claims which I believe, to some extent, diffuse the conflict. First, I would like to claim that whichever point of view one takes, something like the statement of the projection postulate plays a role. Whether it appears as a fundamental statement of the interpretation, or as an approximate and contingent statement which emerges only in the case of decoherent states or histories, the connection to what real observers see can be described in terms of the projection postulate, or something very much like it. Second, I would like to claim that the situation is not different in quantum cosmology then it is in ordinary quantum mechanics. One can make either choice and, in each case, something like the projection postulate must enter when you discuss the results of real observations (at least as long as one is not making quantum observations on the brain of the observer.)

With these preliminaries aside, I will now state how the projection postulate can be phrased so that it applies in the quantum cosmological case:

**Cosmological projection postulate:** *Let $\hat{\mathcal{O}}_I(\tau_0)_{phys}$ be a finite set of physical operators that mutually commute and hence correspond to a set of measurements that can be made simultaneously at the time $\tau_0$. Let us assume that the reference system has been prepared so that the system before the measurements is in the subspace $\mathcal{H}_{phys,T}$. The results of the observations will be a set of eigenvalues $\lambda_I(\tau_0)$. For the purposes of making any further measurements, which would correspond to values of the physical clock field $\tau$ for which $\tau > \tau_0$, the quantum state can be assumed to be projected at the physical clock time $\tau_0$ into the subspace $\mathcal{H}_{phys,T,\lambda(\tau_0)} \subset \mathcal{H}_{phys,T}$ which is spanned by all the eigenvectors of the operators $\hat{\mathcal{O}}_I(\tau_0)_{phys}$ which correspond to the eigenvectors $\lambda_I(t_0)$.*

That is, one is to project the state into the subspace at the time of the measurement, and then continue with the evolution defined by the Hamiltonian constraint.

## 4.3.   Discussion

I would now like to discuss three objections that might be raised concerning the application of the projection postulate to quantum cosmology. Again, let me stress that my goal here is not to argue that one must take Bohr's point of view over that of the other interpretations. I only want to establish that if one is happy with Bohr in ordinary quantum mechanics one can continue to use his point of view in quantum cosmology. The only strong claim I want to make is that the statement sometimes made, that one is required to give up Bohr's point of view when one comes to quantum cosmology, is false.

**First objection:** *Bohr explicitly states that the measurement apparatus must be described classically, which requires that it be outside of the quantum system being studied.* I believe that this represents a misunderstanding of Bohr which, possibly, comes from combining what Bohr did write with an assumption that he did not make, which is that the quantum state is in one to one correspondence with something physically real. For Bohr to have taken such a realistic point of view about the quantum state would have been to directly contradict his fundamental point of view about physics, which is that it does not involve any claim to a realistic correspondence between nature and either the mathematics or the words we use to represent the results of observations we make. Instead, for Bohr, physics is an extension of ordinary language by means of which we describe to each other the results of certain activities we do. Bohr takes it as given that we must use classical language to describe the results of our observations because that is what real experimentalists do. Perhaps the weakest point of Bohr is his claim that it is necessary that we do this, but even if we leave aside his attempts to establish that, we are still left with the fact that up to do this day the only language that we actually do use to communicate with each other what happens when we do experimental physics uses certain classical terms.

Furthermore, rather than insisting that the measuring instrument is outside of the quantum system, Bohr insists repeatedly that the measuring instrument is an inseparable part of the entire system that is described in quantum mechanical terms. He insists that we cannot separate the description of the atomic system from a description of the whole experimental situation, including both the atoms and the apparatus.

Many people do not like this way of talking about physics. My only point here is that there is nothing in this way of speaking that prevents us from doing quantum cosmology. After all, we are in the universe, we are ourselves made of atoms, and we do make observations and describe their results to each other in classical terms. That all these things are true are no more and no less mysterious whether the quantum state is a description of our observations made of the spin of an atom or of the fluctuations in the cosmic black body radiation.

**Second objection:** *It is inconsistent with the idea that the quantum state describes* *the whole universe, including us, to postulate that the result of a measurement that we make is one of the eigenvalues of the measured observable, because that is to employ a classical description, while the whole universe is described by a quantum state.* To say so is, in my opinion, again to misunderstand Bohr and von Neumann and, again, to attempt to combine their way of speaking about physics with some postulate about the reality of the quantum state. To postulate that the result of a measurement is an eigenvalue is to assume that the results of measurements may be *described* using quantities from the language of classical physics. The theory does not, and need not, explain to us why that is the case; that we get definite values for the results of experiments we do is taken as a primitive fact upon which we base the interpretation of the theory.

Furthermore, to assume that the results of measurements are *described* in terms of the language of classical physics is not at all the same thing as to make the (obviously false) claim that the dynamics that governs the physics of either the measuring instruments or ourselves is classical.

**Third objection:** *The fact that we are in the universe might lead to some problem in quantum cosmology because a measurement of a quantum state would involve a measurement of our own state.* There are two replies to this.

First, there is an interpretation of quantum mechanics, suggested by von Neumann[vN] and developed to its logical conclusion by Wigner[EW], that says that all we ever actually do is make observations on our own state. Wigner claims that there is something special about consciousness which is that we can experience only definite things, and not superpositions. This is then taken to be the explanation for why we observe the results of experiments to give definite values. I do not personally believe this point of view, but the fact that it is a logically possible interpretation of quantum mechanics means that there can be no logical problem with including ourselves (and all our measuring instruments and cats, if not friends) in the description of the quantum state.

Second, we can avoid this problem, at least temporarily, if we acknowledge that all measurements we make in quantum cosmology are incomplete measurements. In reality we never determine very much information about the quantum state of the universe when we make a measurement, however we interpret it. We certainly learn very little about our own state when we make a measurement of the gravitational field of the sort I described in section 2.

Thus, as long as we refrain from actually describing experiments in which we make measurements on our own brains, we need not commit ourselves to any claims about the results of making observations on ourselves. Again, the situation here is exactly the same in quantum cosmology as it is in ordinary quantum mechanics-no worse and no better. If there is a possible problem with making observations on ourselves in quantum cosmology, it must occur in ordinary quantum mechanics. And it must be faced there as well, as it cannot matter for the resolution of such a problem whether, besides our brain, Andromeda or the Virgo cluster is also described by the quantum state.

Let me close this section with one comment. Given the measurement postulate above, and the results of the last two sections, we can conclude that in fact the areas of surfaces are quantized in quantum gravity. For, without integrating the evolution equations, we know that $\hat{\mathcal{A}}^I(\tau = 0)_{phys} = \hat{\mathcal{A}}^I(\tau = 0)_{diffeo}$ and the latter operator, from the results of [CR93, LS93] has a discrete spectrum. Thus, quantum gravity makes a physical prediction. Note, further that this result is independent of the form of the Hamiltonian constraint and hence of the dynamics and the matter content of the theory.

## 5. The recovery of conventional quantum field theory

The measurement theory given in the previous section has not required any notion of classical or semiclassical states. One need only assume that it is possible to prepare the fields that describe the spatial and temporal quantum reference frame appropriately so that subsequent measurements are meaningful. One does not need to assume that the gravitational field is in any particular state to do this. Of course, there may be preparations that require some restriction on the state of the gravitational degrees of freedom, but such an assumption is not required in principle. Further, the examples discussed in the last sections show that there are some kinds of physical reference frames whose preparation requires absolutely no restrictions on the gravitational field.

Having said this, we may investigate what happens to the dynamics and the measurement theory if we add the condition that the state is semiclassical in the gravitational degrees of freedom. I will show in this section that by making the assumption that the gravitational field is in a semiclassical state we can recover quantum field theory for the matter fields on a fixed spacetime background. Thus, quantum cosmology, whose dynamics is contained in the quantum constraint equations, and whose interpretation was described in the previous section, does have a limit which reproduces conventional quantum field theory.

Here I will only sketch a version of the demonstration, as my main motive here is to bring out an interesting point regarding a possible role for the zero point energy in the transition from quantum gravity to ordinary quantum field theory. For simplicity, I will also drop in this section the assumption that the state is diffeomorphism invariant, as this will allow me to make use of published results about the semiclassical limit of quantum gravity in the loop representation. However, I will continue to treat the Hamiltonian constraint in the gauge fixed formalism of sections 3.2-3.4

I will make use of the results described in [ARS92] in which it was shown that, given a fixed three metric $q_0^{ab}$, whose curvatures are small in Planck units, we can construct, in the loop representation, a nonperturbative quantum state of the gravitational field that approximates that classical metric up to terms that are small in Planck units.

Such a state can be described as follows[15]. Given the volume element $\sqrt{q_0}$, let me distribute points randomly on $\Sigma$ with a density $1/l_{Planck}^3 \times (2/\pi)^{3/2}$. Let me draw a circle around each point with a radius given by $\sqrt{\frac{\pi}{2}} l_{Planck}$ with an orientation in space that is random, given the metric $q_0^{ab}$. As the curvature is negligible over each of these circles, this is well defined. Let me call the collection of these circles $\Delta = \{\Delta_I\}$, where $I$ labels them.

Let me then define the *weave state associated to* $q_0^{ab}$ as the *characteristic state* of the set of loops $\Delta$. This is denoted $\chi_\Delta$ and, for a non-selfintersecting loop $\Delta$, it is defined so that $\chi_\Delta[\alpha]$ is an eigenstate of the area operator $\mathcal{A}[S]$ which measures the area of the arbitrary surface with eigenvalue $l_{Planck}^2/2$ times the number of times the loop $\Delta$ intersects the surface $S$. The result is that $\chi_\Delta[\alpha]$ is equal to one if $\alpha$ is equivalent to $\Delta$ under the usual rules of equivalence of loops in the loop representation and is equal to zero for most other loops $\alpha$ (including all other distinct non self-intersecting loops.)

I will also assume that the weave is chosen to have no intersections, in which case

$$\lim_{\epsilon \to 0} C_{grav.}^\epsilon(x) \chi_\Delta[\alpha] = 0 \qquad (63)$$

This will simplify our discussion.

I would now like to make the ansatz that the state is of the form

$$\Psi[\gamma, \phi, \tau] = \chi_\Delta[\gamma] \Phi[\gamma, \phi, \tau]. \qquad (64)$$

Note that, as I have dropped for the moment the requirement of diffeomorphism invariance, I have also dropped the dependence on the spatial reference frame field $\mathcal{S}$.

Now, let me assume, as an example, that the matter consists of one scalar field, called $\phi(x)$, with conjugate momenta $\pi(x)$, whose contribution to the regulated Hamiltonian constraint is,

$$\hat{C}_\phi^\epsilon(x) = -\frac{1}{2} \int d^3y \int d^3z\, f^\epsilon(x,y) f^\epsilon(x,z)$$
$$\times \left[ \pi(y)\pi(z) + \hat{T}^{ab}(y,z) \partial_a \phi(y) \partial_b \phi(z) \right]. \qquad (65)$$

Let us focus on the spatial derivative term. Let us assume that the dependence of $\Phi[\gamma, \phi, \tau]$ on the gravitational field loops $\gamma$ can be neglected. (This is exactly to neglect the back reaction and the coupling of gravitons to the matter field.) Let me

---

[15]More details of this construction are given in [ARS92]. See also [AA92, CR91, LS91].

assume also that the support of the state on configurations on which the scalar field is not slowly varying on the Planck scale (relative to $q_0^{ab}$) may be neglected. Then it is not hard to show, following the methods of [ARS92, LS91] that as long as the scalar fields are slowly varying on the Planck scale,

$$
\int d^3y \int d^3z f^\epsilon(x,y) f^\epsilon(x,z) \hat{T}^{ab}(y,z) \partial_a \phi(y) \partial_b \phi(z) \Psi[\gamma,\phi,\tau]
$$

$$
= \left( \int d^3y \int d^3z f^\epsilon(x,y) f^\epsilon(x,z) \hat{T}^{ab}(y,z) \chi_\Delta[\gamma] \right) \partial_a \phi(y) \partial_b \phi(z) \Phi[\gamma,\phi,\tau]
$$

$$
= \sum_I \sum_J \int ds \int dt f^\epsilon(x,\Delta_I(s)) f^\epsilon(x,\Delta_J(t)) \dot{\Delta}_I^a(s) \dot{\Delta}_I^b(t)
$$

$$
\times \left( \sum_{\text{routings}} \chi_\Delta[\gamma \circ \gamma_{x,y}] \right) \partial_a \phi(\Delta_I(s)) \partial_b \phi(\Delta_J(t)) \Phi[\gamma,\phi,\tau]
$$

$$
= det(q_0) q_0^{ab}(x) \partial_a \phi(x) \partial_b \phi(x) \chi_\Delta[\gamma] \Phi[\gamma,\phi,\tau] + O(l_{Planck}^2 \partial_a \phi) \qquad (66)
$$

Putting these results together, we have shown that if we make an ansatz on a state of the form of (64) then, neglecting the dependence of $\Phi$ on $\gamma$, the regulated Hamiltonian constraint (36) is equivalent to

$$
\imath \frac{d\Phi[\gamma,\phi,\tau]}{d\tau} = \sqrt{2\mu} \int d^3x \sqrt{\left[ \frac{1}{2} \hat{\pi}^2(x) + \frac{1}{2} det(q_0) q_0^{ab}(x) \partial_a \phi(x) \partial_b \phi(x) \right]} \Phi[\gamma,\phi,\tau] + \dots
$$
$$(67)$$

This does not yet look like the functional Schroedinger equation for the scalar field. However, we may recall that formally the expression inside the square root is divergent. However, it may not be actually divergent because in computing (66) we have assumed that the scalar field slowly varying on the scale of the weave. If we investigate the action of (65) on states which have support on $\phi(x)$ that are fluctuating on the Planck scale, we can see that in the limit that the regulator is removed the effect of the $T^{ab}$'s is to insure that the the terms in $(\partial_a \phi)^2$ only act at those points which are on the lines of the weave. That is, in the limit of small distances we have a description of a scalar field propagating on a one dimensional subspace of $\Sigma$ picked out by the weave. That is, on scales much smaller than the Planck scale the scalar field is propagating as a $1+1$ dimensional scalar field.

The result must be to cut off the divergence in the zero point energy coming formally from the scalar field Hamiltonian. The effect of this must be the following: If we decompose the scalar field operators into creation and annihilation operators defined with respect to the background metric $q_0^{ab}$ that the weave corresponds to, then the divergent term in the zero point energy must cut off at a scale of $M_{Planck}$. As such, we will have, if we restrict attention to the action of (67) on states that are slowly varying on the Planck scale

$$
\frac{1}{2} \hat{\pi}^2(x) + \frac{1}{2} det(q_0) q_0^{ab}(x) \partial_a \phi(x) \partial_b \phi(x) = \quad a M_{Planck}^4
$$

$$+ : \frac{1}{2}\hat{\pi}^2(x) + \frac{1}{2}det(q_0)q_0^{ab}(x) \qquad \partial_a\phi(x)\partial_b\phi(x) : \qquad (68)$$

where $: ... :$ means normal ordered with respect to the background metric and $a$ is an unknown constant that depends on the short distance structure of the weave.

The reduced Hamiltonian constraint now becomes,

$$\imath\frac{d\Phi[\gamma,\phi,\tau]}{d\tau} = \sqrt{2a\mu}M_{Planck}^2$$

$$\times \int d^3x \quad \sqrt{1 + \frac{1}{aM_{Planck}^4} : \left(\frac{1}{2}\hat{\pi}^2(x) + \frac{1}{2}det(q_0)q_0^{ab}(x)\partial_a\phi(x)\partial_b\phi(x)\right) :}\Phi[\gamma,\phi,$$

$$= \sqrt{2a\mu}M_{Planck}^2 V \quad \Phi[\gamma,\phi,\tau]$$

$$+\frac{\sqrt{\mu}}{M_{Planck}^2\sqrt{2a}} \quad \int d^3x\sqrt{q_0} : \left(\frac{1}{2}\hat{\pi}^2(x) + \frac{1}{2}det(q_0)q_0^{ab}(x)\partial_a\phi(x)\partial_b\phi(x)\right) : \Phi[\gamma,\phi,\tau] +$$

where $V = \int \sqrt{q_0}$ is the volume of space.

Thus, only after taking into account the very large zero point energy do we recover conventional quantum theory for low energy physics.

Before closing this section, I would like to make three comments on this result.

1) Note that the theory we have recovered is Poincare invariant, even if the starting point is not! We may note that the weave state $\chi_\Delta$ is *not* expected to be the vacuum state of quantum gravity because it is a state in which the spatial metric is sharply defined. What we need to describe the vacuum is a Lorentz invariant state which which is some kind of minimal uncertainty wave packet in which the three metric and its conjugate momenta are equally uncertain. A state that has these properties, at least at large wavelengths, can be constructed by dressing $\chi_\Delta$ with a Gaussian distribution of large loops that correspond to a Gaussian distribution of virtual gravitons [ASR91]. It is interesting to note that at the level when we neglect the back-reaction and the coupling to gravitons, the Poincare invariant matter quantum field theory is nevertheless recovered by using the weave state $\chi_\Delta$ as the background. However, before incorporating quantum back-reaction and the coupling to gravitons we must replace $\chi_\Delta$ in (64) with a good approximation to the vacuum state, such as is described in [IR93].

2) Can we add a mass term and self-interaction terms for the scalar field theory? The answer is yes, but to do so we must modify the weave construction in order to add intersections. The reason is that the scalar mass and self interaction is described by the term $\hat{q}V(\hat{\phi})$, where $\hat{q}$ is the operator corresponding to the determinant of the metric. Using results about the volume operator in [LS91], it is easy to see that if we modify the weave construction in order to add intersections, then the effect of this term, after regularization, is to modify $\int d^3x N(x)\mathcal{C}_{matter}\Psi$ by the addition of the

term

$$l^3_{Planck} \sum_i a(i) N(x_i) V(\phi(x_i)) \Phi \tag{70}$$

where the sum is over all intersections involving three or more lines, $x_i$ is the intersection point and $a(i)$ are dimensionless numbers of order one that characterize each intersection. Assuming that there are on the order of $a(i)^{-1}$ intersection points per Planck volume, measured with respect to the volume element $q_0$ (which is consistent with the weave construction described above as that is the approximate number of loops) we arrive at an addition to (69) of the form of $N(x)q(x)V(\phi(x))\Phi$.

Note that once we add intersections that produce volume it is no longer true that the gravitational part of the Hamiltonian constraint is solved by $\chi_\Delta$. This is because there is now a term in the back-reaction of the quantum matter field on the metric coming from the local potential energy of the scalar field. This is telling us that we now cannot neglect the back-reaction of the quantum fields on the background metric to construct solutions of the Hamiltonian constraint.

3) Finally, let me note that the measurement theory of the semiclassical state (64) is already defined because we have a measurement for the full nonperturbative theory. We therefore do not have to supplement the derivation of the equations of quantum mechanics from solutions to the quantum constraint equations of quantum cosmology with the *ab initio* postulation of the standard rules of interpretation of quantum field theory. This is always a suspicious procedure as those rules rely on the background metric that is only a property of a particular state of the form of (64); we cannot then choose inner products or other aspects of the interpretative machinery to fit a particular state.

In this case, since we already have an inner product and a set of rules of interpretation defined for the full theory, what needs to be done is to verify that the usual quantum field theory inner product is recovered from the full physical inner product defined by (46) in the case that the state is of the form of (64) . We have seen in section 3 that this will be the case to the extent to which $\mathcal{W}$ defined by (37) is hermitian. We see that in the approximation that leads to (69), the contribution to $\mathcal{W}$ from the matter fields is hermitian a long as the diffeomorphism inner product implies that the operators for $\phi$ and $\pi$ are also hermitian. In this case, then, the usual inner product of quantum field theory must be recovered.

## 6. Singularities in quantum cosmology

Having established a physical interpretation for quantum cosmology and shown that it leads to the recovery of conventional quantum field theory in appropriate circumstances, we now have tools with which to address what is perhaps the key problem that any quantum theory of quantum gravity must solve, which is what happens to black holes and singularities in the quantum theory.

The main question that must be answered is to what extent the apparent loss of information seen in the semiclassical description of black hole evaporation survives, or is resolved, in the full quantum theory. The key point that must be appreciated to investigate this problem from the fully quantum mechanical point of view is that the problem of loss of information, or of quantum coherence, is a problem about time because the question cannot be asked without assuming that there is a meaningful notion of time with respect to which we can say information or coherence is being lost. If we take an operational approach to the meaning of time in the full quantum theory, along the lines that have been developed here, then the loss of information or coherence, if it exists at the level of the full quantum theory, must show up as a limitation on the possibility of completely specifying the quantum state of the system by measuring the physical observables, $\mathcal{O}(\tau)_{phys}$ for sufficiently late $\tau$.

What I would like to do in this section is to describe how singularities, if they occur in the full quantum theory, will show up in the action of these physical time dependent observables. The result will be the formulation of two conjectures about how singularities may show up in the quantum theory, which I will call the *quantum singularity conjecture* and the *quantum cosmic censorship conjecture*.

While I will not try to prove these conjectures here, I also will argue that there is no evidence that they may not be true. It is possible that quantum effects completely eliminate the singularities of the classical theory as well as the consequent losses of information and coherence in the semiclassical theory. But, it seems to me, it is at least equally possible that the quantum theory does not eliminate the singularities. Rather, given the formulation of the theory along the lines described in this paper, the occurrence of singularities and loss of information, as formulated in these two conjectures, seems to be compatible with both the dynamics and interpretational framework of quantum cosmology.

The loss of information and coherence in the semiclassical theory implies a breakdown in unitarity in any process that describes a black hole forming and completely evaporating. From a naive point of view, this would seem to indicate a breakdown in one of the fundamental principles of quantum mechanics. Hence many discussions about this problem seem to assume that if there is to be a good quantum theory of gravity it must resolve this problem in such a way that unitarity is restored in the full quantum theory. However, the results of the last several sections show that unitarity is not one of the basic principles of quantum cosmology. It is not because unitarity depends on a notion of time that apparently cannot be realized in either classical or quantum cosmology.

To put it most simply, if the concept of time is no longer absolute, but depends for its properties on certain contingent facts about the universe, principally the existence of degrees of freedom that behave as if there is a universal and absolute Newtonian notion of time, then the same must be true for those structures and principles that, in the usual formulation of quantum mechanics, are tied to the absolute background time of the Schroedinger equation. Chief among these are the notions of unitarity evolution and conservation of probability.

As I have argued in detail elsewhere[LS91, LS92c], conventional quantum mechanics, no less than Newtonian mechanics, relies for its interpretation on the assumption of an absolute background time. When we speak of conservation of probability, or unitary evolution in quantum mechanics, we do not have to ask whether something might happen to the clocks that measure time that could make difficulties for our understanding of the operational meaning of these concepts. A single clock could break, but the $t$ in Schroedinger's equation refers to no particular clock but instead to an absolute time that is presumed to exist independently of the both the physical system described the quantum state and of the physical properties of any particular clock. However, neither in classical nor in quantum cosmology does there seem to be available such an absolute notion of time. In any case, if it exists we have not found it. If we then proceed by using an operational notion of time as I have done here then, because that clock must, by diffeomorphism invariance, be a dynamical part of the system under study, we must confront the question. what happens to the notion of time and to all that depends on it if something happens to the clock whose motion is taken as the operational basis of time. Of course, in the theory, as in real life, in most circumstances in which a clock may fail we can imagine constructing a better one. The problem we really have to face as theorists is not the engineering problem of modeling the best possible clock. The problem we have to face is what the implications are for the theory if there are physical effects that can render useless any conceivable clock.

Of course, in classical general relativity there is such an effect; it is called gravitational collapse. Thus, to put the point in the simplest possible way, in quantum cosmology a breakdown of unitarity need not indicate a breakdown in the theory. Rather, it may only indicate a breakdown in the physical conditions that make it possible to speak meaningfully of unitary evolution. This will be the case if there are quantum states in which some of the physical clocks and some of the components of the physical reference frame that make observations in quantum cosmology meaningful cease to exist after certain physical times. This will not prevent us from describing the further evolution of the system in terms of operators whose meaning is tied to the physical clocks that happen to survive the gravitational collapse. But it will prevent us from describing that evolution in terms of the unitary evolution of the initial quantum state.

In this section I proceed in two steps. First, before we describe how singularities may show up in the operator algebra of the quantum theory, we should see how they manifest themselves in terms of the observable algebra of the classical theory. This then provides the basis for the statements of the quantum singularity conjecture and the quantum cosmic censorship conjecture.

## 6.1.   Singularities in the classical observable algebra

In section 3 we found an evolution equations for classical observables in the gauge defined by (19) and (21). We then used this to define both gauge fixed and Dirac observables that correspond to making measurements on the surface defined by $T(x) = \tau$, given that the clocks are synchronized by setting $T(x) = 0$ on the initial surface. In this section I would like to discuss what effects the singularities of classical general relativity have on these observables[16].

Let us begin with a simple point, which is the following: *Given that the matter fields, including the time fields, satisfy the positive energy condition required by the singularity theorems, for any $\tau_0$ there are regions, $\mathcal{R}(\tau_0)$ of the phase space $\Gamma = \{A, E, \phi, ...\}$ such that the future evolution of any data in this region becomes singular before the physical time $\tau_0$ (defined by the gauge conditions (19) and (21)). This means that the evolutions of the data in $\mathcal{R}(\tau_0)$ do not have complete $T(x) = \tau$ surfaces.* This happens because, roughly speaking, some of the clocks that define the $T(x) =$ constant surfaces have encountered spacetime singularities on surfaces with $T(x) < \tau_0$.

Now, let us consider what I will call "quasi-local" observables $\mathcal{O}(\tau)_{phys}$, such as $A^I(\tau)_{phys}$ and $T^I(\tau)_{phys}$, which are associated with more or less local regions of the initial data surface. We may expect that for such observables the following will be the case: For each such $\mathcal{O}(\tau)_{phys}$ and for each $\tau_0$ there will be regions on the phase space $\Gamma$ on which $\mathcal{O}(\tau_0)_{phys}$ is not defined because, for data in that region, the local region measured by that observable (picked out by the spatial reference frame) encountered a curvature singularity at some time $\tau_{sing} < \tau_0$.

Thus, it is clear that the existence of spacetime singularities does limit the operational notion of time tied to a field of clocks. The limitation is that, if we define the evolution of physical quantities in terms of the physical observables $\mathcal{O}(\tau)_{phys}$, only the $\tau = 0$ observables that measure the properties of the initial data surface can be said to give good coordinates for the full space of solutions. If we want a complete description of the full space of solutions (defined as the evolutions of non-singular initial data) we cannot get complete information from the evolving constants of motion for any nonzero $\tau$. What we cannot get is complete information about those solutions for which by $\tau$ some of the clocks have already fallen into singularities.

As far as the classical theory is concerned, this limitation is necessary and entirely unproblematic. The observables become ill defined because we cannot ask any question about what is seen by observers after they have ceased to exist. As general relativity is a local theory, if we choose our observables appropriately, we can still have complete information about all measurements that can be made at a time $\tau$ by local observers who have not yet fallen into a singularity.

---

[16] As in section 4, I will in this section use the notation *phys* to refer either to the gauge fixed or the Dirac formalisms.

I now turn to a consideration of the implications of this for the quantum theory.

## 6.2. Singularities in quantum observables

We have seen how the existence of singularities in classical theory is expressed in terms of the classical observables $\mathcal{O}(\tau)_{phys}$. There are now two question that must be asked: First, in principle can singularities show up in the same way in the physical operators of a quantum cosmological theory? Second, do they actually occur in the physical operator algebra of a realistic theory of quantum gravity coupled to matter fields?

The answer to the first question is yes, as has been shown in two model quantum cosmologies. These are a finite dimensional example, the Bianchi I quantum cosmology[ATU] [17] and an infinite dimensional field theoretic example, the one polarization Gowdy quantum cosmology[VH87]. These are both exactly solvable systems; the first has a physical Hilbert space isomorphic to the state space of a free relativistic particle in $2+1$ dimensional Minkowski spacetime while the Hilbert space of the second is isomorphic to that of a free scalar field theory in $1+1$ dimensions. However, in spite of the existence of these isomorphisms to manifestly non-singular physical systems, they are each singular theories when considered in terms of the operators that represent observables of the corresponding cosmological models. In both cases there is a global notion of time, which is the volume of the universe, $V$ in a homogeneous slicing. One can then construct a physical observable, called $C^2(V)$ which is defined as

$$C^2(V) \equiv \int_{\Sigma(V)} \sqrt{q} g^{\mu\nu} g^{\alpha\beta} C^\rho_{\sigma\mu\alpha} C^\sigma_{\rho\nu\beta} \tag{71}$$

where $\Sigma(V)$ is the three surface defined in the slicing by the condition that the spatial volume is $V$ and $C^\sigma_{\rho\nu\beta}$ is the Weyl curvature. There is also in each case, a $V$-time dependent Hamiltonian that governs the evolution of operators such as $C^2(V)$ through Heisenberg equations of motion.

Now, it is well known that in each of these models the cosmological singularity of the classical theory shows up in the fact that $\lim_{V\to 0} C^2(V)$ is infinite. What is, perhaps, surprising is that the quantum theory is equally singular, in that $\lim_{V\to 0} \hat{C}^2(V)$ diverges. The exact meaning of this is slightly different in the finite dimensional and the quantum field theoretic examples. In the Bianchi I case, it has been shown by Ashtekar, Tate and Uggla[ATU] that for any two normalizable states $|\Psi>$ and $|\Phi>$ in the physical Hilbert space

$$\lim_{V\to 0} < \Psi|\hat{C}^2(V)|\Phi >_{physical} = \infty \tag{72}$$

---

[17]Similar phenomena occur for other Bianchi models[ATU].

In the one polarization Gowdy model, Husain [VH87] has shown that (given a physically reasonable ordering for $\hat{C}^2(V)$) there is a unique state $|0>$ such that (72) holds for all $|\Psi>$ and $|\Phi>$ which are not equal to $|0>$. Furthermore, for all $V$, $<0|\hat{C}^2(V)|0>=0$ so the state $|0>$ represents the vacuum in which no degrees of freedom of the gravitational degrees of freedom are ever excited. As the Gowdy cosmology contains only gravitational radiation, this corresponds to the one point of the classical phase space which is just flat spacetime. For any other states, the quantum cosmology is singular in the sense that the matrix elements of the Weyl curvature squared diverge at the same physical time that the classical singularities occur.

Thus, we see from these examples that there is no principle that prevents spacetime singularities from showing up in quantum cosmological models and that they manifest themselves in the physical quantum operator algebra in the same way they do in the classical observable algebra. It is therefore a dynamical question, rather than a question of principle, whether or not singularities can occur in the full quantum theory. While it is, of course, possible that the singularities are eliminated in every consistent quantum theory of gravity, I think it must be admitted that at present there is little evidence that this is the case. The evidence presently available about the elimination of singularities is the following: a) Cosmological singularities are not eliminated in the semiclassical approximation of quantum cosmology [HH]. b) There are exact solutions of string theory which are singular [GH92]. c) In $1+1$ models of quantum gravity, singularities are sometimes, but not always, eliminated, at the semiclassical level[EW91, GH92, HS92, JP92].

Furthermore, there does not seem to be any reason why quantum cosmology requires the removal of the singularities. Both the mathematical structure and the physical interpretation of the quantum theory are, just like those of the classical theory, robust enough to survive the occurrence of singularities.

Given this situation, it is perhaps reasonable to ask whether and how singularities may appear in full quantum gravity. As a step towards answering this question, we may postulate the following conjecture:

**Quantum singularity conjecture** *There exist, in the Hilbert space $\mathcal{H}_{phys}$ of quantum gravity, normalizable states $|\Psi>$ such that:*

a) *the expectation values $<\Psi|\hat{O}_I(0)_{phys}|\Psi>_{phys}$ are finite for all $I$, so that at the initial time $\tau=0$, all physical observables are finite.*

b) *There is a subset of the physical operators, $\hat{O}_I(\tau)_{phys}$, whose expectation values in the state $|\Psi>$ develops singularities under evolution in the physical time $\tau$. That is, for each such $|\Psi>$ and for each $\hat{O}_I(\tau)_{phys}$ in this subset there is a finite time $\tau_{sing}$*

*such that*

$$\lim_{\tau \to \tau_{sing}} < \Psi | \hat{\mathcal{O}}^I(\tau)_{phys} | \Psi >_{phys} = \infty \qquad (73)$$

This means that if we want to predict what observers in the universe described by the state $|\Psi >$ will see at some time $\tau_0$, they may measure only the $\hat{\mathcal{O}}_I(\tau_0)_{phys}$ which do not go singular in this sense by the time $\tau_0$. This means that they may be able to recover less information about the state $|\Psi >$ by making measurements at that time than was available to observers at the time $\tau = 0$. Thus, the occurrence of singularities in the solutions to the operator evolution equations (43) (or (51)) means that real loss of information happens in the full quantum theory.

The possibility of this happening can be captured by a conjecture that I will call the *quantum cosmic censorship conjecture*. The name is motivated by analogy to the classical conjecture: if there is censorship then there is missing information. This is the content of the following:

**Quantum cosmic censorship conjecture:** a) *There exists states $|\Psi >$ in $\mathcal{H}_{phys}$ which are singular for at least one observable $\mathcal{O}_{sing}(\tau)_{phys}$ at some time $\tau_{sing}$, but for which there are a countably infinite number of other observables $\mathcal{O}'_I(\tau)$ such that $< \Psi | \mathcal{O}_I(\tau)_{phys} | \Psi >$ are well defined and are finite for some open interval of times $\tau > \tau_{sing}$.*

b) *Let $|\Psi >$ be a state which satisfies these conditions. Then for every $\tau$ in this open interval there is a proper density matrix $\rho_\Psi(\tau)$ such that, for every $\hat{\mathcal{O}}_I(\tau)$, for which $< \Psi | \hat{\mathcal{O}}_I(\tau) | \Psi >$ is finite, then*

$$< \Psi | \hat{\mathcal{O}}_I(\tau) | \Psi >_{phys} = Tr \rho_\Psi(\tau) \hat{\mathcal{O}}_I(\tau). \qquad (74)$$

Here the trace is to be defined with respect to the physical inner product a proper density matrix is one that corresponds to no pure state.

The first part of the conjecture means that there are states which describe what we might want to call black holes in the sense that while some observables become singular at some time $\tau_{sing}$ there are other observables which remain nonsingular for later times. The second part means that there exists a density matrix that contains all the information about the quantum state that is relevant for physical times $\tau$ after the time of the first occurrence of singularity. Because a density matrix contains all the information that could be gotten by measuring the pure state, we may say that loss of information has occurred.

Finally, we may note that for any state $|\Psi >$ there may be a finite time $\tau_{final}$ such that, for every observable $\mathcal{O}_I(\tau)_{phys}$, $< \Psi | \mathcal{O}_I(\tau)_{phys} | \Psi >_{phys}$ is undefined, divergent or zero for every $\tau > \tau_{final}$. This would correspond to a quantum mechanical version of a final singularity.

Suppose these conjectures can be proven for quantum general relativity, or some other quantum theory of gravity. Would this mean that the theory would be inadequate for a description of nature? While someone may want to argue that it may be preferable to have a quantum theory of gravity without singularities, I do not think an argument can be made that such a theory must be either incomplete, inconsistent or in disagreement with anything we know about nature. What self-consistency and consistency with observation require of a quantum theory of cosmology is much less. The following may be taken to be a statement of the minimum that we may require of a quantum theory of cosmology:

**Postulate of adequacy**. A quantum theory of cosmology, constructed within the framework described in this paper, may be called *adequate* if,

a) *For every $\tau > 0$ there exists a physical state $|\Psi > \in \mathcal{H}_{phys}$ and a countable set of operators $\hat{O}_I(\tau)$ such that the $< \Psi|\hat{O}_I(\tau)_{phys}|\Psi >_{phys}$ are finite.*

b) *The theory has a flat limit, which is quantum field theory on Minkowski spacetime. This means that there exists physical states and operators whose expectation values are equal to those of quantum field theory on Minkowski spacetime for large regions of space and time, up to errors which are small in Planck units.*

c) *The theory has a classical limit, which is general relativity coupled to some matter fields. This means that there exists physical states and operators whose physical expectation values are equal to the values of the corresponding classical observables evaluated in a classical solution to general relativity, up to terms that are small in Planck units.*

If a theory satisfies these conditions, we would have a great deal of trouble saying it was not a satisfactory quantum theory of gravity. Thus, just like in the classical theory, the presence of singularities and loss of information cannot in principle prevent a quantum theory of cosmology from providing a meaningful and adequate description of nature. Whether there is an adequate quantum theory of cosmology that eliminates singularities and preserves information is a dynamical question, and whether that theory, rather than another adequate theory for which the quantum singularity and quantum cosmic censorship conjectures hold, is the correct description of nature is, in the end, an empirical question.

## 7.  Conclusions

The purpose of this paper has been to explore the implications of taking a completely pragmatic approach to the problem of time in quantum cosmology. The main conclusion of the developments described here is that such an approach may be possible at the nonperturbative level. This may allow the theory to address problems such as the effect of quantum effects on singularities which most likely require a nonperturbative treatment, while remaining within the framework of a coherent interpretation.

I would like to close this paper by discussing two questions. First, are there ways in which the ideas described here may be tested? Second, is it possible that there is a more fundamental solution to the problem of time in quantum gravity which avoids the obvious limitations of this pragmatic approach?

## 7.1. Suggestions for future work

The proposals and conjectures described here are only meaningful to the extent that they can be realized in the context of a full quantization of general relativity or some other quantum theory of gravity. In order to do this, the key technical problem that must be resolved is, as we saw in section 3, the construction of the operator $\mathcal{W}$. Given that we know rather a lot about both the kernel and the action of the Hamiltonian constraint, I believe that this is a solvable problem.

Beyond this, it would be very interesting to test these ideas and conjectures in the context of certain model systems. Among those that could be interesting are 1) The Bianchi IX model 2) The full two polarization Gowdy models 3) Models of spherically symmetric general relativity coupled to matter[18] 4) Other $1 + 1$ dimensional models of quantum gravity coupled to matter such as the dilaton theories that have recently received some attention [CGHS, GH92, HS92, B92, ST92, SWH92]. 5) the chiral $G \rightarrow 0$ limit of the theory[LS92d] and finally, 6) $2 + 1$ general relativity coupled to matter[EW88, AHRSS]. Each of these are systems that have not yet been solved, and in which the difficulty of finding the physical observables has, as in the full theory, blocked progress.

## 7.2. Is there an alternative framework for quantum cosmology not based on such an operational notion of time?

We are now, if the above is correct, faced with the following situation. Taking an operational approach to the meaning of time we have been able to provide a complete physical interpretation for quantum cosmology that reduces to quantum field theory in a suitable limit. However, the quantum field theory that is reproduced may turn out to be unitary only in the approximation in which we can neglect the possibility that some of the clocks that define operationally surfaces of simultaneity become engulfed in black hole singularities. This need not be disturbing; it says that we cannot count on probability conservation in time if the notion of time we are thinking of is based

---

[18]For the complete quantum theory without matter, see [TT].

on the existence of a certain field of dynamical clocks and there is finite probability that these clocks themselves cease to exist. However, if there is no other notion of time with respect to which probability conservation can be maintained, so that this is the best that can be done, it is still a bit disturbing.

We seem at this juncture to have two choices. It may indeed be that we cannot do better than this, so that we must accept that the Hilbert space structure that forms the basis of our interpretation of quantum cosmology is tied to the existence of certain physical frames of reference and that, as the existence of the conditions that define these frames of reference is contingent, we cannot ascribe any further meaning to unitarity. If this is the case then we have to accept a further "relativization" of the laws of physics, in which different Hilbert structures, with different inner products, are associated to observations made by different observers. This means, roughly, that not only are the actualities (to use a distinction advocated by Shimony[Sh]) in quantum mechanics dependent on the physical conditions of the observer, so are the potentialities. This point of view has been advocated by Finkelstein[F] and developed mathematically in a very interesting recent paper of Crane[LC92].

On the other hand, it is possible to imagine that there is some meaning to what the possibilities are for actualization that is independent of the conditions of the observer. If such a level of the theory existed, it could be used to deduce the relationships between what could be, and what is, seen by different observers in the same universe.

Barbour has recently made a proposal about the role of time in quantum cosmology, which I think can be understood along these lines[JBB92]. I would now like to sketch it, as it may serve as a prototype for all such proposals in which the probability interpretation of quantum cosmology is not relativized so as to make the inner produce dependent on the conditions of the observer.

Barbour's proposal is at once a new point of view about time and a new proposal for an interpretation of quantum cosmology. He posits that time actually does not exist, so that our impressions of the existence and passage of time are illusions caused by certain properties of the classical limit of quantum cosmology. More precisely, he proposes that an interpretation can be given entirely in the context of the diffeomorphism invariant theory.

Barbour's fundamental postulate, to which he gives the colorful name of the "heap hypothesis", is as follows: *The world consists of a timeless real ensemble of configurations, called "the heap".* The probability for any given spatially diffeomorphism invariant property to occur in "the heap" is governed by a quantum state, $|\Psi >$ which is assumed to satisfy all the constraints of quantum gravity. This probability is considered to be an actual ensemble average. Given a particular diffeomorphism invariant observable $\hat{\mathcal{O}}_I$, the ensemble expectation value in the heap is given by

$$< \Psi|\hat{\mathcal{O}}_I|\Psi >_{complete \; diffeo} . \tag{75}$$

Here the inner product is required to be the spatially diffeomorphism invariant inner product for the whole system, including any clocks that may be around. Thus, this proposal is different than the one made in section 3 in which the inner product proposed in (46) is the spatially diffeomorphism invariant inner product for a specifically reduced system in which the clock has been removed.

There are several comments that must be made about this proposal. First, this is the complete statement of the interpretation of the theory. Quantum cosmology is understood as giving a statistical description of a real ensemble of configurations, or moments. There is no time. The fact that we have an impression of time's passage is entirely to be explained by certain properties of the quantum state of the universe. In particular, Barbour wants to claim that our experience of each complete moment is, so to speak, a world unto itself. It is only because we have memories that we have an impression in this moment that there have been previous moments. It is only because the quantum state of the universe is close to a semiclassical state in which the laws of classical physics approximately hold that the world we experience at this moment gives us the strong impression of causal connections to the other moments.

Second, the probabilities given by (75) are not quantities that are necessarily or directly accessible to observers like ourselves who live inside the universe. Only an observer who is somehow able to look at the whole ensemble is able to directly measure the probabilities given by (75). Of course, we are not in that situation. Barbour must then explain how the probabilities for observations that we make are related to the probability distribution for elements of the heap to have different properties. In order to do this, the key thing that he must do is show how the probabilities defined by the heap ensemble (75) are related to what we measure.

One way in which this may happen is that if one considers only states in which some variables corresponding to a particular clock are semiclassical, then Barbour's inner product (75) may reduce to the inner product defined by (46) in which the clock degrees of freedom have been removed. From Barbour's point of view, the inner product proposed in this paper could only be an approximation to the true ensemble probabilities (75) that holds in the case that the quantum state of the universe is semiclassical in the degrees of freedom of a particular physical clock.

This discussion of Barbour's proposal brings us back to the choice I mentioned above and points up what I think is a paradox that must confront any quantum theory of cosmology. The two possibilities we must, it seems, choose between can be described as follows: a) The inner product and the resulting probabilities are tied in an operational sense to what may be seen by a physical observer inside the universe. In this case, as we have seen here, the notions of unitarity and conservation of probability can only be as good as the clocks carried by a particular observer may be reliable. We are then in danger of the kind of relativization of the interpretation mentioned above. b) The basic statement of the interpretation refers neither to a particular set of observers nor to time as measured by clocks that they carry. In this case the relativization can be avoided, but at the cost that the fundamental quantities of the theory do not in general refer to any observations made by observers living in the universe. In this case the probabilities seen by any observers living in the universe can only approximate the true probabilities for particular semiclassical configurations. Furthermore, if there is a finite probability that any physical clock may in its future encounter a spacetime singularity, however small, then it is difficult to see how a breakdown of unitarity evolution can be avoided, if by evolution we mean anything tied to the readings of a physical clock.

This situation brings us back to the problem of what happens to the information inside of an evaporating black hole. I think that the minimum that can be deduced from the considerations of section 6 is that, at the nonperturbative level, this question cannot be resolved without resolving the dynamical question of what happens to the singularities inside classical black holes. As pointed out a long time ago by Wheeler[JAW], this problem challenges all of our ideas about short distance physics and its relation to cosmology[19]. Unfortunately, it must be admitted that the quantum theory of gravity still has little to say about this problem. What I hope to have shown here is that there may be a language which allows the problem to be addressed by a nonperturbative formulation of the quantum theory. Whether it can be answered by such a formulation remains a problem for the future.

## ACKNOWLEDGEMENTS

This work had its origins in my attempts over the last several years to understand and resolve issues that arose in collaborations and discussions with Abhay Ashtekar, Julian Barbour, Louis Crane, Ted Jacobson and Carlo Rovelli. I am grateful to them for continual stimulation, criticism and company on this long road to quantum gravity. I am in addition indebted to Carlo Rovelli for pointing out an error in a previous version of this paper. I am also very grateful to a number of other people who have provided important stimulus or criticisms of these ideas, including Berndt Bruegmann, John Dell, David Finkelstein, James Hartle, Chris Isham, Alejandra Kandus, Karel Kuchar, Don Marolf, Roger Penrose, Jorge Pullin, Rafael Sorkin, Rajneet Tate and John Wheeler. This work was supported by the National Science Foundation under grants PHY90-16733 and INT88-15209 and by research funds provided by Syracuse University.

## References

[AAn93] A. Anderson *Thawing the frozen formalism: the difference between observables and what we observe* in this volume.

---

[19]One speculative proposal about this is in [LS92a].

[YaA80] Ya. Aref'feva, Theor. and Math. Phys. 43 (1980) 353 (Teor.i Mat. Fiz. 43 (1980) 111.

[AA86] A. V. Ashtekar, Physical Review Letters 57, 2244–2247 (1986) ; Phys. Rev. D 36 (1987) 1587.

[AA91] A. Ashtekar , *Non-perturbative canonical gravity.* Lecture notes prepared in collaboration with Ranjeet S. Tate. (World Scientific Books, Singapore,1991).

[AA92] A. Ashtekar , in the Procedings of the 1992 Les Houches lectures.

[AHRSS] A. Ashtekar, V. Husain, C. Rovelli, J. Samuel and L. Smolin *2+1 quantum gravity as a toy model for the 3+1 theory* Class. and Quantum Grav. L185-L193 (1989)

[AI92] A. Ashtekar and C. J. Isham, *Inequivalent observer algebras: A new ambiguity in field quantization* Phys. Lett. B 274 (1992) 393-398;*Representations of the holonomy algegbra of gravity and non-abelian gauge theories,* Class. and Quant. Grav. 9 (1992) 1433-67.

[AR91] A. Ashtekar and C. Rovelli, *Quantum Faraday lines: Loop representation of the Maxwell theory,* Class. Quan. Grav. 9 (1992) 1121-1150.

[ASR91] A. Ashtekar, C. Rovelli and L. Smolin, *Gravitons and Loops,* Phys. Rev. D 44 (1991) 1740-1755; J.Iwasaki and C. Rovelli, *Gravitons as embroidery on the weave,* Pittsburgh and Trento preprint (1992); J. Zegwaard, *Gravitons in loop quantum gravity,* Nucl. Phys. B378 (1992) 288-308.

[ARS92] A. Ashtekar, C. Rovelli and L. Smolin, Physical Review Letters 69 (1992) 237-240.

[AS91] A. Ashtekar and J. Stachel, ed. *Conceptual Problems in Quantum Gravity,* (Birkhauser,Boston,1991).

[ATU] A. Ashtekar, R. Tate and C. Uggla, —it Minisuperspaces: observables and quantization Syracuse preprint, 1992.

[JB92] J. Baez, University of California, Riverside preprint (1992), to appear in Class. and Quant. Grav.

[B92] T. Banks, A Dabholkar, M. R. Douglas and M. O'Loughlin, Phys. Rev. D45 (1992) 3607; T. Banks and M. O'Loughlin, Rutgers preprints RU-92-14, RU-92-61 (1992); T. Banks, A. Strominger and M. O'Loughlin, Rutgers preprints RU-92-40 hepth/9211030.

[JBB92] J. B. Barbour,to appear in the *Procedings of the NATO Meeting on the Physical Origins of Time Asymmetry* eds. J. J. Halliwell, J. Perez-Mercader and W. H. Zurek (Cambridge University Press, Cambridge, 1992); *On the origin of structure in the universe* in *Proc. of the Third Workshop on Physical and Philosophical Aspects of our Understanding of Space and Time* ed. I. O. Stamatescu (Klett Cotta); *Time and the interpretation of quantum gravity* Syracuse University Preprint, April 1992.

[Bl]  M. Blencowe, Nuclear Physics B 341 (1990) 213.

[B]  N. Bohr, *Quantum theory and the description of nature* (Cambridge University Press, 1934); *Atomic Physics and Human Knowledge* (Wiley,New York, 1958); Phys. Rev. 48 (1935) 696-702.

[BP92] B. Bruegmann and J. Pullin, *On the Constraints of Quantum Gravity in the Loop Representation* Syracuse preprint (1992), to appear in Nuclear Physics B.

[DB92] D. Brill, Phys. Rev. D46 (1992) 1560.

[CGHS] C. G. Callen, S. B. Giddings, J. A. Harvey, and A. Strominger, Phys. Rev. D45 (1992) R1005.

[SC92] S. Carlip, Phys. Rev. D 42 (1990) 2647; D 45 (1992) 3584; UC Davis preprint UCD-92-23.

[LC92] L. Crane, *Categorical Physics*, Kansas State Univesity preprint, (1992).

[BD62] B. S. DeWitt, in *Gravitation, An Introduction to Current Research* ed. L. Witten (Wiley, New York,1962).

[AE]  A. Einstein, in *Relativity, the special and the general theory* (Dover,New York).

[HE57] H. Everett III, Rev. Mod. Phys. 29 (1957) 454; in B.S. DeWitt and N. Graham, editors *The Many Worlds Interpretation of Quantum Mechanics*(Princeton University Press,1973); J.A. Wheeler, Rev. Mod. Phys. 29 (1957) 463; R. Geroch, Nous 18 (1984) 617.

[F]  D. Finkelstein, *Q*, to appear.

[GT81] R. Gambini and A. Trias, Phys. Rev. D23 (1981) 553, Lett. al Nuovo Cimento 38 (1983) 497; Phys. Rev. Lett. 53 (1984) 2359; Nucl. Phys. B278 (1986) 436; R. Gambini, L. Leal and A. Trias, Phys. Rev. D39 (1989) 3127.

[RG86] R. Gambini, Phys. Lett. B 171 (1986) 251; P. J. Arias, C. Di Bartolo, X. Fustero, R. Gambini and A. Trias, Int. J. Mod. Phys. A 7 (1991) 737.

[RG90] R. Gambini, *Loop space representation of quantum general relativity and the group of loops*, preprint University of Montevideo 1990, Physics Letters B 255 (1991) 180. R. Gambini and L. Leal, *Loop space coordinates, linear representations of the diffeomorphism group and knot invariants* preprint, University of Montevideo, 1991; M. Blencowe, Nuclear Physics B 341 (1990) 213.; B. Bruegmann, R. Gambini and J. Pullin, Phys. Rev. Lett. 68 (1992) 431-434; *Knot invariants as nondegenerate staes of four dimensional quantum gravity* Syracuse University Preprint (1991), to appear in the proceedings of the XXth International Conference on Differential Geometric Methods in Physics, ed. by S. Catto and A. Rocha (World Scientific,Singapore,in press).

[GMH] M. Gell-Mann and J. Hartle, *Alternative decohering histories in quantum mechanics* in K. Phua and Y. Yamaguchi, eds. "Proceedings of the 25th International Conference on High Energy Physics, Singapore 1990", (World Scientific,Singapore,1990); *Quantum mechanics in the light of quantum cosmology* in S. Kobayashi, H. Ezawa, Y. Murayama and S. Nomura eds. "Proceedings of the Third International Symposium on the Foundations of Quantum Mechanics in the Light of New Technology", Physical Society of Japan, Tokyo, pp. 321-343, and in W. Zurek, ed. "Complexity, Entropy and the Physics of Information, SFI Studies in the Science of Complexity, Vol VIII, Addison-Wesley, Reading pp. 425-458; J. Hartle, Phys. Rev. D38 (1988) 2985-2999; *The quantum mechanics of cosmology* in S. Coleman, J. Hartle, T. Piran and S. Weinberg, eds. "Quantum Cosmology and Baby Universes", (World Scientific, Singapore, 1991).

[HPMZ] J. J. Halliwell, J. Perez-Mercader and W. H. Zurek, eds. *Procedings of the NATO Meeting on the Physical Origins of Time Asymmetry* (Cambridge University Press, Cambridge, 1992) and references contained theirin.

[HH] J. Hartle and B.-l. Hu, Phys. Rev. D 20 (1979) 1757; D 21 (1980) 2756.

[HS92] J. Harvey and A. Strominger, *Quantum aspects of black holes* hep-th/9209055, to appear in the proceedings of the 1992 Trieste Spring School on String Theory and Quantum Gravity.

[SWH75] S. W. Hawking, Commun. Math. Phys. 43 (1975) 199; Phys. Rev. D13 (1976) 191; D14 (1976) 2460.

[SWH92] S. W. Hawking, Phys. Rev. Lett. 69 (1992) 406.

['tH85] G. 't Hooft, Nucl. Phys. B256 (1985) 727; B225 (1990) 138.

[GH92] G. T. Horowitz *The dark side of string theory: black holes and black strings* Santa Barbara preprint UCSBTH-92-32, hep-th/9210119, to appear in the proceedings of the 1992 Trieste Spring School on String Theory and Quantum Gravity; G. Horowitz and A. Steif, Phys. Rev. Lett. 64 (1990) 260; Phys. Rev. D42 (1990) 1950.

[VH87]  V. Husain, Class. Quant. Grav. 4 (1987) 1587.

[VH93]  V. Husain, U. of Alberta preprint (1993).

[I92]  C. J. Isham, *Canonical quantum gravity and the problem of time*, Imperial College preprint Inperial/TP/91-92/25, to appear in the procedings of the NATO advanced study institute "Recent Problems in Mathematical Physics", Salamanca (1992).

[IR93]  J.Iwasaki and C. Rovelli, Pittsburgh and Trento preprint in preparation (1993).

[JS88]  T. Jacobson and L. Smolin, Nucl. Phys. B 299 (1988).

[KK92]  K. Kuchar, *Time and interpretations of quantum gravity* in the *Proceedings of the 4th Canadian Conference on General Relativity and Relativistic Astrophysics*, eds. G. Kunstatter, D. Vincent and J. Williams (World Scientific, Singapore,1992).

[KT91]  K. Kuchar and C. Torre, Phys. Rev. D43 (1991) 419-441; D44 (1991) 3116-3123.

[MS93]  M. Miller and L. Smolin *A new discretization of classical and quantum gravity* Syracuse preprint 1993.

[RP]  R. Penrose, *The Emperor's New Mind* (Oxford University Press,Oxford).

[JP92]  J. Preskill, *Do black holes destroy information?* Cal Tech preprint CALT-68-1819, hep-th/9209058, to appear in the proceedings of the International Symposium on Black Holes, Membranes, Wormholes and Superstrings, The Woodlands, Texas, January 1992.

[CR91]  C. Rovelli, Classical and Quantum Gravity, 8 (1991) 1613-1676.

[CRT91]  C. Rovelli, Phys. Rev. D 42 (1991) 2638; 43 (1991) 442; in *Conceptual Problems of Quantum Gravity* ed. A. Ashtekar and J. Stachel, (Birkhauser,Boston,1991).

[CRM91]  C. Rovelli, Class. and Quant. Grav. 8 (1991) 297,317.

[CR93]  C. Rovelli, *A generally covariant quantum field theory* Trento and Pittsburgh preprint (1992).

[RS88]  C. Rovelli and L. Smolin, *Knot theory and quantum gravity* Phys. Rev. Lett. **61**, 1155 (1988); *Loop representation for quantum General Relativity*, Nucl. Phys. B133 (1990) 80.

[Sh] A. Shimony, in *Quantum concepts in space and time* ed. R. Penrose and C. J. Isham (Clarendon Press, Oxford, 1986).

[LS84] L. Smolin, *On quantum gravity and the many worlds interpretation of quantum mechanics* in *Quantum theory of gravity* (the DeWitt Feschrift) ed. Steven Christensen (Adam Hilger, Bristol, 1984).

[LS91] L. Smolin *Space and time in the quantum universe* in the proceedings of the Osgood Hill conference on *Conceptual Problems in Quantum Gravity* ed. A. Ashtekar and J. Stachel, (Birkhauser,Boston,1991).

[LS91] L. Smolin, *Recent developments in nonperturbative quantum gravity* in the Proceedings of the 1991 GIFT International Seminar on Theoretical Physics: *Quantum Gravity and Cosmology*, held in Saint Feliu de Guixols, Catalonia, Spain (World Scientific, Singapore,in press).

[LS92a] L. Smolin, *Did the Universe evolve?* Classical and Quantum Gravity 9 (1992) 173-191.

[LS92b] *What can we learn from the study of non-perturbative quantum general relativity?* to appear in the procedings of GR13, IOP publishers, and in the procedings of the Lou Witten Feshchrift, World Scientific (1993).

[LS92c] L. Smolin, *Time, structure and evolution* Syracuse Preprint (1992); to appear (in Italian translation) in the Procedings of a conference on *Time in Science and Philosophy* held at the Istituto Suor Orsola Benincasa, in Naples (1992) ed. E. Agazzi. Syracuse preprint, October 1992.

[LS92d] L. Smolin, *The $G_{Newton} \to 0$ limit of Euclidean quantum gravity*, Classical and Quantum Gravity, 9 (1992) 883-893.

[LS93] L. Smolin *Diffeomorphism invariant observables from coupling gravity to a dynamical theory of surfaces* Syracuse preprint, January (1993).

[ST92] L. Susskind and L. Thorlacius, Nucl. Phys. B382 (1992) 123; J. Russo, L. Susskind and L. Thorlacius, Phys. Lett. B292 (1992) 13; A. Peet, L. Susskind and L. Thorlacius, Stanford preprint SU-ITP-92-16.

[RT92] R. S. Tate, *Constrained systems and quantization, Lectures at the Advanced Institute for Gravitation Theory*, December 1991, Cochin University, Syracuse Univesity Preprint SU-GP-92/1-4; *An algebraic approach to the quantization of constrained systems: finite dimensional examples* Ph.D. Dissertation (1992) Syracuse University preprint SU-GP-92/8-1.

[TT] T. Thiemann and H. Kastrup, Nucl. Phys. B (to appear).

[vN] J. von Neumann *Mathematical Foundations of Quantum Mechanics* (Princeton University Press, 1955).

[JAW] J. A. Wheeler, *Geometrodynamics and the issue of the final state*, in DeWitt and DeWitt, eds. *Relativity, Groups and Topology* (Gordon and Breach, New York, 1964); *Superspace and the nature of quantum geometrodynamics*, in De-Witt and Wheeler, eds. ;*Battelle Recoontres: 1967 Lectures in Mathematics and Physics* (W. A. Benjamin,New York,1968); in H. Woolf *Some strangeness in the proportion: A centanary symposium to celebrate the achievement of Albert Einstein* (Addison -Wesley,Reading MA, 1980).

[EW] E. Wigner, *Symmetries and Reflections* (Indiana University Press, 1967); Am. J. Phys. 31 (1963) 6-13.

[EW88] E. Witten, Nucl. Physics B311 (1988) 46.

[EW91] E. Witten, Phys. Rev. D44 (1991) 314.

# Impossible Measurements on Quantum Fields

## RAFAEL D. SORKIN

Department of Physics, Syracuse University, Syracuse NY 13244-1130

*Abstract*

It is shown that the attempt to extend the notion of ideal measurement to quantum field theory leads to a conflict with locality, because (for most observables) the state vector reduction associated with an ideal measurement acts to transmit information faster than light. Two examples of such information-transfer are given, first in the quantum mechanics of a pair of coupled subsystems, and then for the free scalar field in flat spacetime. It is argued that this problem leaves the Hilbert space formulation of quantum field theory with no definite measurement theory, removing whatever advantages it may have seemed to possess vis a vis the sum-over-histories approach, and reinforcing the view that a sum-over-histories framework is the most promising one for quantum gravity.

## 1. INTRODUCTION: IDEAL MEASUREMENTS AND QUANTUM FIELD THEORY

Whatever may be its philosophical limitations, the textbook interpretation of non-relativistic quantum mechanics is probably adequate to provide the quantum formalism with all the predictive power required for laboratory applications. It is also self-consistent in the sense that there exist idealized models of measurements which allow the system-observer boundary to be displaced arbitrarily far in the direction of the observer. And the associated "transformation theory" possesses a certain formal beauty, seemingly realizing the "complementarity principle" in terms of the unitary equivalence of all orthonormal bases. It is therefore natural to try to generalize this semantic framework to relativistic quantum field theory in the hope of learning something new, either from the success or failure of the attempt.

In fact we will see that the attempt fails in a certain sense; and the way it fails suggests that the familiar apparatus of states and observables must give way to a more spacetime-oriented framework, in which the new physical symmetry implied by the "transformation theory" is lost, and the role of measurement as a fundamental concept is transformed or eliminated. Although such a renunciation of the usual measurement formalism might seem a step backwards for quantum field theory, it can also be viewed as a promising development for quantum gravity. It means that some of the conceptual problems which are normally thought of as peculiar to quantum gravity or quantum cosmology are already present in flat space, where their analysis and resolution may be easier.

The framework we will attempt to generalize is based on the notion of an *ideal measurement*, by which I mean one for which (*i*) the possible outcomes are the eigenvalues of the corresponding operator, realized with probabilities given by the usual trace-rule, and (*ii*) the standard *projection postulate* correctly describes the effect of the measurement on the subsequent quantum state. Such a "minimally disturbing measurement" is not only a "detection" but is simultaneously a "preparation" as well and, if one follows the textbook interpretation, it is precisely from this dual character of measurement that the predictive power of the quantum mechanical formalism derives. It is also from this dual aspect that the difficulty in relativistic generalization will arise—a difficulty with so-called superluminal signaling.

Now in non-relativistic quantum mechanics, measurements are idealized as occurring at a single moment of time. Correspondingly the interpretive rules for quantum field theory are often stated in terms of ideal measurements which take place on Cauchy hypersurfaces. However, in the interests of dealing with well-defined operators, one usually thickens the hypersurface, and in fact the most general formulations of quantum field theory assume that there corresponds to any open region of spacetime an algebra of observables which—presumably—can be measured by procedures occurring entirely within that region. (Unlike for the non-relativistic case, however, no fully quantum models of such field measurements have been given, as far as I am aware.)

The statement that "one can measure" a single observable $A$ associated to a spacetime region $O$ is fine as far as it goes, and an obvious generalization of the projection postulate can be adopted as part of the definition of such a measurement. But a potential confusion arises as soon as we think of two or more separate measurements being made. In the non-relativistic theory, measurements carry a definite temporal order from which the logical sequence of the associated state-vector reductions is derived; but in Minkowski space, the temporal relationships among regions can be more complicated, and the rules for "collapsing" the state are not necessarily evident. Nonetheless, I claim that a natural set of rules exists, which directly generalizes the prescription of the non-relativistic theory.

## 2. A RELATIVISTIC PROJECTION POSTULATE

The problem is that these "obvious" rules fail to be consistent with established ideas of causality/locality. Hence, the kind of measurements they envisage can presumably not be accomplished. It would of course be very interesting to try to construct models within quantum field theory, to see what goes wrong, but I will not attempt such a von-Neumann-like analysis here. Also, I will restrict myself to flat-space for definiteness, although nothing would be changed by going over to an arbitrary globally hyperbolic spacetime. Finally it seems most convenient to work in the Heisenberg picture, since the association of field operators to regions of spacetime is most direct in that picture.

With these choices made, let us envisage a (finite) collection of ideal measurements to be performed on some quantum field $\Phi$. We are then faced with a collection of regions, $O_k$ in Minkowski space, and corresponding to each region, we are given an observable $A_k$, formed from the restriction of $\Phi$ to $O_k$. Given all this and an initial state $\rho_0$ specified to the past of all of the $O_k$, we may ask for the probability of obtaining any specified set of eigenvalues $\alpha_k$ of the $A_k$ as measurement outcomes.

Non-relativistically, we would determine these probabilities by ordering the $A_k$ in time (say with $A_1$ preceding $A_2$ preceding $A_3$ ...), then using $\rho_0$ to compute probabilities for the earliest observable $A_1$, then "reducing" $\rho$ conditioned on the eigenvalue $\alpha_1$, then using this reduced state to compute probabilities for $A_2$, etc. In the special case where each of the $A_k$ is a projection or "question" $E_k$, this procedure results, as is well known, in the remarkably simple expression

$$\langle E_1 E_2 \ldots E_{n-1} E_n E_{n-1} \ldots E_2 E_1 \rangle \tag{1}$$

for the probability of 'yes' answers to all the questions, where expectation in the initial state $\rho_0$ has been denoted simply by angle brackets, a practice I will adhere to henceforth. (That is, $\langle A \rangle \equiv \text{trace}(\rho_0 A)$.)

Now let us return to the relativistic situation. In the special case where the regions $O_k$ are non-intersecting Cauchy surfaces (or slight thickenings thereof), their time-ordering permits a unique labeling, and the generalization of the above non-relativistic procedure is immediate. For a more general set of regions we may try to foliate the spacetime in such a way that the $O_k$ *acquire* a well-defined temporal ordering (each $O_k$ being separated from its predecessor by one of the leaves of the foliation). Obviously not every labeling of the regions can arise from a foliation by Cauchy surfaces, since no region which comes later with respect to such a foliation can intersect the causal past of one which comes earlier. Indeed this restriction makes sense independent of any choice of foliation, and merely says that if a measurement made in one region can possibly influence the outcome of a measurement made in a second region, then the second measurement should be regarded as taking place "later" than the first.

The labelings of the regions which respect this causal restriction can be described systematically in terms of an order relation "$\prec$" reflecting the possibilities of causal influence among the regions. To define $\prec$ we merely specify that $O_j \prec O_k$ iff some point of $O_j$ causally precedes some point of $O_k$. A labeling — or equivalently a linear ordering — of the regions is then *compatible with* $\prec$ iff $O_j \prec O_k \Rightarrow j \leq k$. Of course, it can happen that no such labeling exists, in which case the rules described above admit of no natural generalization. To express the exclusion of such cases in a systematic manner, we may take the *transitive closure* of $\prec$, obtaining thereby an extended relation, for which I will use the same symbol $\prec$. The condition that compatible labelings exist is then that this extended $\prec$ be what is called a partial order, that is that it never happen that both $O_j \prec O_k$ and $O_k \prec O_j$ for some $j \neq k$. When, on the contrary, this does happen, we have the analog of two non-relativistic measurements being simultaneous, which, even non-relativistically, leads to no well-defined probabilities unless the corresponding operators happen to commute. We may exclude such cases, by simply requiring that the regions $O_k$ be disposed so that no such circularity occurs. In particular overlapping regions are thereby excluded by fiat.

Given the partial order $\prec$, it is easy to state the natural generalization of the non-relativistic rules for forming probabilities: We simply extend $\prec$ to a linear order (as can always be done) and use the rules precisely as they were stated earlier. Unless $\prec$ happens to be already a linear order, the particular choice of linear extension (or equivalently labeling) is not unique, but this ambiguity is harmless *as long as* the field $\Phi$ satisfies "local commutativity", i.e. as long as field observables belonging to spacelike separated regions commute. In particular, when the $A_k$ are all projection-operators, the formula (1) holds exactly as given above, for any labeling of the regions which is compatible with $\prec$.

(It is often objected that the idea of state-vector reduction cannot be Lorentz-invariant, since "collapse" will occur along different hypersurfaces in different rest-frames. However we have just seen that well-defined probability rules can be given without associating the successive collapses to any particular hypersurface. Thus the objection is unfounded to the extent that one regards the projection postulate as nothing more than a rule for computing probabilities. Of course if one takes the state-vector (or density operator?) itself to be physically real, then the puzzle about "where" it collapses might remain.)

Abstractly considered, the scheme we have just taken the trouble to construct seems impeccable, but in fact it has a problem foreshadowed by our need to take a transitive closure in defining $\prec$. The problem is that the state-vector reduction implied by an ideal measurement is non-local in such a way as to transmit observable effects faster than light, something like an EPR experiment gone haywire. If we want to reject such superluminal effects, then we will be forced to exclude the possibility of ideal measurements of most of the spacetime observables we have been contemplating.

## 3. THE CONTRADICTION WITH LOCALITY

In speaking of superluminal effects, the situation I have in mind concerns three regions $O_1$, $O_2$ and $O_3$, situated so that some points of the first precede points of the second, and some points of the second precede points of the third, but all points of the first and third are spacelike separated. In fact, let us specialize $O_2$ to be a thickened spacelike hyperplane, with $O_1$ and $O_3$ being bounded regions to its past and future respectively. The corresponding observables $A_1$, $A_2$, $A_3$, I will call '$A$', '$B$' and '$C$' in order to save writing of subscripts. The non-locality (or "acausality") in question then shows up in the fact that, for generic choices of the observables $A$, $B$, $C$ and of initial state $\rho_0$, *the results obtained by measuring $C$ depend on whether or not $A$ was measured*, even though $A$ is spacelike to $C$. By arranging beforehand that $B$ will certainly be measured, someone at $O_1$ could clearly use this dependence of $C$ on $A$ to transmit information "superluminally" to a friend at $O_3$.[1]

### A Simple Example with Coupled Systems

To see this effect at its simplest it may help to retreat from quantum field theory to a more elementary situation, namely a pair of coupled quantum systems together with three observables: $A$ belonging to the first subsystem, $C$ belonging to the second, and $B$ being a joint observable of the combined system. As in the field-theory case, the effect of measuring $B$ will be to make a prior intervention on the first subsystem felt by the second.

To simplify the analysis further we can make a change which is actually a generalization. Instead of considering specifically a measurement at $O_1$ [respectively, a measurement on the first subsystem] we consider an arbitrary intervention, implemented mathematically by a unitary operator $U$ formed (like $A$ itself) from the restriction of $\Phi$ to $O_1$ [respectively a unitary element of the observable algebra of the first sub-

---

[1] This effect is reminiscent of the acausality of [1], but in that case, information is transmitted directly into the past, rather than over spacelike separations, as here. Also, the acausality there depends on the assumption that one can directly "observe" spacetime properties of a history which need not correspond to any traditionally defined operator in Hilbert space. Here, in contrast, even the traditionally defined observables give trouble.

system]. That a measurement is effectively a special case of this kind of intervention follows from the observation that the effect of a measurement of the observable $A$ on the density-operator $\rho$ is to convert it into the $\lambda$-average of $\exp(-i\lambda A)\rho\exp(i\lambda A)$, as a one-line calculation will confirm. (Here, of course, I mean the effect of the measurement before the value of the result is taken into account.) Thus, measuring $A$ is equivalent to applying $U = \exp(-i\lambda A)$ with a random value of the parameter $\lambda$, and in this sense is a special kind of "unitary intervention". It follows in particular that if unitary intervention cannot transmit information, then neither can any measurement.

Incidentally, the subsuming of measurement under unitary intervention in this way leads to a more unified criterion for a theory to be "local": it is local if interventions confined to some region can affect only the future of that region. But it is interesting that, although it no longer mentions measurement directly, this criterion is still expressed in terms of intervention by an external agent. As far as I know there is no way to directly express the idea of locality in the context of a completely self-contained system.

A specific example of the non-local influence in question is now easy to obtain. Let the two subsystems be spin-1/2 objects, and let their initial state be $|dd\rangle$, where the two spin-states are $u$ and $d$. At time $t_1$ let us "kick" the first subsystem (or more politely "intervene") by exchanging $u$ with $d$ (we apply the unitary operator $\sigma_1$). At time $t_2$ we measure the operator $B$ of orthogonal projection onto the state $(|uu\rangle + |dd\rangle)/\sqrt{2}$; and at $t_3$ we measure an arbitrary observable $C$ of the second subsystem.

It is straightforward to work out the density-operator which governs this final measurement of $C$ by "tracing out" the first subsystem from the state resulting from the $B$-measurement. The result is that the second subsystem ends up in the pure spin-down state, $|d\rangle\langle d|$, and the expectation-value of $C$ is accordingly $\langle d|C|d\rangle$. In contrast, without the kick, we would have obtained an entirely different effective state, namely the totally random density operator $(1/2)(|u\rangle\langle u| + |d\rangle\langle d|)$, corresponding to a $C$-expectation of $(1/2)\mathrm{tr}C$. Thus, the detection of spin-up for the second subsystem would be an unambiguous signal that the first subsystem had not been kicked.

If you feel uneasy about using a kick rather than an actual measurement, you can replace the intervention at $O_1$ with a measurement of $A = \sigma_1$ (we apply $\exp(i\sigma_1\lambda)$ with a random value of $\lambda$ instead of with $\lambda = \pi/2$). The computation is not much longer, and yields for the effective state of subsystem 2, the density-operator $(1/4)|u\rangle\langle u| + (3/4)|d\rangle\langle d|$. This again differs from the totally random state, though not so strikingly as before.

## An Example in Quantum Field Theory

In a sense, the two subsystem example just given is all we need, since one would expect to be able to embed it in any quantum field theory which is sufficiently general to be realistic. Still, one might worry about non-localities having snuck in in connection with the particle concept on which the identification of the subsystems would probably be based in such an embedding; so it seems best to present an example couched directly in terms of a quantum field and its observables.

The example will follow the lines of that just given, but the computation is a bit more involved, and I will present it in slightly more detail, working for convenience in the interaction picture, for which the field $\phi$ evolves independently of the intervention, while the state $\rho$ get "kicked" from its initial value $\rho_0$ to $U\rho_0 U^*$, $U$ being the unitary operator which implements the kick. The quantum field will be a free scalar field $\phi(x)$ initially in its vacuum state, and the three spacetime regions will be those introduced above. The kicking operator will be taken to be $U = \exp(i\lambda\phi(y))$, where $y \in O_1$; and the observable measured in $O_3$ will be $C = \phi(x)$, where $x \in O_3$. (Really the fields should be smeared, but it will be clear that this would make no difference.) Finally, the observable $B$ measured in $O_2$ (which can be chosen as any operator at all, since $O_2$ includes the whole spacetime in its domain of dependence) will be orthogonal projection onto the state-vector

$$|b\rangle = \alpha|0\rangle + \beta|1\rangle, \tag{2}$$

where $|0\rangle$ is the vacuum and $|1\rangle$ is some convenient one-particle state; thus $B = |b\rangle\langle b|$.

Denoting vacuum expectation values simply by $\langle\cdot\rangle$, we may express the mean value predicted for $C$ as

$$\langle U^*BCBU\rangle + \langle U^*(1-B)C(1-B)U\rangle, \tag{3}$$

whose two terms correspond respectively to the outcomes 1 and 0 for the $B$-measurement. [To derive this expression, we may begin with the state $\rho = U\rho_0 U^*$, as it is after the kick but before the measurement of the projection $B$. The probability of the outcome $B = 1$ is then $\mathrm{tr}\rho B = \mathrm{tr}\rho_0 U^*BU = \langle U^*BU\rangle$, and that of the outcome $B = 0$ is $\langle U^*(1-B)U\rangle$. In the former case, the projection postulate yields for the consequent (normalized) state,

$$\sigma = \frac{B\rho B}{\mathrm{tr}B\rho B} = \frac{B\rho B}{\mathrm{tr}\rho B},$$

and therefore for the consequent expectation value of $C$,

$$\mathrm{Exp}(C|B=1) = \mathrm{tr}\,\sigma C = \frac{\mathrm{tr}\rho BCB}{\mathrm{tr}\rho B}.$$

When weighted with the probability $\mathrm{tr}\rho B$ of actually obtaining $B = 1$, this expression becomes the contribution of the $B = 1$ outcome to the final expectation-value of $C$:

$$\mathrm{Exp}(C, B = 1) = \mathrm{tr}\rho BCB = \langle U^* BCBU \rangle.$$

Finally, adding in the contribution of the $B = 0$ outcome yields (3), as desired.]

In order that the mean value (3) not depend on the magnitude of the kick, it is necessary in particular that its derivative with respect to $\lambda$ vanish at $\lambda = 0$ (or in other words that an infinitesimal kick have no effect). Now, this derivative may easily be computed, and turns out to be (twice) the imaginary part of

$$\langle \phi(y)(C + 2BCB - BC - CB) \rangle,$$

which therefore must be purely real in order that locality be respected. However the first and last terms are separately real; the former equals $\langle \phi(y)\phi(x) \rangle$, which is real because $x \natural y$ (a notation meaning that $x$ is spacelike to $y$) whence $\phi(x)\natural\phi(y)$ (a notation meaning that $\phi(x)$ and $\phi(y)$ commute), and the latter reduces to $|\alpha|^2$ times the former when the definition of $B$ is used. With the aid of the notation, $\psi(x) \equiv \langle 0|\phi(x)|1\rangle$, the results of combining the two remaining terms can be written as

$$2(\alpha^* \beta)^2\, \psi(x)\psi(y) + (2|\alpha|^2 - 1)|\beta|^2\, \psi(x)^*\psi(y). \tag{4}$$

Here the star denotes complex conjugation, and the fact that $\phi(x)$ changes the particle number by $\pm 1$ has been used in places to eliminate some terms which would have been present had $\phi(x)$ not been a free field.

To show that (4) need not be real, we can, for example, eliminate its first term by taking $|\alpha|^2 = |\beta|^2 = 1/2$. What remains can then be given any desired phase by an appropriate choice of the relative phase of $\alpha$ and $\beta$, unless it happens that $\psi(x)$ or $\psi(y)$ vanishes. Avoiding this possibility in our choice of $\psi$ (indeed would be difficult not to avoid it, since $\psi$ is purely positive frequency!), we arrive at the conclusion announced earlier. Notice, incidentally, that we could have arranged $\psi(\cdot)$ to manufacture a problem even with $\alpha = 0$, but the superposition with the vacuum state allows us to control the phase of (4) for an arbitrary choice of the one-particle wave-function $\psi$. On the other hand, setting $\alpha = 0$ in (2) does have the advantage of lending a particularly simple physical meaning to the measurement $B$: it merely asks whether or not there is precisely one particle present and if so whether that particle is in the specific state, $|1\rangle$.

## 4. POSSIBLE IMPLICATIONS FOR QUANTUM FIELD THEORY AND QUANTUM GRAVITY

In a way it is no surprise that a measurement such as of $B$, which occupies an entire hypersurface, should entail a physical non-locality; but surprising or not, the implications seem far from trivial. Unless one admits the possibility of superluminal signaling, the entire interpretive framework constructed above for the quantum field formalism must be rejected as it stands. What then remains of the apparatus of states and observables, on which the interpretation of quantum mechanics is traditionally based?

A possible way to salvage our framework would be to further restrict the allowed measurement-regions $O_j$ in such a manner that the transitive closure we took in defining $\prec$ would be redundant. For example, we could require that for each pair of regions $O_j$, $O_k$, all pairs of points $x \in O_j$ and $y \in O_k$ be related in the same way (i.e. either $x \natural y$ in all cases, or $x < y$ in all cases, or $x > y$). Such a restriction would block the kind of example just presented, but it would also be a very severe limitation on the allowed measurements (excluding measurements on Cauchy surfaces, for example.) Also it is difficult to see how the ability to perform a measurement in a given region—or the effect of that measurement on future probabilities—could be sensitive to whether some other measurement was located totally to its past, or only partly to its past and partly spacelike to it.[2]

Another way out might be to select the allowed measurements on some more ad hoc basis than that which was set up in Section 2. For example it can be shown, in the situation with the coupled subsystems, that information transfer never occurs when $B$ is a *sum* of observables, one belonging to each subsystem. This suggests that one might allow Cauchy-surface observables which were integrals of local operators, even though other Cauchy-surface observables would still have to be excluded. Spatially smeared fields have this additive character, for example (though they might be very singular as operators), but fields which are also smeared in time do not. Similarly, the most obvious gauge-invariant hypersurface observables in nonabelian gauge theories like QCD, i.e. holonomies, are functions of the commuting set of local variables $A_j(x)$, but they are not linear functions. This puts them in jeopardy because, in the two subsystem situation, nonlinear combinations of pairs of observables do in general lead

---

[2] On the other hand, there exists a purely formal consideration which suggests that in fact there might be some difference between the two cases. If one demands an answer to "where does the collapse occur?", the only viable response would seem to be "along the past light cone", and that would indeed appear as an influence of a later measurement on an earlier one, when the latter is partly spacelike to the former.

to information transfer. (For example, even something as simple as the product of two projections has this difficulty).[3] This also portends trouble for diffeomorphism-invariant hypersurface observables in quantum gravity.

This may an appropriate place to comment on one of the few attempts I know of to design concrete models of field measurements, namely that of Bohr and Rosenfeld [2]. Their idealized apparatus is designed to measure averaged field values in an arbitrary pair of spacetime regions (even overlapping ones!). When their regions are related as we required in Section 2, their results are consistent with the conclusion that their procedure does indeed furnish ideal measurements, as we defined this term earlier. Specifically, the apparatus interacts with the field only in the two specified regions, and the uncontrollable disturbance exerted by the earlier measurement on the later one is no greater than required by the commutator of the corresponding operators. It would thus seem important to extend (or reinterpret) their essentially classical treatment of the apparatus to a quantum one, in order to learn how close they come to actually fulfilling the requirements for an ideal measurement. Specifically, one can ask whether they actually measure the field averages they claim to, and whether the probabilities of the different possible outcomes are those predicted by the quantum formalism (with special reference to the use of the projection postulate after the first measurement, since its effect could *only* be seen in a full quantum treatment). Such an analysis would be especially interesting (even though Bohr and Rosenfeld only treat a free field) because there is no obvious formal reason why their temporally smeared observables should not suffer from the type of non-locality we have been discussing here.

However such an analysis would turn out, though (and it doesn't look that hard to do), there remains the more general fact that the need to resort to a case-by-case analysis would still leave us without any clear formal criterion for which "observables" can be ideally-measured, and which cannot; and we might also be left without any general rule to take the place of the projection postulate. Moreover the charm of the "transformation theory" would be lost as well, since different orthonormal bases would no longer be equal before the Law of Locality (in the case of hypersurface measurements, for example).

Now, as we are all aware, the question of what is the best dynamical framework for quantum gravity is not one which everyone will answer in the same way. It is very possible that some as yet unknown framework will be needed, but among existing

---

[3] An interesting problem would be to characterize which joint observables of a pair of subsystems potentially lead to "information transfer", and which don't.

interpretations of quantum mechanics, I have long felt that the sum-over-histories is the most promising, both for philosophical reasons and for practical ones. It has the great advantage that it deals with spacetime rather than just space, so that what is usually called the "problem of time" hardly makes an appearance. In particular, notions like horizon-area are well-defined, and quantum cosmology can investigate the early history of the universe because the universe really does have a history.[4]

In the non-relativistic context, however, the sum-over-histories has the disadvantage of allowing you to ask "too many" questions, including ones whose answering seems to lead to causality violations similar to those of Section 3 [1]. And in the face of such difficulties, one lacks a well-defined criterion to know in advance what measurements are possible, or even what interactions should count as a measurement. In these ways, however, the sum-over-histories would now appear to be no worse off than what survives of the state-and-observable framework when one tries to extend it to flat space quantum field theory (not to mention quantum gravity!)[5]

With the formal notion of measurement compromised as it seems to be already in quantum field theory, the greatest advantage of the sum-over-histories may be that it does not employ measurement as a basic concept. Instead it operates with the idea of a *partition* (or "coarse-graining") of the set of all histories, and assigns probabilities directly to the members of a given partition, using what I would call the quantum replacement for the classical probability calculus.

Actually, there are (at least) two variants of this idea. In the way I like to think about it [3], the partition is "implemented" by a designated subsystem which gains information about the rest of the universe. In this way of thinking, the basic idea would be that of probability *relative to* such a subsystem, and the difficulty is that it is not fully clear under what conditions information is actually obtained. (In practice, though, no problem is apparent in either of the two extreme cases which the history of science has brought us so far — laboratory experiments and astronomical observations.) In the other approach [4], the partition is given a priori, but subjected

---

[4] Another approach which deserves mention here is that of David Finkelstein, who views dynamics in terms of networks of elementary processes of input/output or creation/annihilation, and correspondingly draws a fundamental distinction between preparations and detections. Clearly this is relevant to the examples of Section 2, since our reasoning there required an ideal measurement to be both a detection and a preparation at once.

[5] The sum-over-histories also suffers esthetically from its inability to incorporate the unitary symmetry of the so called transformation theory. But we have seen that the Hilbert-space framework is now no better off in this respect either.

to a condition of decoherence. In this approach, the classical probability calculus applies, and one effectively returns to a classical (but stochastic) dynamical law. Some drawbacks of this variant are that one must do a very difficult analysis even to authorize the use of probability, and (more fundamentally) that decoherence is always *provisional*, since it applies to entire histories and is therefore always in danger of being overturned by future activities.

Let us hope that future activities by all of us will clarify some of these issues, and bring out the new insights which further study of the "quantum measurement problem" undoubtedly has to offer. After all, the dialectical inter-penetration of all existing things is at the heart of the interpretation which Bohr wanted to give to the new quantum formalism. In the textbook formulation this inseparability shows up in the impossibility of observing without also disturbing. It will be interesting to see how it shows up in the modified dynamical framework which will have to be developed for quantum gravity, and, as we see now, perhaps even for quantum field theory.

In concluding I would like to dedicate this article to Dieter Brill, in honor of his sixtieth birthday and in recollection of the many happy hours I have spent discussing physics with him (and once in a while playing sonatas together). Happy Birthday, Dieter!

I would also like to thank John Friedman, Josh Goldberg, Jim Hartle and David Malament for discussions and correspondence on the topic of this paper. This work was supported in part by NSF Grant No. PHY-9005790.

## REFERENCES

[1] Sorkin, R.D., "Problems with Causality in the Sum-over-histories Framework for Quantum Mechanics", in A. Ashtekar and J. Stachel (eds.), *Conceptual Problems of Quantum Gravity* (Proceedings of the conference of the same name, held Osgood Hill, Mass., May 1988), 217–227 (Boston, Birkhäuser, 1991).

[2] Bohr, N. and L. Rosenfeld, "Zur Frage der Messbarkeit der Elektromagnetis-chen Feldgrössen", *Det Kgl. Danske Videnskabernes Selskab., Mathematisk-fysiske Meddelelser*, **12**, No. 8 (1933); "Field and Charge Measurements in Quantum Electrodynamics", *Phys. Rev.* **78**:794-798 (1950).

[3] Sorkin, R.D., "On the Role of Time in the Sum-over-histories Framework for Gravity", paper presented to the conference on The History of Modern Gauge Theories, held Logan, Utah, July, 1987, to be published in *Int. J. Theor. Phys.* (1993, to

appear); Sinha, Sukanya and R.D. Sorkin, "A Sum-Over-Histories-Account of an EPR(B) Experiment" *Found. of Phys. Lett.*, **4**, 303-335, (1991).

[4] For a review see: Hartle, J.B., "The Quantum Mechanics of Cosmology", in *Quantum Cosmology and Baby Universes: Proceedings of the 1989 Jerusalem Winter School for Theoretical Physics*, eds. S. Coleman et al. (World Scientific, Singapore, 1991)

# A New Condition Implying
# The Existence of a
# Constant Mean Curvature Foliation

*Frank J. Tipler* [*]

### Abstract

It is shown that if a non-flat spacetime $(M, g)$ whose future c-boundary is a single point satisfies $R_{ab}V^a V^b \geq 0$ for all timelike vectors $V^a$, equality holding only if $R_{ab} = 0$, then sufficiently close to the future c-boundary the spacetime can be uniquely foliated by constant mean curvature compact hypersurfaces. The uniqueness proof uses a variational method developed by Brill and Flaherty to establish the uniqueness of maximal hypersurfaces.

In 1976 Dieter Brill and Frank Flaherty (1976) published an extremely important paper[1], "Isolated Maximal Hypersurfaces in Spacetime", establishing that maximal hypersurfaces are unique in closed universes with attractive gravity everywhere. That is, there is only one such hypersurface, if it exists at all. In an earlier paper, Brill had established that in three-torus universes, only suitably identified flat space possessed a maximal hypersurface, so the existence of a maximal hypersurface is not guaranteed. These results by Brill are important because maximal hypersurfaces are very convenient spacelike hypersurfaces upon which to impose initial data; on such hypersurfaces the constraint equations are enormously simplified. Furthermore, in asymptotically flat space, foliations of spacetime by maximal hypersurfaces often exist, and the simplifications of the constraint equations on such a foliation make it

---

[*]Bitnet address: TIPLER@MATH.TULANE.EDU, Institut für Theoretische Physik, Universität Wien, Boltzmanngasse 5, A-1090 Wien, AUSTRIA. Permanent Address: Department of Mathematics and Department of Physics, Tulane University, New Orleans, Louisiana 70118 USA

[1]I regard this paper of Dieter's as important because I've used its results in about ten of my own papers. But the relativity community also finds this paper important. According to the Science Citation Index, it has been cited 6 times in the period January 1989 through August of 1992, thus gathering about 6 per cent of Dieter's total citation count of 102 for this period. Most papers are never cited ten years after publication, so this citation frequency is quite impressive.

easy to numerically solve[2] the full four-dimensional vacuum Einstein equations for physically interesting situations.

Constant mean curvature foliations give similar simplifications, and such foliations often exist in closed universes. As Brill and Flaherty realized, their method can be generalized to show that if such a foliation exists, then it is unique. What I shall do in this paper is establish the existence of such a foliation near the final singularity in the case that the singularity is an "omega point". I shall conclude this paper with a discussion on the connection between Penrose's Weyl Curvature Hypothesis and the existence of a foliation of the entire spacetime by constant mean curvature hypersurfaces.

Let me begin with a

*Definition.* A spacetime (M,g) will be said to terminate in an *omega point* if its future c-boundary consists of a single point.

I have discussed elsewhere (Barrow & Tipler 1986, Tipler 1986, 1989, 1992) reasons for believing that the actual universe may terminate in an omega point. Misner's Mixmaster universe (Misner 1967) was specifically constructed to have a point c-boundary in the past, though it is not known if in fact there is a vacuum Bianchi type IX universe with such a c-boundary. Doroshkevich *et al* (1971) established (using different terminology) that if such a vacuum solution exists, it is of very small measure in the vacuum Bianchi type IX vacuum initial data. Vacuum solutions to the Einstein equations which terminate in an omega point are known (Löbell 1931; Hawking & Ellis 1973, p. 120 & p. 205; Budic & Sachs 1976), but they are all locally flat.

Let me give two examples of $S^3$ spatial topology Friedmann universes which terminate in omega points. Recall that the $S^3$ Friedmann metric is

$$ds^2 = -dt^2 + a^2(t)[d\chi^2 + \sin^2 \chi(d\theta^2 + \sin^2 \theta d\phi^2)] \tag{1}$$

where $0 \leq \chi \leq \pi$, $0 \leq \theta \leq \pi$, and $0 \leq \phi < 2\pi$. In the Friedmann universe, all null geodesics are radial, with comoving coordinates given by $ds^2 = 0 = -dt^2 + a^2(t)d\chi^2$, which upon integration yields Rindler's equation:

$$\chi_f - \chi_i = \pm \int_{t_i}^{t_f} \frac{dt}{a(t)} \tag{2}$$

---

[2]I leave this split infinitive in for Dieter's amusement. When I was his graduate student in the early 1970's, he was always finding them in my drafts of papers. (As a German, he naturally had a better command of English than a native American.) But just before I received my Ph.D. from him in 1976, he circulated a note announcing that, since he had just found a large number of split infinitives in the *Congressional Record*, he would henceforth regard split infinitives as officially correct American English.

The future c-boundary is a single point if and only if the integral in equation (2) diverges as proper time approaches its future limit $t_f = t_{max}$, since only in this case will the event horizons disappear: light rays circumnavigate the universe an infinity of times no matter how close to the c-boundary one is.

*example 1:* If $a(t) =$ constant, the metric (1) represents the Einstein static universe. Since the integral (2) diverges in this case because $t_{max} = +\infty$ and $t_{min} = -\infty$, both the future and past c-boundaries are single points.

*example 2:* Let $a(t) = \sin t$. Then the integral (2) is $\ln\left|\frac{\tan(t_f/2)}{\tan(t_i/2)}\right|$. There are s.p. curvature singularities (Hawking & Ellis 1973) at $t_f = \pi$ and at $t_i = 0$. At either of these limits, the integral (2) diverges, so this example has the same c-boundary structure as the Einstein static universe: both the future and past c-boundaries are each single points.

The Friedmann universe of example 2 does not satisfy the Einstein equations with any standard equation of state. However, this example is worth analysis for two reasons: first, because this $a(t)$ can be smoothly joined to a Friedmann universe which is matter and/or radiation dominated to the future of $10^{-15}$ seconds (before which $a(t) = \sin t$ may be appropriate, for who knows what the stress-energy tensor is like at extremely high densities), and second, because it nevertheless obeys all the usual energy conditions, thus showing that even in the case of the closed Friedmann universe, one need not violate the energy conditions to get the future and past c-boundaries to be single points. (This example is thus a counter-example to a conjecture of Budic and Sachs (1976), that to have a single point as its c-boundary, "... a cosmological model may have to 'coast into the [singularity] so slowly it almost bounces' corresponding to a 'near violation' of the timelike convergence condition. (Budic & Sachs 1976, p. 28)". But the metric of example 2 does not "nearly violate" the timelike convergence condition.

To see this, let us compute the stress-energy tensor for the metric of example 2. The mass density is

$$\mu \equiv T_{\hat{t}\hat{t}} \equiv \frac{1}{8\pi} G_{\hat{t}\hat{t}} = \frac{3}{8\pi}\left(\frac{a'^2+1}{a^2}\right) = \frac{3}{8\pi}\left(\frac{\cos^2 t + 1}{\sin^2 t}\right) \geq \frac{3}{8\pi}$$

The principal pressure is

$$p \equiv T_{\hat{x}\hat{x}} \equiv \frac{1}{8\pi} G_{\hat{x}\hat{x}} = -\left(\frac{1}{8\pi}\right)\frac{2aa'' + a'^2 + 1}{a^2} = \frac{1}{8\pi}\left(1 - 2\cot^2 t\right)$$

which is negative for $|\cot t| > \frac{1}{\sqrt{2}}$ — that is, near the singularities — and $p \to -\infty$ as $t \to 0$ or $\pi$. But we have

$$\mu + p = \frac{1}{8\pi}\left(\frac{4}{\sin^2 t}\right) > \frac{1}{2\pi} \quad , \quad \mu + 3p = \frac{6}{8\pi}$$

Since the weak energy condition requires (Hawking & Ellis 1973) $\mu \geq 0$ and $\mu + p \geq 0$, the weak energy condition is satisfied. Since the strong energy condition (here, also the timelike convergence condition) requires $\mu + p \geq 0$ and $\mu + 3p \geq 0$, the strong energy condition is satisfied. Furthermore, since both $\mu + p$ and $\mu + 3p$ are bounded well away from zero at *all* times, the timelike convergence condition is never "nearly violated". The dominant energy condition (Hawking & Ellis, 1973) requires $\mu \geq 0$ and $-\mu \leq p \leq +\mu$, so the dominant energy condition is satisfied. Finally it is easily checked that the generic condition is satisfied. The Ricci scalar is $R = 6(aa'' + a^2 + 1)/a^2 = 6\sin^{-2} t$, so the single c-boundary points are true s.p. curvature singularities at $t = 0$ and at $t = \pi$.

Note that examples 1 and 2 collectively suggest that if the closed Friedmann universe is not the Einstein static universe, then negative pressures are required in order for the c-boundary to be a single point. This can in fact be proven, but I shall omit the proof.

Budic and Sachs (1976) were motivated by Misner's model to prove some general theorems on spacetimes with either the future or the past c-boundaries being single points. Their theorems can be applied to either the past or the future c-boundary, though they stated their theorems in terms of a single point c-boundary in the past (since they were thinking of Misner's model). Similarly, though I shall state the theorems below in terms of an omega point — I shall be thinking of the final rather than the initial singularity — the theorems can be trivially modified to apply to the initial singularity.

Requiring that a spacetime end in an omega point imposes very powerful constraints on the spacetime. For example,

*Theorem* (Seifert 1971): a spacetime which terminates in an omega point and which satisfies the chronology condition has a compact Cauchy hypersurface.

This theorem was first stated by Seifert (1971), but unfortunately his proof is defective (In his Theorem 6.3, Seifert claims that the existence of an omega point in both the past and future directions is equivalent to the existence of a compact Cauchy surface). Budic and Sachs (1976) have stated that the existence of an omega point in a future and past distinguishing spacetime (Hawking & Ellis 1973) implies the existence of a compact Cauchy surface. It is easy to check that Seifert's Theorem holds if the spacetime is stably causal, so I've stated his Theorem with this causality condition.

As a converse to Seifert's Theorem, we have

*Theorem 1*: If the future c-boundary of a stably causal spacetime consists of an omega point, then for all points $q$ sufficiently close to the future c-boundary, $\partial I^-(q)$ is also a Cauchy surface.

*Proof:* By Seifert's Theorem, the spacetime admits a compact Cauchy surface. Since the spacetime has a compact Cauchy surface, Geroch's Theorem (Proposition 6.6.8 of Hawking & Ellis 1973), page 212), all Cauchy surfaces in the spacetime have the same topology, and further, the spacetime can be foliated by compact diffeomorphic spacelike Cauchy surfaces. Let $S(t)$ represent such a foliation, where $t$ increases in the future direction, and let $V^a(\vec{x}, t)$ represent the timelike future-directed unit vector field which is everywhere normal to S(t). Let $\lambda(t)$ be any flow line of this vector field. I claim that there exists $t_\lambda$ such that $\partial I^-(\lambda(t_\lambda))$ is a Cauchy surface. Suppose not. Then there would exist another flow line $\mu(t)$ of $V^a(\vec{x}, t)$ which never intersects $\partial I^-(\lambda(t))$, for any $t$. But then the flow line $\mu(t)$ would define a different future c-boundary point than $\lambda(t)$, contrary to the fact that there is only one c-boundary point. Thus for each $\lambda(t)$ in $V^a(\vec{x}, t)$, there is a time $t_\lambda$ for which $\partial I^-(\lambda(t))$ is a Cauchy surface, for all $t > t_\lambda$. Since the leaves of the foliation $S(t)$ are compact, $sup[t_\lambda] \equiv t_C$ is achieved in the spacetime. Then $\partial I^-(q)$ will be a Cauchy surface provided $q$ is any event to the future of $S(t_C)$; i.e., $q \in I^+(S(t_C))$. QED

We thus know that $\partial I^-(q)$ is a Cauchy surface for $q$ sufficiently close to the omega point, so in principle, all information is available at $q$. This property allows us to show that a foliation of spacetime by constant mean curvature hypersurfaces exists, at least sufficiently near the omega point.

*Theorem 2*: If a non-flat stably causal spacetime $(M, g)$ satisfies $R_{ab}V^aV^b \geq 0$ for all timelike vectors $V^a$, equality holding only if $R_{ab} = 0$, and $(M, g)$ has an omega point, then there exists a point $p \in M$ such that through $p$ there passes a $C^{2,\alpha}$ Cauchy surface $S$ with constant mean curvature, and further, $I^+(S)$ can be uniquely foliated by $C^{2,\alpha}$ Cauchy surfaces with constant mean curvature.

That is, a spacetime which satisfies the timelike convergence condition and which ends in an omega point has sufficiently near the omega point a foliation by compact Cauchy surfaces with constant mean curvature. However, the entire spacetime might not have such a foliation; the foliation is guaranteed to exist only for that part of spacetime sufficiently close to the omega point. The meaning of "sufficiently close" is made precise in the proof of Theorem 1 above. (A $C^{2,\alpha}$ Cauchy surface (Bartnik 1984) is one which is $C^2$ with these second derivatives being Hölder continuous of order $\alpha$.)

*Proof:* Bartnik (1988) has shown that if for any point $p$ in $(M, g)$, the set $M - I^+(p) \cup I^-(p)$ is compact, then there is a spacelike $C^{2,\alpha}$ constant mean curvature Cauchy surface through $p$. I shall need two Lemmas to combine with Bartnik's result:

*Lemma 1*: If the future c-boundary of a spacetime $(M, g)$ which satisfies the chronology condition is an omega point, then the achronal boundary $\partial I^+(p)$ is a Cauchy surface for *any* point $p$ in the spacetime.

*Proof:* Suppose not. Then there is a future- and past-endless timelike curve $\gamma$ which never intersects $\partial I^+(p)$, which, since the chronology condition holds, is non-empty and is generated by null geodesic segments at least some of which intersect $p$. If (1) $\gamma \cap I^-(\partial I^+(p)) \neq \emptyset$, or (2) $\gamma \cap I^-(\partial I^+(p)) = \emptyset$ and $\gamma \cap I^+(\partial I^+(p)) = \emptyset$, then $I^-(\gamma)$ would not intersect $I^+(p)$, so $I^-(\gamma)$ defines a different c-boundary point than does a future-endless timelike curve which eventually enters $I^+(p)$. Thus there are at least two distinct c-boundary points, contradicting the hypothesis that there is just one future c-boundary point.

The other possibility, which we now eliminate, is $\gamma \cap I^+(p) \neq \emptyset$, but $\gamma \cap \partial I^+(p) = \emptyset$. Since $\gamma \cap I^+(p) \neq \emptyset$, there exists a timelike curve $\beta_q$ from $p$ to *some* point $q \in \gamma$. Consider the sequence of timelike curves $\beta_{q_i}$ as the point $q$ moves into the past along $\gamma$ through a sequence of points $q_i$. This sequence defines a subsequence which converges to some causal curve $\hat{\beta}$ in $\overline{I^+(p)}$ (since $\overline{I^+(p)}$ is closed). However, $\hat{\beta}$ must be disconnected since if it were connected, $\gamma \cup \hat{\beta}$ would be a connected curve, contrary to the assumption that $\gamma$ is past-endless. The connected subset of $\hat{\beta}$ — call it $\hat{\beta}_p$ — which ends in the point $p$ is thus future-endless, and since $\overline{I^+(\gamma)} \cap I^-(\hat{\beta}_p) = \emptyset$, the causal curves $\gamma$ and $\hat{\beta}_p$ define different TIPs, contrary to the assumption that there is just once TIP in $(M, g)$. QED.

*Lemma 2:* If the future c-boundary of $M, g)$ is a single point and the chronology condition holds, then $\partial I^+(p)$ is non-empty and compact for every event $p$ in the spacetime.

*Proof:* If the chronology condition holds, then $p \in \partial I^+(p)$. By the remarks on page 188 of Hawking and Ellis (1973), $\partial I^+(p)$ is generated by null geodesic segments which either have no endpoints or have endpoints at $p$. Thus all the null geodesics from $p$ into the future are generators of $\partial I^+(p)$. If every null geodesic generator of $\partial I^+(p)$ from p leaves $\partial I^+(p)$ in the future, then $\partial I^+(p)$ is compact, since one can put on the collection of null geodesic generators of $\partial I^+(p)$ an affine parameterization such that the length of the segment of the null geodesic in $\partial I^+(p)$ from $p$ varies continuously with the null direction into the future from $p$, and the collection of null directions at $p$ is compact (actually, a 2-sphere). Thus the only way that part of $\partial I^+(p)$ for which $\partial I^+(p) \cap \{p\} \neq \emptyset$ could fail to be compact is for there to exist a null geodesic $\gamma$ of $\partial I^+(p)$ which never leaves $\partial I^+(p)$.

But then $I^-(\gamma)$ would define a TIP which is distinct from a TIP generated by any future-inextendible timelike curve which crossed $\gamma$ from $I^-(\gamma)$ into $I^+(p)$. But this would mean more than one TIP, contrary to assumption, so that part of $\partial I^+(p)$ for which $\partial I^+(p) \cap \{p\} \neq \emptyset$ is compact for all p, and also all null geodesic generators of $\partial I^+(p)$ from $p$ must eventually leave $\partial I^+(p)$.

We now eliminate the possibility that $\partial I^+(p)$ has a null geodesic generator $\beta$ which does not intersect $p$. Suppose it does, and let $q$ be a point of $\beta$ with normal neighborhood $N$. Then there is a timelike curve from $p$ to any point in $N \cap I^+(p)$, (which is non-empty since $\beta \subset \partial I^+(p)$.) Consider a sequence of points $q_i$ in $N \cap I^+(p)$ converging to $q$. This sequence defines a sequence of timelike curves $\beta_i$ from $p$ to $q_i$. If this sequence of timelike curves converged to a single connected causal curve, it would have to be a null geodesic with past endpoint at $p$, which is impossible by definition of $\beta$. Since locally (in any convex normal neighbourhood) the sequence converges, it must converge globally to at least two (possibly more) distinct disconnected causal curves, the one terminating at p being future-endless. This future-endless curve, call it $\hat{\beta}$, defines a TIP which is different from at least one TIP defined by some future-endless timelike curve in $I^+(\beta)$, since by construction $I^-(\hat{\beta}) \cap I^+(\beta) = \emptyset$. QED

To continue the proof of Theorem 2, recall that by Theorem 1 above, $\partial I^-(p)$ is a compact Cauchy surface for all points $p$ sufficiently close to the omega point. Together these imply that $M - [I^+(p) \cup I^-(p)]$ is compact for $p$ sufficiently close to the omega point. (The set $M - [I^+(p) \cup I^-(p)]$ is closed since both $I^+(p)$ and $I^-(p)$ are open). Also, for any foliation of $(M, g)$ by spacelike hypersurfaces $S(t)$, there will times $t_1$ and $t_0$ with $t_1 > t_0$ such that $\partial I^+(p) \subset I^-(S(t_1))$ and $\partial I^-(p) \subset I^+(S(t_0))$. Hence, the closed set $M - [I^+(p) \cup I^-(p)]$ is contained in the compact set $M - [I^+(S(t_1)) \cup I^- S(t_0))] \approx S(t) \times [0,1]$, for any fixed $t$, and so is compact.) Thus through every point sufficiently close to the omega point, there passes a spacelike $C^{2,\alpha}$ constant mean curvature Cauchy surface. Brill and Flaherty (1976), and Marsden and Tipler (1980) have modified a theorem by Brill and Flaherty (1976) to show that any constant mean curvature compact Cauchy surface on which the constant mean curvature $\chi^a{}_a$ is non-zero, is unique if the timelike convergence condition holds. Following Geroch (see Hawking and Ellis 1973, p. 274), Marsden and Tipler (1980) have shown that in all non-flat spacetimes with $R_{ab} V^a V^b \geq 0$, equality holding only if $R_{ab} = 0$, compact Cauchy surfaces with $\chi^a{}_a = 0$ are also unique. Hence, there exists a point $p$ in M such that through $p$ there passes a $C^{2,\alpha}$ Cauchy surface $S$ with constant mean curvature, and further, $I^+(S)$ can be uniquely foliated by Cauchy surfaces with constant mean curvature. QED.

The non-flatness and $R_{ab} V^a V^b = 0$ only when $R_{ab} = 0$ assumptions were only needed for uniqueness of the maximal hypersurface (if in fact any exists). The existence of a constant mean curvature compact Cauchy surface foliation follows merely from the timelike convergence condition and the existence of the omega point. If both the future and past c-boundaries are single points — as in examples 1 and 2 — then the proof of Theorem 2 shows that the entire spacetime is foliated by constant mean curvature compact Cauchy surfaces.

Budic and Sachs (1976) have shown that if the total spacetime volume $\int \sqrt{-g}\, d^4x$ of an omega point spacetime is finite (as it would be in example 2, for instance), then there is another natural foliation $S_{BS}(t)$ of $(M, g)$ by spacelike hypersurfaces, namely for a given $t$, the value of $\int_{I^+(p)} \sqrt{-g}\, d^4x$ is the same for each point $p \in S_{BS}(t)$. Budic and Sachs show that this foliation is $C^1$, and a modification of the proof of Theorem 2 shows that sufficiently close to the omega point, the hypersurfaces $S_{BS}(t)$ will be compact Cauchy surfaces. The question then arises, what is the relationship — if any — between these two natural spacelike foliations of $(M, g)$? In example 2, the two foliations are exactly the same, but in general this will not be the case. For instance, if $(M, g)$ is the spacetime of example 2, then $M - J^-(p)$ for any point $p \in M$ is a spacetime with an omega point which can be foliated with constant mean curvature Cauchy surfaces only to the future of $p$, while $S_{BS}(t)$ foliate the entire spacetime (though with Cauchy surfaces only to the future of $p$).

Budic and Sachs (1976) show that $\overline{M}$, the spacetime with its c-boundary, is second-countable and metrizable, so some constraints are imposed on the initial singularity by the requirement that the final singularity is an omega point. I conjecture that if we require that the entire spacetime be foliated by constant mean curvature Cauchy surfaces which everywhere coincide with the $S_{BS}(t)$ hypersurfaces, then the spacetime must be spatially homogeneous.

Penrose's Weyl Curvature Hypothesis (Penrose 1979), namely that time is defined so that a physical spacetime's "initial" singularity is characterized by the vanishing of the Weyl curvature as one approaches the initial singularity (and the "final" singularity is characterized by the dominance of the Weyl curvature over the Ricci curvature) is another proposal to connect the initial and final singularities. Tod (1990) conjectured and Newman (1991) proved (at least for the $\gamma = \frac{4}{3}$ case) that if the Weyl curvature vanished at a singularity (which is "conformally compactifiable"), then the spacetime was necessarily Friedmann everywhere. Goode et al (1985, 1991, 1992) have restated the Weyl Curvature Hypothesis to mean that

$$\lim_{T \to 0^+} \frac{C_{abcd}C^{abcd}}{R_{ab}R^{ab}} = 0 \qquad (3)$$

at an "initial" singularity. Goode et al (1992) have shown that many of the standard Cosmological Problems (flatness problem, horizon problem, etc.) can be solved if one imposes this modified Weyl Curvature Hypothesis. However, they do not propose strongly believable reasons why the Weyl Curvature Hypothesis should be true.

Perhaps by connecting these two approaches to connecting the initial and final singu-laries a strongly believeable reason can be found. Goode et al and Tod, in their defini-tions of "conformally comfactifiable" or "isotropic" singularity, require the existence, near the initial singularity, of a foliation of spacetime by spacelike hypersurfaces, but they do not require that the foliation be one of the "natural" ones discussed above.

However, suppose we require that globally, the Second Law of Thermodynamics must always hold: the total entropy of the universe at time $t_i$ must always be greater than or equal to the total entropy at time $t_j$ whenever $t_i \geq t_j$. Clearly, this inequality cannot hold globally for all foliations, since locally we can always decrease the entropy at the expense of an even greater entropy increase at another spatial position, and we can use this fact to construct a foliation of spacetime by spacelike hypersurfaces in which the above entropy inequality was violated, at least for a short time. But it conceivably *might* be true for one (or both) of the natural foliations described above — if the modified Penrose Weyl Curvature Hypothesis holds. If the entropy inequalities do not hold for *some* natural foliation, then we would be forced to admit that the Second Law of Thermodynamics simply does not always hold globally (or is inconsistent with general relativity), an admission we should be loath to make.

The modified Penrose Weyl Curvature Hypothesis would have to hold for two reasons. First, to ensure that the purely gravitation degrees of freedom — gravitational waves — when degraded into heat, do not by themselves violate the Second Law of Thermodynamics. Second, to ensure the global existence of both of the above foliations: I conjecture that if the initial singularity is "isotropic" in the sense of Goode *at al* and "conformally compactifiable" in the sense of Tod, then the foliation of constant mean curvature Cauchy surfaces and the Budic-Sachs foliation by Cauchy surfaces — which must exist near an omega point — can be extended globally to the entire spacetime.

If so, then the modified Penrose Weyl Curvature Hypothesis would be equivalent to requiring the global validity of the Second Law of Thermodynamics. Here would be a strongly believable reason for accepting the Weyl Curvature Hypothesis and its resolution of the Cosmological Problems! This modified Penrose Curvature Hypothesis also gives another reason for studying constant mean curvature foliations, a research topic to which Dieter Rolf Brill has contributed so much.

It is a pleasure to thank P.C. Aichelburg, J. D. Barrow, R. Beig, H. Kühnelt, H. Narnhofer, R. Penrose, and H. Urbantke for helpful discussions. My research was supported in part by the University of Vienna, by the BMWF of Austria under grant number GZ30.401/1-23/92 and by Fundacion Federico.

# REFERENCES

Barrow, J.D. & Tipler, F.J. (1986). *The Anthropic Cosmological Principle*. Oxford University Press: Oxford.

Bartnik, R. (1984). *Commun. Math. Phys.*, **94**, 155.

Bartnik, R. (1988). *Commun. Math. Phys.*, **117**, 615.

Brill, D.R. & Flaherty, F. (1976). *Comm. Math. Phys.*, **50**, 157.

Budic, R. & Sachs, R.K. (1976). *Gen. Rel. Grav.*, **7**, 21.

Doroshkevich, A.G. & Novikov, I.D. (1971). *Sov. Astron. AJ*, **14**, 763.

Doroshkevich, A.G., Lukash, V.N., & Novikov, I.D. (1971). *Sov. Phys. JETP*, **33**, 649.

Goode, S.W. (1991). *Class. Quantum Grav.*, **8**, L1.

Goode, S.W., Coley, A.A. & Wainwright, J. (1992). *Class. Quantum Grav.*, **9**, 445.

Goode, S.W. & Wainwright, J. (1985). *Class. Quantum Grav.*, **2**, 99.

S.W. Hawking, S.W. & Ellis, G.F.R. (1973). *The Large- Scale Structure of Space-Time*. (Cambridge University Press: Cambridge.

Lifshitz, E.M., Lifshitz, I.M. & Khalatnikov, I.M. (1971). *Sov. Phys. JETP*, **32**, 173.

Löbell, F. (1931). *Ber. Verhandl. Sächs. Akad. Wiss. Leipzig, Math. Phys. Kl.*, **83**, 167.

Marsden, J.E. & Tipler, F.J. (1980). *Phys. Rep.*, **66**, 109.

Misner, C.W. (1967). *Nature*, **214**, 40.

Newman, R.P.A.C. (1991). *Twistor Newsletter*, **33**, 11.

Penrose, R. (1979) in *General Relativity: An Einstein Centenary Survey*, ed. S.W. Hawking & W. Israel. Cambridge Univeristy Press: Cambridge.

Seifert, H.J. (1971). *Gen. Rel. Grav.*, **1**, 247.

Tipler, F.J. (1986). *Int. J. Theor. Phys.*, **25**, 617.

Tipler, F.J. (1989). *Zygon*, **24**, 217.

Tipler, F.J. (1992). *Phys. Lett. B*, **286**, 36.

Tod, K.P. (1990). *Class. Quantum Grav.*, **7**, L13.

# Maximal Slices in Stationary Spacetimes with Ergoregions

*Robert M. Wald* *

## Abstract

We describe some recent results (obtained in collaboration with Piotr Chruściel) which establish existence of a maximal slice in a class of stationary spacetimes which contain an ergoregion but no black or white hole. No use of Einstein's equation or energy conditions is made in the proof. The result enables one to prove that all stationary solutions to the Einstein-Yang-Mills equations which have vanishing electric charge and do not contain a black or white hole must be static. Similar results for the case where a black and white hole with bifurcate horizon are present are briefly described.

A *maximal slice* in a spacetime $(M, g_{ab})$ is a closed, embedded, spacelike, submanifold of co-dimension one whose trace, $K = K^a{}_a$, of extrinsic curvature vanishes. The issue of whether maximal slices exist in certain classes of spacetimes in general relativity has arisen in many analyses. One of the most prominent early examples of the relevance of this issue occurs in the positive energy argument given by Dieter Brill in collaboration with Deser [5], where the existence of a maximal slice in asymptotically flat spacetimes was needed in order to assure positivity of the "kinetic terms" in the Hamiltonian constraint equation. The existence and properties of maximal slices has remained a strong research interest of Brill, and he has made a number of important contributions to the subject. In particular, he provided an important example of a globally hyperbolic, asymptotically flat solution to Einstein's equation with physically reasonable matter which fails to possess a maximal Cauchy surface [6]. In collaboration with Flaherty, he also obtained some uniqueness results on maximal slices [7].

The issue of existence of maximal slices recently arose again in an investigation of solutions to the Einstein-Yang-Mills equations [9] – which includes, as special cases, solutions to the vacuum Einstein equations and the Einstein-Maxwell equations. In

*Enrico Fermi Institute and Department of Physics, University of Chicago, 5640 S. Ellis Avenue, Chicago, IL 60637, USA. This research was supported in part by the National Science Foundation under Grant No. PHY89-18388.

[9] the following results were obtained: (1) If an asymptotically flat solution to the Einstein-Yang-Mills equations admits a Killing field $X$ which approaches a time translation at infinity and also admits an asymptotically flat maximal slice $\Sigma$ with "compact interior" which is asymptotically orthogonal to $X$, and if, furthermore, the Yang-Mills electric charge, $Q$, or the electrostatic potential, $V$, of the solution vanishes at infinity, then the solution must be static and the electric field must vanish. (2) Similarly, if an asymptotically flat solution to the Einstein-Yang-Mills equations admits a Killing field $X$ which approaches a time translation at infinity and which vanishes on the bifurcation surface, $S$, of a bifurcate Killing horizon, if the spacetime admits an asymptotically flat maximal hypersurface, $\Sigma$, with boundary $S$ and "compact interior" which is asymptotically orthogonal to $X$, and if, furthermore, $Q$ or $V$ vanishes at infinity, then the solution must be static and the electric field must vanish. I have outlined the salient features of the proof of these results in the accompanying volume [10].

The case of greatest interest in the above two results is where $X$ fails to be globally timelike, i.e, where "ergoregions" occur in the spacetime. In particular, in that case, as discussed further in [9] and [10], result (2) closed a gap that had remained open for the past twenty years in the proof of the black hole uniqueness theorems in the vacuum and Einstein-Maxwell cases. However, the results (1) and (2) require the existence of of a maximal slice or, respectively, a maximal hypersurface with boundary $S$, with the stated properties. Thus, it is imperative to determine whether such a slice or hypersurface exists in the spacetimes of interest.

Some important advances have been made in the past decade toward obtaining sufficient conditions for the existence of maximal slices in asymptotically flat spacetimes, due mainly to the work of Bartnik [1], [2], wherein the existence of a time function satisfying a "uniform interior condition" on the spacetime was shown to be sufficient. More recently, the asymptotic conditions assumed in [1] were considerably weakened, and the basic method was used to establish existence of maximal slices for asymptotically flat stationary spacetimes in which the Killing field $X$ is strictly timelike everywhere [4]. In spacetimes where $X$ is strictly timelike, a time function, $t$, can be defined as follows: Start with any smooth, acausal, asymptotically flat slice, $\Sigma$, which has compact interior and which is asymptotically orthogonal to $X$. Since $X$ is timelike, it is automatically transverse to $\Sigma$. We define $t$, as the orbital parameter along $X$ starting from $\Sigma$. It then follows immediately that eqs. (4.2) and (4.3) of [4] hold. Furthermore, since $X$ is timelike, a bound can be put on the "slopes of the light cones" [4] which establishes that $t$ satisfies Bartnik's interior condition. Existence of a maximal slice then follows [4].

There are two basic difficulties which arise if one attempts to generalize the results of [4] to the case where "ergoregions" may be present, i.e., where $X$ is only assumed to be timelike "near infinity". First, there need not exist a spacelike slice $\Sigma$ transverse to $X$, so it is not obvious how to define a time function analogous to the above function

$t$. Indeed, for the case of interest for result (2) above, $X$ cannot be transverse to $\Sigma$ at the bifurcation surface $S$, so a precise analog of $t$ cannot exist. Second, when $X$ fails to be timelike one loses control over the "slopes of the light cones", so even if an analog of $t$ could be defined, Bartnik's interior condition need not be satisfied.

That the existence of maximal slices satisfying the desired conditions may fail when $X$ is not globally timelike can be seen from the following example due to Dieter Brill (private communication). Start with maximally extended Schwarzschild spacetime. This spacetime, of course, admits maximal slices with "compact interior" which are asymptotically flat and are asymptotically orthogonal to the timelike Killing field at both "ends". However, from uniqueness results, it can be shown that all such maximal slices are spherically symmetric. Now remove from the spacetime a future-directed "radial" null geodesic starting at the "north pole" ($\theta = 0$) of the bifurcation 2-sphere at $r = 2m$, and remove a past-directed "radial" null geodesic starting at the "south pole" ($\theta = \pi/2$) of the same bifurcation 2-sphere. The resulting spacetime continues to possess a one-parameter group of isometries (since only complete Killing orbits were removed) and continues to possess slices with compact interior which are asymptotically flat and are asymptotically orthogonal to the timelike Killing field. (Furthermore, the domain of dependence of such a slice yields a globally hyperbolic spacetime with these properties.) However, in this modified spacetime, there are no spherically symmetric, asymptotically flat slices with compact interior, and, hence, no asymptotically flat, maximal slices with compact interior which are asymptotically orthogonal to the timelike Killing field.

What further conditions on a stationary spacetime possessing an ergoregion will ensure that maximal slices and/or hypersurfaces of the desired type will exist? It turns out that for the case relevant to result (1) above, the condition that no black or white hole be present suffices. Similarly, for the case relevant to result (2), it suffices that there not be any black or white hole "exterior to $S$". These existence results have been proven very recently in work done in collaboration with Piotr Chruściel [8]. The details of the proofs are quite technical, and I refer the reader to [8] for the precise statement and proofs of our existence theorems on maximal hypersurfaces. However, the key ideas which enter the proof for the case relevant to result (1) above are fairly simple. I now shall give a sketch of these key ideas in this case; the modifications needed for the case relevant to result (2) then will be briefly indicated.

We consider a strongly causal spacetime, $(M, g_{ab})$, which possesses an acausal, asymptotically flat slice $\Sigma$ (with a finite number of asymptotically flat "ends", $\Sigma_i$), whose "interior region" is compact. (The fall-off conditions imposed upon the metric are quite weak; the reader is referred to [8] for the precise definition of asymptotic flatness we use.) Furthermore, we require that $(M, g_{ab})$ possess a one-parameter group of isometries generated by a Killing field $X$ which is timelike (with norm asymptotically bounded away from zero) at each of the ends. However, we emphasize that Einstein's equation is *not* imposed. In such a spacetime, we define $M_i$ to be the orbit of $\Sigma_i$ under these isometries, and we define $M_{ext}$ to be the union of the $M_i$. We define the *black hole* region, $B$ of $M$, to consist of those events which do not lie in the past of $M_{ext}$. The *white hole* region, $W$ is defined similarly. We then prove the following result: *If the interior of the domain of dependence of $\Sigma$ does not contain any events lying in $B$ or $W$, then there exists a smooth, asymptotically flat maximal slice $\hat{\Sigma}$ which is asymptotically orthogonal to $X$ and is a Cauchy surface for $\mathrm{int}D(\Sigma)$.*

As already stated above, the failure of $X$ to be timelike poses two fundamental difficulties for proving existence of a maximal slice via the methods of [4]. Remarkably, the absence of a black or white hole in $\mathrm{int}D(\Sigma)$ suffices to overcome both of these difficulties. I now shall indicate how this occurs.

The first difficulty – arising from the fact that $X$ need not be transverse to $\Sigma$ – concerns the definition of a suitable time function $t$ which agrees with Killing parameter along the orbits of $X$. To treat this situation, we note, first, that the condition that $[B \cup W] \cap \mathrm{int}D(\Sigma) = \emptyset$ implies that $X$ must have the same (future or past) orientation at each "end" (see Lemma 3.2 of [8]). Reversing the orientation of $X$ if necessary, we take $X$ to be future-directed at each end. Next, if $p \notin W$, it follows that for large positive $t$, the orbit of $X$ through $p$ must enter and remain in the future of at least one end, $\Sigma_i$, of $\Sigma$. To prove this, we note that $p \notin W$ implies that $p$ lies in the future of a point $q$ in some $M_i$. However, there then exists $T \in \mathbb{R}$ such that the events at parameter $t \geq T$ on the orbit through $q$ lie to the future of $\Sigma_i$. It then follows immediately that the portion $t \geq T$ of the orbit through $p$ lies in the future of $\Sigma_i$. Similarly, if $p \notin B$, for large negative $t$, the orbit of $X$ through $p$ must enter and remain in the past of $\Sigma_j$ for some $j$. Finally, if $p \in \mathrm{int}D(\Sigma)$, it can be shown that the orbit through $p$ is contained in $\mathrm{int}D(\Sigma)$ (see Proposition 3.1 of [8]). Hence, since this orbit enters both $I^+(\Sigma)$ and $I^-(\Sigma)$, it must intersect $\Sigma$. Thus, although $X$ need not be transverse to $\Sigma$, it follows that when $[B \cup W] \cap \mathrm{int}D(\Sigma) = \emptyset$ the intersection with $\Sigma$ of an orbit of $X$ through any $p \in \mathrm{int}D(\Sigma)$ must be both non-empty and compact.

The desired time function $t$ now may be constructed as follows: Let $\tau$ be a smooth function such that $\tau = 0$ outside the future of a small neighborhood, $\mathcal{O}$, of $\Sigma$, $\tau = 1$ outside the past of $\mathcal{O}$, and the gradient of $\tau$ is timelike within $\mathcal{O}$. (An explicit construction of such a $\tau$ can be found in [8].) For $p \in \mathrm{int}D(\Sigma)$, define

$$t(p) = \int_{-\infty}^{0} \tau(\phi_s(p))ds + \int_{0}^{\infty} (\tau - 1)(\phi_s(p))ds$$

where $\phi_s$ denotes the one-parameter group of isometries generated by $X$. It then can be shown that $t$ is a global time function on $\mathrm{int}D(\Sigma)$ which agrees with Killing parameter along the orbits of $X$. Indeed the level surfaces of $t$ are Cauchy surfaces for $\mathrm{int}D(\Sigma)$ which are everywhere transverse to $X$. Thus, although the failure of $X$ to be globally timelike may result in $X$ failing to be transverse to the original slice $\Sigma$, if no black or white hole exists in $\mathrm{int}D(\Sigma)$, then there exist other slices to which $X$ will be transverse, and a time function, $t$, can be constructed with the same essential properties as in the case where $X$ is globally timelike.

As indicated above, the second difficulty that arises when $X$ fails to be globally timelike is the loss of control over the "slopes of the light cones" with respect to local coordinates with $x^0 = t$ which are "comoving" with $X$. The key property needed to ensure that Bartnik's interior condition holds is that the domain of dependence of the "interior portion" of $\Sigma$ be compact. When $X$ is globally timelike, bounds on the light cone slope imply that the set $D(\Sigma_{\text{int}})$ must be contained within a bounded range of $t$. Compactness of $D(\Sigma_{\text{int}})$ then follows directly. Nevertheless, in the case of interest here, despite the loss of bounds upon the light cone slope, compactness of $D(\Sigma_{\text{int}})$ can be proven as follows: Consider a sequence $\{x_i\}$ in $D(\Sigma_{\text{int}})$, and let $\gamma_i$ denote the orbit of $X$ through $x_i$. Then, as shown above, $\gamma_i$ must intersect $\Sigma$ (possibly in more than one point). Let $y_i$ denote an intersection point of $\gamma_i$ with $\Sigma$. Clearly, $y_i \in \Sigma_{\text{int}}$, and since $\Sigma_{\text{int}}$ is compact, the sequence $\{y_i\}$ has a subsequence $\{z_i\}$ which converges to a point $z \in \Sigma_{\text{int}}$. By the above arguments, for $t$ large and positive, the orbit, $\gamma$, through $z$ must enter and remain in the future of the "exterior portion", $\Sigma_{\text{ext}}$, of $\Sigma$, whereas for $t$ large negative, $\gamma$ must enter and remain in the past of $\Sigma_{\text{ext}}$. From this, it follows that the sequence $\{t_i\}$ defined by $\phi_{t_i}(z_i) = x_i$ is bounded, and thus has an accumumlation point $T$. It then follows that the point $x$ lying at parameter $T$ along $\gamma$ is an accumulation point of the original sequence $\{x_i\}$. This establishes compactness of $D(\Sigma_{\text{int}})$. The satisfaction of Bartnik's interior condition and the existence of maximal slices then can be proven in close parallel with the results of [4].

Remarkably, it turns out that the nonexistence of black or white holes in $\text{int}D(\Sigma)$ in stationary spacetimes with ergoregions is not only sufficient to guarantee existence of maximal slices, but it also appears to be *necessary* in order to prove existence via the basic methods of [4]. In theorem 3.1 of [8], it is proven that the condition $[\mathcal{B} \cup \mathcal{W}] \cap \text{int}D(\Sigma) = \emptyset$ actually is equivalent to the following two conditions: (i) $X$ has the same orientation at each end. (ii) $D(\Sigma_{\text{int}})$ is compact. However, if condition (i) fails, then no analog of the time function $t$ exists. On the other hand, if condition (ii) fails, then even if an analog of $t$ can be constructed, there is no reason to expect it to satisfy Bartnik's interior condition. Of course, the example of Brill described above shows that existence of suitable maximal slices will fail, in general, when $[\mathcal{B} \cup \mathcal{W}] \cap \text{int}D(\Sigma) \neq \emptyset$.

In order to extend this analysis to cover the case relevant to result (2) above, we again consider a strongly causal spacetime, $(M, g_{ab})$, but now assume the presence of an acausal, asymptotically flat hypersurface $\Sigma$ with compact boundary, $S$, (again, with a finite number of asymptotically flat "ends"), whose "interior region" is compact. Again, we require $(M, g_{ab})$ to possess a one-parameter group of isometries generated by a Killing field $X$ which is timelike at each of the ends. In addition, we require $X$ to be tangent to $S$, so that $S$ is mapped into itself under the isometries. These conditions are appropriate for the case where both a black and white hole are present in the spacetime, and their horizons intersect on $S$. In this case, we prove the following: If the interior of the domain of dependence of $\Sigma$ does not contain any events lying within a black or white hole (i.e., if there is no black or white hole "exterior to $S$"), then there exists a smooth asymptotically flat maximal hypersurface $\hat{\Sigma}$ with boundary $S$ and compact interior, which is asymptotically orthogonal to $X$. Furthermore, $\Sigma \setminus S$ is a Cauchy surface for $\mathrm{int} D(\Sigma \setminus S)$. In order to prove this result, we must alter the definition of the time function, $t$, in a neighborhood of $S$ so that it does not agree with Killing parameter there. Further technical complications arise in the construction of $t$ and in the proof of compactness of $D(\Sigma_{\mathrm{int}})$. In addition, further arguments must be given in the existence proof on account of the fact that the desired maximal hypersurface has $S$ as its interior boundary. However, otherwise the proof proceeds in close parallel with the previous case. We refer the reader to [8] for further details.

The above results, of course, are applicable only to the case of spacetimes possessing a one-parameter group of isometries whose orbits are timelike near infinity. However, the key role played by the condition that no black or white hole be present in the interior of $D(\Sigma)$ suggests that this condition may be relevant to the issue of existence of maximal slices in more general classes of asymptotically flat spacetimes. This idea may be closely related to the "no hidden infinities" conjecture of Bartnik (see conjecture 3 of [3]).

As mentioned above, all of the results reported here were obtained in collaboration with Piotr Chruściel and are presented in detail in [8].

# References

[1] Bartnik R. *Commun. Math. Phys.* **94**, 155 (1984).

[2] Bartnik R. *Acta Math.* **161**, 145 (1988).

[3] Bartnik R. *Commun. Math. Phys.* **117**, 615 (1988).

[4] Bartnik R., Chruściel P.T., O'Murchadha N. *Commun. Math. Phys.* **130**, 95 (1990).

[5] Brill D, Deser S. *Ann. Phys.* **50** 548 (1968).

[6] Brill D. *On spacetimes without maximal surfaces.* Proceedings of the Third Marcel Grossman Conference, Ning H. (ed.), North Holland, Amsterdam 1991.

[7] Brill D., Flaherty F. *Commun. Math. Phys.* **50** 157 (1976).

[8] Chruściel P.T., Wald R.M. *Maximal hypersurfaces in asymptotically stationary spacetimes.* (preprint, 1992).

[9] Sudarsky D., Wald R.M. *Phys. Rev.* **D46**, 1453 (1992).

[10] Wald R.M. *The first law of black hole mechanics* in *Misner Festschrift*, Vol. 1 of *Directions in General Relativity*, eds. B. L. Hu, M. P. Ryan and C. V. Vishveshwara (Cambridge University Press, Cambridge, 1993)

# (1+1)-Dimensional Methods for General Relativity

*Jong-Hyuk Yoon* *

## Abstract

We present the (1+1)-dimensional method for studying general relativity of 4-dimensions. We first discuss the general formalism, and subsequently draw attention to the algebraically special class of space-times, following the Petrov classification. It is shown that this class of space-times can be described by the (1+1)-dimensional Yang-Mills action interacting with matter fields, with the spacial diffeomorphisms of the 2-surface as the gauge symmetry. The (Hamiltonian) constraint appears polynomial in part, whereas the non-polynomial part is a non-linear sigma model type in (1+1)-dimensions. It is also shown that the representations of $w_\infty$-gravity appear naturally as special cases of this description, and we discuss briefly the $w_\infty$-geometry in term of the fibre bundle.

## 1. Introduction

For past years many 2-dimensional field theories have been intensively studied as laboratories for many theoretical issues, due to great mathematical simplicities that often exist in 2-dimensional systems. Recently these 2-dimensional field theories have received considerable attention, for different reasons, in connection with general relativistic systems of 4-dimensions, such as self-dual spaces [1] and the black-hole space-times [2, 3]. These 2-dimensional formulations of self-dual spaces and black-hole space-times of allow, in principle, many 2-dimensional field theoretic methods developed in the past relevant for the description of the physics of 4-dimensions. This raises an intriguing question as to whether it is also possible to describe general relativity itself as a 2-dimensional field theory. Recently we have shown that such a description is indeed possible, and obtained, at least formally, the corresponding (1+1)-dimensional action principle based on the (2+2)-decomposition of general space-times[1][4]. In particular, the algebraically special class of space-times (the

---

*Center for Theoretical Physics and Department of Physics, Seoul National University, Seoul 151-742, Korea. e-mail address: SNU00162@KRSNUCC1.Bitnet. This work was partially supported by the Ministry of Education and by the Korea Science and Engineering Foundation.

[1]Here we are viewing space-times of 4-dimensions as locally fibrated, $M_{1+1} \times N_2$, with $M_{1+1}$ as the base manifold of signature $(-, +)$ and $N_2$ as the 2-dimensional fibre space of signature $(+, +)$.

Petrov type II), following the Petrov classification [5], was studied as an illustration from this perspective, and the (1+1)-dimensional action principle and the constraints for this class were identified [6]. In this (2+2)-decomposition general relativity shows up as a (1+1)-dimensional gauge theory interacting with (1+1)-dimensional matter fields, with the *minimal* coupling to the gauge fields, where the gauge symmetry is the diffeomorphisms of the fibre of 2-spacial dimensions. In this article we shall review our recent attempts of the (1+1)-dimensional formulation of general relativity. This article is organized as follows. In section 2, we present the general formalism of (2+2)-decomposition of general relativity, and establish the corresponding (1+1)-dimensional action principle.

In section 3, we draw attention to the algebraically special class of space-times [5, 7], following the Petrov classification, and present the (1+1)-dimensional action principle for this entire class of space-times. We shall show that the spacial diffeomorphisms of the 2-surface becomes the gauge fixing condition in this description. The (Hamiltonian) constraint is polynomial in part, whereas the non-polynomial term is a non-linear sigma model type in (1+1)-dimensions. As such, this formulation might render the problem of the constraints of general relativity manageable, at least formally.

In section 4, we discuss the realizations of the so-called $w_\infty$-gravity as special cases of this description. We find the fibre bundle as the natural framework for the geometric description of $w_\infty$-gravity, whose geometric understanding was lacking so far [8, 9]. In this picture the local gauge fields for $w_\infty$-gravity are identified as the connections valued in the infinite dimensional Lie algebra associated with the area-preserving diffeomorphisms of the 2-dimensional fibre. Due to this picture of $w_\infty$-geometry, we are able to construct field theoretic realizations of $w_\infty$-gravity in a straightforward way. In section 5, we summarize this review and discuss a few problems for the future investigations.

## 2.  $(2+2)$-decomposition of general relativity

Consider a 4-dimensional manifold $P_4 \simeq M_{1+1} \times N_2$, equipped with a metric $g_{AB}$ $(A, B, \cdots = 0, 1, 2, 3)^2$. Let $\partial_\mu = \partial/\partial x^\mu$ $(\mu, \nu, \cdots = 0, 1)$ and $\partial_a = \partial/\partial y^a$ $(a, b, \cdots = 2, 3)$ be a coordinate basis of $M_{1+1}$ and $N_2$, respectively, and choose $\partial_A = (\partial_\mu, \partial_a)$ as a coordinate basis of $P_4$. In this basis the most general metric on $P_4$ can be written as [10]

$$ds^2 = \phi_{ab}dy^a dy^b + \left(\gamma_{\mu\nu} + \phi_{ab}A_\mu{}^a A_\nu{}^b\right)dx^\mu dx^\nu + 2\phi_{ab}A_\mu{}^b dx^\mu dy^a. \tag{2.1}$$

---

[2]From here on, we shall distinguish the two manifolds by their signatures to avoid confusion. Namely, $M_{1+1}$ shall be referred to as the (1+1)-dimensional manifold and $N_2$ as 2-dimensional manifold.

Formally this is quite similar to the 'dimensional reduction' in Kaluza-Klein theory, where $N_2$ is regarded as the 'internal' fibre [3] and $M_{1+1}$ as the 'space-time'. In the standard Kaluza-Klein reduction one assumes a restriction on the metric, namely, an isometry condition, to make $A_\mu{}^a$ a gauge field associated with the isometry group. Here, however, we do not assume any isometry condition, and allow all the fields to depend arbitrarily on both $x^\mu$ and $y^a$. Nevertheless $A_\mu{}^a(x, y)$ can still be identified as a connection, but now associated with an infinite dimensional diffeomorphism group diff$N_2$. To show this, let us consider the following diffeomorphism of $N_2$,

$$y'^a = y'^a(y^b, x^\mu), \qquad x'^\mu = x^\mu. \tag{2.2}$$

Under these transformations, we find

$$\gamma'_{\mu\nu}(y', x) = \gamma_{\mu\nu}(y, x), \tag{2.3a}$$

$$\phi'_{ab}(y', x) = \frac{\partial y^c}{\partial y'^a} \frac{\partial y^d}{\partial y'^b} \phi_{cd}(y, x), \tag{2.3b}$$

$$A_\mu{}'^a(y', x) = \frac{\partial y'^a}{\partial y^c} A_\mu{}^c(y, x) - \partial_\mu y'^a. \tag{2.3c}$$

For the corresponding infinitesimal variations such that

$$\delta y^a = \xi^a(y^b, x^\mu), \qquad \delta x^\mu = 0, \tag{2.4}$$

(2.3) become

$$\delta\gamma_{\mu\nu} = -[\xi, \gamma_{\mu\nu}] = -\mathcal{L}_\xi \gamma_{\mu\nu} = -\xi^c \partial_c \gamma_{\mu\nu}, \tag{2.5a}$$

$$\delta\phi_{ab} = -[\xi, \phi]_{ab} = -\mathcal{L}_\xi \phi_{ab}$$

$$= -\xi^c \partial_c \phi_{ab} - (\partial_a \xi^c)\phi_{cb} - (\partial_b \xi^c)\phi_{ac}, \tag{2.5b}$$

$$\delta A_\mu{}^a = -\partial_\mu \xi^a + [A_\mu, \xi]^a = -\partial_\mu \xi^a + \mathcal{L}_{A_\mu} \xi^a$$

$$= -\partial_\mu \xi^a + (A_\mu^c \partial_c \xi^a - \xi^c \partial_c A_\mu{}^a), \tag{2.5c}$$

where $\mathcal{L}_\xi$ represents the Lie derivative along the vector fields $\xi = \xi^a \partial_a$, and acts only on the 'internal' indices $a, b$, etc. Notice that the Lie derivative, an *infinite* dimensional generalization of the finite dimensional matrix commutators, appears naturally. Clearly (2.4) defines a gauge transformation which leaves the line element (2.1) invariant. Associated with this gauge transformation, the *covariant* derivative $D_\mu$ is defined by

$$D_\mu = \partial_\mu - \mathcal{L}_{A_\mu}, \tag{2.6}$$

where the Lie derivative is taken along the vector field $A_\mu = A_\mu{}^a \partial_a$. With this definition, we have

$$\delta A_\mu{}^a = -D_\mu \xi^a, \tag{2.7}$$

---

[3]For the algebraically special class of space-times we shall consider in section 3, the fibre space $N_2$ may be interpreted as the physical transverse wave-surface [5].

which clearly indicates that $A_\mu{}^a$ is the gauge field valued in the infinite dimensional Lie algebra associated with the diffeomorphisms of $N_2$. Moreover the transformation properties (2.3a) and (2.3b) show that $\gamma_{\mu\nu}$ and $\phi_{ab}$ are a scalar and tensor field, respectively, under diff $N_2$. The field strength $F_{\mu\nu}{}^a$ corresponding to $A_\mu{}^a$ can now be defined as

$$[D_\mu, D_\nu] = -F_{\mu\nu}{}^a \partial_a = -\{\partial_\mu A_\nu{}^a - \partial_\nu A_\mu{}^a - [A_\mu, A_\nu]^a\}\partial_a. \qquad (2.8)$$

Notice that the field strength transforms covariantly under the infinitesimal transformation (2.4),

$$\delta F_{\mu\nu}{}^a = -[\xi, F_{\mu\nu}]^a = -\pounds_\xi F_{\mu\nu}{}^a. \qquad (2.9)$$

To find the (1+1)-dimensional action principle of general relativity, we must compute the scalar curvature of space-times in the (2+2)-decomposition. For this purpose it is convenient to introduce the following non-coordinate basis $\hat{\partial}_A = (\hat{\partial}_\mu, \hat{\partial}_a)$ where [11]

$$\hat{\partial}_\mu \equiv \partial_\mu - A_\mu{}^a(x, y)\partial_a, \qquad \hat{\partial}_a \equiv \partial_a . \qquad (2.10)$$

From the definition we have

$$[\hat{\partial}_A, \hat{\partial}_B] = f_{AB}{}^C(x, y)\hat{\partial}_C, \qquad (2.11)$$

where the structure coefficients $f_{AB}{}^C$ are given by

$$\begin{aligned} f_{\mu\nu}{}^a &= -F_{\mu\nu}{}^a, \\ f_{\mu a}{}^b &= -f_{a\mu}{}^b = \partial_a A_\mu{}^b, \\ f_{AB}{}^C &= 0, \qquad \text{otherwise.} \end{aligned} \qquad (2.12)$$

The virtue of this basis is that it brings the metric (2.1) into a block diagonal form

$$g_{AB} = \begin{pmatrix} \gamma_{\mu\nu} & 0 \\ 0 & \phi_{ab} \end{pmatrix}, \qquad (2.13)$$

which drastically simplifies the computation of the scalar curvature. In this basis the Levi-Civita connections are given by

$$\Gamma_{AB}{}^C = \frac{1}{2}g^{CD}(\hat{\partial}_A g_{BD} + \hat{\partial}_B g_{AD} - \hat{\partial}_D g_{AB}) + \frac{1}{2}g^{CD}(f_{ABD} - f_{BDA} - f_{ADB}), \qquad (2.14)$$

where $f_{ABC} = g_{CD}f_{AB}{}^D$. For completeness, we present the connection coefficients in components,

$$\begin{aligned} \Gamma_{\mu\nu}{}^\alpha &= \frac{1}{2}\gamma^{\alpha\beta}\left(\hat{\partial}_\mu \gamma_{\nu\beta} + \hat{\partial}_\nu \gamma_{\mu\beta} - \hat{\partial}_\beta \gamma_{\mu\nu}\right), \\ \Gamma_{\mu\nu}{}^a &= -\frac{1}{2}\phi^{ab}\partial_b \gamma_{\mu\nu} - \frac{1}{2}F_{\mu\nu}{}^a, \end{aligned}$$

$$\Gamma_{\mu a}{}^{\nu} = \Gamma_{a\mu}{}^{\nu} = \frac{1}{2}\gamma^{\nu\alpha}\partial_a\gamma_{\mu\alpha} + \frac{1}{2}\gamma^{\nu\alpha}\phi_{ab}F_{\mu\alpha}{}^{b},$$

$$\Gamma_{\mu a}{}^{b} = \frac{1}{2}\phi^{bc}\hat{\partial}_\mu\phi_{ac} + \frac{1}{2}\partial_a A_\mu{}^{b} - \frac{1}{2}\phi^{bc}\phi_{ae}\partial_c A_\mu{}^{e},$$

$$\Gamma_{a\mu}{}^{b} = \frac{1}{2}\phi^{bc}\hat{\partial}_\mu\phi_{ac} - \frac{1}{2}\partial_a A_\mu{}^{b} - \frac{1}{2}\phi^{bc}\phi_{ae}\partial_c A_\mu{}^{e},$$

$$\Gamma_{ab}{}^{\mu} = -\frac{1}{2}\gamma^{\mu\nu}\hat{\partial}_\nu\phi_{ab} + \frac{1}{2}\gamma^{\mu\nu}\phi_{ac}\partial_b A_\nu{}^{c} + \frac{1}{2}\gamma^{\mu\nu}\phi_{bc}\partial_a A_\nu{}^{c},$$

$$\Gamma_{ab}{}^{c} = \frac{1}{2}\phi^{cd}\left(\partial_a\phi_{bd} + \partial_b\phi_{ad} - \partial_d\phi_{ab}\right). \tag{2.15}$$

For later purposes it is useful to have the following identities,

$$\Gamma_{\alpha\mu}{}^{\alpha} = \frac{1}{2}\gamma^{\alpha\beta}\hat{\partial}_\mu\gamma_{\alpha\beta}, \qquad \Gamma_{a\mu}{}^{a} = \frac{1}{2}\phi^{ab}\hat{\partial}_\mu\phi_{ab} - \partial_a A_\mu{}^{a}, \tag{2.16a}$$

$$\Gamma_{\beta a}{}^{\beta} = \frac{1}{2}\gamma^{\alpha\beta}\partial_a\gamma_{\alpha\beta}, \qquad \Gamma_{ba}{}^{b} = \frac{1}{2}\phi^{bc}\partial_a\phi_{bc}. \tag{2.16b}$$

The curvature tensors are defined as

$$R_{ABC}{}^{D} = \hat{\partial}_A\Gamma_{BC}{}^{D} - \hat{\partial}_B\Gamma_{AC}{}^{D} + \Gamma_{AE}{}^{D}\Gamma_{BC}{}^{E} - \Gamma_{BE}{}^{D}\Gamma_{AC}{}^{E} - f_{AB}{}^{E}\Gamma_{EC}{}^{D},$$
$$R_{AC} = R_{ABC}{}^{B}, \qquad R = g^{AC}R_{AC}. \tag{2.17}$$

Explicitly, the scalar curvature $R$ is given by

$$R = \gamma^{\mu\nu}(R_{\mu\alpha\nu}{}^{\alpha} + R_{\mu a\nu}{}^{a}) + \phi^{ab}(R_{acb}{}^{c} + R_{a\mu b}{}^{\mu}), \tag{2.18}$$

which becomes, after a lengthy computation,

$$\begin{aligned}
R =\ & \gamma^{\mu\nu}R_{\mu\nu} + \phi^{ac}R_{ac} + \frac{1}{4}\phi_{ab}\gamma^{\mu\nu}\gamma^{\alpha\beta}F_{\mu\alpha}{}^{a}F_{\nu\beta}{}^{b} \\
& + \frac{1}{4}\gamma^{\mu\nu}\phi^{ab}\phi^{cd}\left\{(D_\mu\phi_{ac})(D_\nu\phi_{bd}) - (D_\mu\phi_{ab})(D_\nu\phi_{cd})\right\} \\
& + \frac{1}{4}\phi^{ab}\gamma^{\mu\nu}\gamma^{\alpha\beta}\left\{(\partial_a\gamma_{\mu\alpha})(\partial_b\gamma_{\nu\beta}) - (\partial_a\gamma_{\mu\nu})(\partial_b\gamma_{\alpha\beta})\right\} + \nabla_A j^{A}, \tag{2.19}
\end{aligned}$$

where $R_{\mu\nu}$ and $R_{ac}$ are defined by

$$R_{\mu\nu} = \hat{\partial}_\mu\Gamma_{\alpha\nu}{}^{\alpha} - \hat{\partial}_\alpha\Gamma_{\mu\nu}{}^{\alpha} + \Gamma_{\mu\beta}{}^{\alpha}\Gamma_{\alpha\nu}{}^{\beta} - \Gamma_{\beta\alpha}{}^{\beta}\Gamma_{\mu\nu}{}^{\alpha}, \tag{2.20a}$$

$$R_{ac} = \partial_a\Gamma_{bc}{}^{b} - \partial_b\Gamma_{ac}{}^{b} + \Gamma_{ad}{}^{b}\Gamma_{bc}{}^{d} - \Gamma_{db}{}^{d}\Gamma_{ac}{}^{b}. \tag{2.20b}$$

The last term in (2.19) is given by

$$\nabla_A j^{A} = \nabla_\mu j^{\mu} + \nabla_a j^{a}, \tag{2.21a}$$

$$\nabla_\mu j^{\mu} = \left(\hat{\partial}_\mu + \Gamma_{\alpha\mu}{}^{\alpha} + \Gamma_{c\mu}{}^{c}\right)j^{\mu}, \tag{2.21b}$$

$$\nabla_a j^{a} = \left(\partial_a + \Gamma_{ca}{}^{c} + \Gamma_{\alpha a}{}^{\alpha}\right)j^{a}, \tag{2.21c}$$

where $j^\mu$ and $j^a$ are given by

$$j^\mu = \gamma^{\mu\nu}\left(\phi^{ab}\hat{\partial}_\nu\phi_{ab} - 2\partial_a A_\nu{}^a\right), \qquad j^a = \phi^{ab}\gamma^{\mu\nu}\partial_b\gamma_{\mu\nu}. \tag{2.22}$$

That $\nabla_A j^A$ is a surface term in the action integral can be seen easily, using (2.16). For instance let us show that $\sqrt{-\gamma}\sqrt{\phi}\nabla_\mu j^\mu$ is a surface term, where $\gamma = \det\gamma_{\mu\nu}$ and $\phi = \det\phi_{ab}$. From (2.21b) we have

$$\sqrt{-\gamma}\sqrt{\phi}\nabla_\mu j^\mu = \sqrt{-\gamma}\sqrt{\phi}\left[\partial_\mu j^\mu - A_\mu{}^a\partial_a j^\mu + \left(\Gamma_{\alpha\mu}{}^\alpha + \Gamma_{c\mu}{}^c\right)j^\mu\right]. \tag{2.23}$$

The first term in the r.h.s. of (2.23) can be written as

$$\sqrt{-\gamma}\sqrt{\phi}\,\partial_\mu j^\mu = -\frac{1}{2}\sqrt{-\gamma}\sqrt{\phi}\left(\gamma^{\alpha\beta}\partial_\mu\gamma_{\alpha\beta} + \phi^{ab}\partial_\mu\phi_{ab}\right)j^\mu + \partial_\mu\left(\sqrt{-\gamma}\sqrt{\phi}j^\mu\right), \tag{2.24}$$

and for the second term, we have

$$\sqrt{-\gamma}\sqrt{\phi}A_\mu{}^a\partial_a j^\mu = -\sqrt{-\gamma}\sqrt{\phi}\left[\left\{A_\mu{}^a(\Gamma_{\alpha a}{}^\alpha + \Gamma_{ba}{}^b) + \partial_a A_\mu{}^a\right\}j^\mu\right] + \partial_a\left(\sqrt{-\gamma}\sqrt{\phi}A_\mu{}^a j^\mu\right). \tag{2.25}$$

The last two terms in the r. h. s. of (2.23) becomes, using (2.16),

$$\sqrt{-\gamma}\sqrt{\phi}\left(\Gamma_{\alpha\mu}{}^\alpha + \Gamma_{c\mu}{}^c\right)j^\mu = \sqrt{-\gamma}\sqrt{\phi}\Big(\frac{1}{2}\gamma^{\alpha\beta}\partial_\mu\gamma_{\alpha\beta} - A_\mu{}^a\Gamma_{\alpha a}{}^\alpha + \frac{1}{2}\phi^{ab}\partial_\mu\phi_{ab}$$
$$-A_\mu{}^a\Gamma_{ba}{}^b - \partial_a A_\mu{}^a\Big)j^\mu. \tag{2.26}$$

Putting (2.24), (2.25), and (2.26) into (2.23), we find that it is a total divergence term,

$$\sqrt{-\gamma}\sqrt{\phi}\nabla_\mu j^\mu = \partial_\mu\left(\sqrt{-\gamma}\sqrt{\phi}j^\mu\right) - \partial_a\left(\sqrt{-\gamma}\sqrt{\phi}A_\mu{}^a j^a\right), \tag{2.27}$$

which we may ignore. Similarly, $\sqrt{-\gamma}\sqrt{\phi}\nabla_a j^a$ is also a surface term. This altogether shows that $\sqrt{-\gamma}\sqrt{\phi}\nabla_A j^A$ is indeed a total divergence term.

At this point it is important to notice the followings. First, $D_\mu\phi_{ab}$, written as

$$\begin{aligned} D_\mu\phi_{ab} &= \partial_\mu\phi_{ab} - \mathcal{L}_{A_\mu}\phi_{ab} \\ &= \partial_\mu\phi_{ab} - \left\{A_\mu{}^c(\partial_c\phi_{ab}) + (\partial_a A_\mu{}^c)\phi_{cb} + (\partial_b A_\mu{}^c)\phi_{ac}\right\} \end{aligned} \tag{2.28}$$

indeed transforms covariantly under the infinitesimal diffeomorphism (2.4),

$$\delta(D_\mu\phi_{ab}) = -\mathcal{L}_\xi(D_\mu\phi_{ab}) = -[\xi, D_\mu\phi]_{ab}. \tag{2.29}$$

Second, the derivative $\hat{\partial}_\mu$, when applied to $\gamma_{\mu\nu}$, becomes the covariant derivative

$$\hat{\partial}_\mu\gamma_{\alpha\beta} = \partial_\mu\gamma_{\alpha\beta} - \mathcal{L}_{A_\mu}\gamma_{\alpha\beta} = D_\mu\gamma_{\alpha\beta}, \tag{2.30}$$

so that $\hat{\partial}_\mu \gamma_{\alpha\beta}$ transforms covariantly

$$\delta(\hat{\partial}_\mu \gamma_{\alpha\beta}) = -\pounds_\xi(D_\mu \gamma_{\alpha\beta}) = -[\xi, D_\mu \gamma_{\alpha\beta}]. \tag{2.31}$$

These observations play an important role when we discuss the gauge invariance of the theory under diff$N_2$. It is worth mentioning here that, from (2.20a) and (2.30), $R_{\mu\nu}$ becomes the 'covariantized' Ricci tensor

$$R_{\mu\nu} = D_\mu \Gamma_{\alpha\nu}{}^\alpha - D_\alpha \Gamma_{\mu\nu}{}^\alpha + \Gamma_{\mu\beta}{}^\alpha \Gamma_{\alpha\nu}{}^\beta - \Gamma_{\beta\alpha}{}^\beta \Gamma_{\mu\nu}{}^\alpha, \tag{2.32}$$

as $\Gamma_{\mu\nu}{}^\alpha$'s do not involve the 'internal' indices $a, b$, etc. Thus we might call $\gamma^{\mu\nu} R_{\mu\nu}$ as the 'gauged' gravity action in (1+1)-dimensions [12].

With the scalar curvature at hand, one can easily write down the lagrangian for the Einstein-Hilbert action on $P_4$. From (2.19) we have

$$\begin{aligned}
\mathcal{L}_2 = {} & -\sqrt{-\gamma}\sqrt{\phi}\Big[\gamma^{\mu\nu} R_{\mu\nu} + \phi^{ab} R_{ab} + \frac{1}{4}\phi_{ab} F_{\mu\nu}{}^a F^{\mu\nu b} \\
& + \frac{1}{4}\gamma^{\mu\nu}\phi^{ab}\phi^{cd}\big\{(D_\mu \phi_{ac})(D_\nu \phi_{bd}) - (D_\mu \phi_{ab})(D_\nu \phi_{cd})\big\} \\
& + \frac{1}{4}\phi^{ab}\gamma^{\mu\nu}\gamma^{\alpha\beta}\big\{(\partial_a \gamma_{\mu\alpha})(\partial_b \gamma_{\nu\beta}) - (\partial_a \gamma_{\mu\nu})(\partial_b \gamma_{\alpha\beta})\big\}\Big],
\end{aligned} \tag{2.33}$$

neglecting the total divergence term (2.27). Clearly the action principle describes a (1+1)-dimensional field theory which is invariant under the gauge transformation of diff$N_2$, as the gauge field $A_\mu{}^a$ couples *minimally* to both $\gamma_{\mu\nu}$ and $\phi_{ab}$. Therefore each term in (2.33) is invariant under diff$N_2$. To understand the physical contents of the theory we notice the followings. First, unlike the ordinary gravity, the metric $\gamma_{\mu\nu}$ of $M_{1+1}$ here is 'charged', because it couples to $A_\mu{}^a$ (with the coupling constant 1). Second, the metric $\phi_{ab}$ of $N_2$ can be identified as a non-linear sigma field, whose self-interaction potential is determined by the scalar curvature $\phi^{ab} R_{ab}$ of $N_2$. The theory therefore describes a gauge theory of diff$N_2$ interacting with the 'gauged' gravity and the non-linear sigma field on $M_{1+1}$.

## 3.   Algebraically special class of space-times

In contrast to the cases of the self-dual spaces and black-hole space-times, the (1+1)-dimensional action principle for general space-times, as we derived in the previous section, appears to be rather formal and consequently, of little practical use. In this section we therefore draw attention following the Petrov classification to a specific class of space-times, namely, the algebraically special class, and interpret the entire class from the (1+1)-dimensional point of view. It turns out that space-times of this class can be formulated as (1+1)-dimensional field theory in a remarkably simple form.

Let us consider a class of space-times that contain a twist-free null vector field $k^A$. These space-times belong to the algebraically special class of space-times, according to the Petrov classification. This class of space-times is rather broad, since most of the known exact solutions of the Einstein's equations are algebraically special. Being twist-free, the null vector field may be chosen to be a gradient field, so that $k_A = \partial_A u$ for some function $u$. The null hypersurface $N_2$ defined by $u = $ constant spans the 2-dimensional subspace for which we introduce two space-like coordinates $y^a$. The general line element for this class has the form [5, 7]

$$ds^2 = \phi_{ab}dy^a dy^b - 2du(dv + m_a dy^a + Hdu), \tag{3.1}$$

where $v$ is the affine parameter, and $\phi_{ab}$, $m_a$ and $H$ are functions of all of the four coordinates $(u, v, y^a)$, as we assume no Killing vector fields.

For the class of space-times (3.1), we shall find the (1+1)-dimensional action principle defined on the $(u, v)$-surface. For this purpose let us first introduce the 'light-cone' coordinates $(u, v)$ such that

$$u = \frac{1}{\sqrt{2}}(x^0 + x^1), \qquad v = \frac{1}{\sqrt{2}}(x^0 - x^1), \tag{3.2}$$

and define $A_u{}^a$ and $A_v{}^a$

$$A_u{}^a = \frac{1}{\sqrt{2}}(A_0{}^a + A_1{}^a), \qquad A_v{}^a = \frac{1}{\sqrt{2}}(A_0{}^a - A_1{}^a). \tag{3.3}$$

For $\gamma_{\mu\nu}$, we assume the Polyakov ansatz [13]

$$\gamma_{\mu\nu} = \begin{pmatrix} -2h & -1 \\ -1 & 0 \end{pmatrix}, \qquad \gamma^{\mu\nu} = \begin{pmatrix} 0 & -1 \\ -1 & 2h \end{pmatrix}, \qquad (\det\gamma_{\mu\nu} = -1), \tag{3.4}$$

in the $(u, v)$-coordinates. Then the line element (2.1) becomes

$$\begin{aligned} ds^2 &= \phi_{ab}dy^a dy^b - 2dudv - 2h(du)^2 + \phi_{ab}(A_u{}^a du + A_v{}^a dv)(A_u{}^b du + A_v{}^b dv) \\ &\quad + 2\phi_{ab}(A_u{}^a du + A_v{}^a dv)dy^b. \end{aligned} \tag{3.5}$$

If we choose the 'light-cone' gauge [4] $A_v{}^a = 0$, then this becomes

$$ds^2 = \phi_{ab}dy^a dy^b - 2\,du\left[dv - \phi_{ab}A_u{}^b dy^a + \left(h - \frac{1}{2}\phi_{ab}A_u{}^a A_u{}^b\right)du\right]. \tag{3.6}$$

A comparison of (3.1) and (3.6) tells us that if the following identifications

$$m_a = -\phi_{ab}A_u{}^b, \qquad H = h - \frac{1}{2}\phi_{ab}A_u{}^a A_u{}^b \tag{3.7}$$

---

[4]Here we are referring to the disposable gauge degrees of freedom in the action. There could be topological obstruction against globalizing this choice, as the general coordinate transformation of $N_2$ corresponds to the gauge transformation.

are made, then the two line elements are the same. This shows that the Polyakov ansatz (3.4) amounts to the restriction (modulo the gauge choice $A_v{}^a = 0$) to the algebraically special class of space-times that contain a twist-free null vector field.

Let us now examine the transformation properties of $h$, $\phi_{ab}$, $A_u{}^a$, and $A_v{}^a$ under the diffeomorphism of $N_2$,

$$y'^a = y'^a(y^b, u, v), \qquad u' = u, \qquad v' = v. \tag{3.8}$$

Under these transformations, we find that

$$h'(y', u, v) = h(y, u, v), \tag{3.9a}$$

$$\phi'_{ab}(y', u, v) = \frac{\partial y^c}{\partial y'^a} \frac{\partial y^d}{\partial y'^b} \phi_{cd}(y, u, v), \tag{3.9b}$$

$$A_u'^a(y', u, v) = \frac{\partial y'^a}{\partial y^c} A_u{}^c(y, u, v) - \partial_u y'^a, \tag{3.9c}$$

$$A_v'^a(y', u, v) = -\partial_v y'^a, \tag{3.9d}$$

which become, under the infinitesimal variations, $\delta y^a = \xi^a(y, u, v)$ and $\delta x^\mu = 0$,

$$\delta h = -[\xi, h] = -\xi^a \partial_a h, \tag{3.10a}$$

$$\delta \phi_{ab} = -[\xi, \phi]_{ab} = -\xi^c \partial_c \phi_{ab} - (\partial_a \xi^c)\phi_{cb} - (\partial_b \xi^c)\phi_{ac}, \tag{3.10b}$$

$$\delta A_u{}^a = -D_u \xi^a = -\partial_u \xi^a + [A_u, \xi]^a, \tag{3.10c}$$

$$\delta A_v{}^a = -\partial_v \xi^a. \tag{3.10d}$$

This shows that $h$ and $\phi_{ab}$ are a scalar and tensor field, respectively, and $A_u{}^a$ and $A_v{}^a$ are the gauge fields valued in the infinite dimensional Lie algebra associated with the group of diffeomorphisms of $N_2$. That $A_v{}^a$ is a pure gauge is clear, as it depends on the gauge function $\xi^a$ only. Therefore it can be always set to zero, at least locally, by a suitable coordinate transformation (3.8). To maintain the explicit gauge invariance, however, we shall work with the line element (3.5) in the following, with the understanding that $A_v{}^a$ is a pure gauge.

Let us now proceed to write down the action principle for (3.5) in terms of the fields $h$, $\phi_{ab}$, $A_u{}^a$, and $A_v{}^a$. For this purpose, it is convenient to decompose the 2-dimensional metric $\phi_{ab}$ into the conformal classes

$$\phi_{ab} = \Omega \rho_{ab}, \qquad (\Omega > 0 \text{ and } \det \rho_{ab} = 1). \tag{3.11}$$

The kinetic term $K$ of $\phi_{ab}$ in (2.33) then becomes

$$
\begin{aligned}
K &\equiv \frac{1}{4} \sqrt{-\gamma} \sqrt{\phi} \gamma^{\mu\nu} \phi^{ab} \phi^{cd} \left\{ (D_\mu \phi_{ac})(D_\nu \phi_{bd}) - (D_\mu \phi_{ab})(D_\nu \phi_{cd}) \right\} \\
&= -\frac{(D_\mu \Omega)^2}{2\Omega} + \frac{1}{4} \Omega \gamma^{\mu\nu} \rho^{ab} \rho^{cd}(D_\mu \rho_{ac})(D_\nu \rho_{bd}) \\
&= -\frac{1}{2} e^\sigma (D_\mu \sigma)^2 + \frac{1}{4} e^\sigma \gamma^{\mu\nu} \rho^{ab} \rho^{cd}(D_\mu \rho_{ac})(D_\nu \rho_{bd}), \tag{3.12}
\end{aligned}
$$

where we defined $\sigma$ by $\sigma = \ln\Omega$, and the covariant derivatives $D_\mu\Omega$, $D_\mu\rho_{ab}$, and $D_\mu\sigma$ are

$$D_\mu\Omega = \partial_\mu\Omega - A_\mu{}^a\partial_a\Omega - (\partial_a A_\mu{}^a)\Omega, \tag{3.13a}$$
$$D_\mu\rho_{ab} = \partial_\mu\rho_{ab} - [A_\mu, \rho]_{ab} + (\partial_c A_\mu{}^c)\rho_{ab}, \tag{3.13b}$$
$$D_\mu\sigma = \partial_\mu\sigma - A_\mu{}^a\partial_a\sigma - \partial_a A_\mu{}^a, \tag{3.13c}$$

respectively, where $[A_\mu, \rho]_{ab}$ is given by

$$[A_\mu, \rho]_{ab} = A_\mu{}^c\partial_c\rho_{ab} + (\partial_a A_\mu{}^c)\rho_{cb} + (\partial_b A_\mu{}^c)\rho_{ac}. \tag{3.14}$$

The inclusion of the divergence term $\partial_a A_\mu{}^a$ in (3.13) is necessary to ensure (3.13) transform covariantly (as the tensor fields) under diff$N_2$, since $\Omega$ and $\rho_{ab}$ are the tensor densities of weight $-1$ and $+1$, respectively. Using the ansatz (3.4), the kinetic term (3.12) becomes

$$\begin{aligned} K = {} & e^\sigma(D_+\sigma)(D_-\sigma) - \frac{1}{2}e^\sigma\rho^{ab}\rho^{cd}(D_+\rho_{ac})(D_-\rho_{bd}) \\ & -he^\sigma\left\{(D_-\sigma)^2 - \frac{1}{2}\rho^{ab}\rho^{cd}(D_-\rho_{ac})(D_-\rho_{bd})\right\}, \end{aligned} \tag{3.15}$$

where $+(-)$ stands for $u(v)$. The Polyakov ansatz (3.4) simplifies enormously the remaining terms in the action (2.33), as we now show. Let us first notice that $\det\gamma_{\mu\nu} = -1$. Therefore the term

$$\sqrt{-\gamma}\sqrt{\phi}\phi^{ac}R_{ac} = \sqrt{\phi}\phi^{ac}R_{ac} \tag{3.16}$$

can be removed from the action being a surface term. Moreover, since we have

$$\gamma^{\mu\nu}\partial_a\gamma_{\mu\nu} = \frac{2}{\sqrt{-\gamma}}\partial_a\sqrt{-\gamma} = 0, \tag{3.17}$$

the last term in the action (2.33) vanishes. Furthermore, one can easily verify that

$$\begin{aligned} \phi^{ab}\gamma^{\mu\nu}\gamma^{\alpha\beta}(\partial_a\gamma_{\mu\alpha})(\partial_b\gamma_{\nu\beta}) &= \phi^{ab}(\partial_a\gamma_{++})\gamma^{+-}(\partial_b\gamma_{-\alpha})\gamma^{\alpha+} \\ &= 0, \end{aligned} \tag{3.18}$$

since $\partial_b\gamma_{-\alpha} = 0$. The only remaining terms that contribute to the action (2.33) are thus the (1+1)-dimensional Yang-Mills action and the 'gauged' gravity action. The Yang-Mills action becomes

$$\frac{1}{4}\phi_{ab}F_{\mu\nu}{}^a F^{\mu\nu b} = -\frac{1}{2}e^\sigma\rho_{ab}F_{+-}{}^a F_{+-}{}^b. \tag{3.19}$$

To express the 'gauged' Ricci scalar $\gamma^{\mu\nu}R_{\mu\nu}$ in terms of $h$ and $A_v{}^a$, etc., we have to compute the Levi-Civita connections first. They are given by

$$\begin{aligned} \Gamma_{++}^+ &= -D_-h, & \Gamma_{++}^- &= D_+h + 2hD_-h, \\ \Gamma_{+-}^- &= \Gamma_{-+}^- = D_-h, \end{aligned} \tag{3.20}$$

and vanishing otherwise. Thus the 'gauged' Ricci tensor becomes

$$R_{+-} = R_{-+} = -D_-^2 h, \qquad R_{--} = 0. \tag{3.21}$$

From (3.4) and (3.21), the 'gauged' Ricci scalar $\gamma^{\mu\nu} R_{\mu\nu}$ is given by

$$\gamma^{\mu\nu} R_{\mu\nu} = 2\gamma^{+-} R_{+-} = 2D_-^2 h, \tag{3.22}$$

since $\gamma^{++} = R_{--} = 0$. Putting together (3.15), (3.19), and (3.22) into (2.33), the action becomes

$$
\begin{aligned}
\mathcal{L}_2 = \; & -\frac{1}{2} e^{2\sigma} \rho_{ab} F_{+-}{}^a F_{+-}{}^b + e^\sigma (D_+\sigma)(D_-\sigma) - \frac{1}{2} e^\sigma \rho^{ab} \rho^{cd} (D_+\rho_{ac})(D_-\rho_{bd}) \\
& + h e^\sigma \left\{ \frac{1}{2} \rho^{ab} \rho^{cd} (D_-\rho_{ac})(D_-\rho_{bd}) - (D_-\sigma)^2 \right\} + 2 e^\sigma D_-^2 h.
\end{aligned}
\tag{3.23}
$$

The last term in (3.23) can be expressed as

$$
\begin{aligned}
e^\sigma D_-^2 h \; = \; & e^\sigma \left( \partial_- - A_-^b \partial_b \right) \left( \partial_- h - A_-^a \partial_a h \right) \\
= \; & e^\sigma \left\{ \partial_-^2 h - \partial_- \left( A_-^a \partial_a h \right) - A_-^a \partial_a (D_- h) \right\} \\
= \; & -(\partial_- e^\sigma)(\partial_- h) + (\partial_- e^\sigma)\left( A_-^a \partial_a h \right) + \partial_a \left( e^\sigma A_-^a \right)(D_- h) \\
& + \partial_- \left( e^\sigma \partial_- h \right) - \partial_- \left( e^\sigma A_-^a \partial_a h \right) - \partial_a \left( e^\sigma A_-^a D_- h \right) \\
\simeq \; & -e^\sigma (\partial_-\sigma)(D_- h) + e^\sigma A_-^a (\partial_a \sigma)(D_- h) + e^\sigma (\partial_a A_-^a)(D_- h) \\
= \; & -e^\sigma (D_-\sigma)(D_- h),
\end{aligned}
\tag{3.24}
$$

where we dropped the surface term and used (3.13c). This can be written as

$$
\begin{aligned}
e^\sigma (D_-\sigma)(D_- h) \; = \; & e^\sigma (D_-\sigma)\left( \partial_- h - A_-^a \partial_a h \right) \\
= \; & -h \partial_- \left( e^\sigma D_-\sigma \right) + h \partial_a \left( e^\sigma A_-^a D_-\sigma \right) + \partial_- \left( h e^\sigma D_-\sigma \right) \\
& - \partial_a \left( h e^\sigma A_-^a D_-\sigma \right) \\
\simeq \; & -h e^\sigma (\partial_-\sigma)(D_-\sigma) - h e^\sigma \partial_- (D_-\sigma) + h e^\sigma A_-^a (\partial_a \sigma)(D_-\sigma) \\
& + h e^\sigma (\partial_a A_-^a)(D_-\sigma) + h e^\sigma A_-^a \partial_a (D_-\sigma) \\
= \; & -h e^\sigma \left\{ D_-^2 \sigma + (D_-\sigma)^2 \right\}.
\end{aligned}
\tag{3.25}
$$

We therefore have

$$e^\sigma D_-^2 h \simeq h e^\sigma \left\{ D_-^2 \sigma + (D_-\sigma)^2 \right\}, \tag{3.26}$$

neglecting the surface terms. The resulting (1+1)-dimensional action principle therefore becomes

$$
\begin{aligned}
\mathcal{L}_2 = \; & -\frac{1}{2} e^{2\sigma} \rho_{ab} F_{+-}{}^a F_{+-}{}^b + e^\sigma (D_+\sigma)(D_-\sigma) - \frac{1}{2} e^\sigma \rho^{ab} \rho^{cd} (D_+\rho_{ac})(D_-\rho_{bd}) \\
& + h e^\sigma \left\{ 2 D_-^2 \sigma + (D_-\sigma)^2 + \frac{1}{2} \rho^{ab} \rho^{cd} (D_-\rho_{ac})(D_-\rho_{bd}) \right\},
\end{aligned}
\tag{3.27}
$$

up to the surface terms. Notice that $h$ is a Lagrange multiplier, whose variation yields the constraint

$$H_0 = D_-^2 \sigma + \frac{1}{2}(D_-\sigma)^2 + \frac{1}{4}\rho^{ab}\rho^{cd}(D_-\rho_{ac})(D_-\rho_{bd}) \approx 0. \qquad (3.28)$$

From this (1+1)-dimensional point of view, $h$ is the lapse function (or a pure gauge) that prescribes how to 'move forward in the $u$-time', carrying the surface $N_2$ at each point of the section $u = $ constant. The (Hamiltonian) constraint, $H_0 \approx 0$, is *polynomial* in $\sigma$ and $A_-^a$, and contains a non-polynomial term of the non-linear sigma model type but in (1+1)-dimensions, where such models often admit exact solutions. This allows us to view the problem of the constraints of general relativity [14] from a new perspective.

We now have the (1+1)-dimensional action principle for the algebraically special class of space-times that contain a twist-free null vector field. It is described by the Yang-Mills action, interacting with the fields $\sigma$ and $\rho_{ab}$ on the 'flat' (1+1)-dimensional surface, which however must satisfy the (Hamiltonian) constraint $H_0 \approx 0$. (The flatness of the (1+1)-dimensional surface can be seen from the fact that the lapse function, $h$, can be chosen as zero, provided that $H_0 \approx 0$ holds.) The infinite dimensional group of the diffeomorphisms of $N_2$ is *built-in* as the local gauge symmetry, via the minimal couplings to the gauge fields.

Having formulated the algebraically special class of space-times as a gauge theory on (1+1)-dimensions, we may wish to apply varieties of field theoretic methods developed in (1+1)-dimensions. For instance, the action (3.27) can be viewed as the bosonized form [15] of *some* version of the (1+1)-dimensional QCD in the infinite dimensional limit of the gauge group [16]. For small fluctuations of $\sigma$, the action (3.27) becomes

$$\mathcal{L}_2 = -\frac{1}{2}\rho_{ab}F_{+-}^{\ a}F_{+-}^{\ b} + (D_+\sigma)(D_-\sigma) - \frac{1}{2}\rho^{ab}\rho^{cd}(D_+\rho_{ac})(D_-\rho_{bd}), \qquad (3.29)$$

modulo the constraint $H_0 \approx 0$. It is beyond the scope of this article to investigate these theories as (1+1)-dimensional quantum field theories. However, this formulation raises many intriguing questions such as: would there be any phase transition in quantum gravity as viewed as the (1+1)-dimensional quantum field theories? If it does, then what does that mean in quantum geometrical terms? Thus, general relativity, as viewed from the (1+1)-dimensional perspective, renders itself to be studied as a gauge theory in full sense [17], at least for the class of space-times discussed here.

## 4.    $w_\infty$-gravity as special cases

In the previous section we derived the action principle on (1+1)-dimensions as the vantage point of studying general relativity for this algebraically special class of space-times. We now ask different but related questions: what kinds of other (1+1)-dimensional field theories related to this problem can we study? For these, let us consider the case where the local gauge symmetry is replaced by the area-preserving diffeomorphisms of $N_2$. (For these varieties of field theories, we shall drop the constraint (3.28) for the moment. It is at this point that we are departing from general relativity.) This class of field theories naturally realizes the so-called $w_\infty$-gravity [8, 9] in a linear and geometric way, as we now describe.

The area-preserving diffeomorphisms are generated by the vector fields $\xi^a$, tangent to the surface $N_2$ and divergence-free,

$$\partial_a \xi^a = 0. \tag{4.1}$$

Let us find the gauge fields $A_\pm^a$ compatible with the divergence-free condition (4.1). Taking the divergence of both sides of (3.10c) and (3.10d), we have

$$\partial_a \delta A_\pm^a = -\partial_\pm(\partial_a \xi^a) + \partial_a[A_\pm, \xi]^a. \tag{4.2}$$

This shows that the condition $\partial_a A_\pm^a = 0$ is invariant under the area-preserving diffeomorphisms, and characterizes a special subclass of the gauge fields, compatible with the condition (4.1). Moreover, when $\partial_a A_\pm^a = 0$, the fields $\rho_{ab}$ and $\sigma$ behave under the area-preserving diffeomorphisms as a tensor and a scalar field, respectively, as (3.13b) and (3.13c) suggest. Indeed, the Jacobian for the area-preserving diffeomorphisms is just 1, disregarding the distinction between the tensor fields and the tensor densities. The (1+1)-dimensional action principle now becomes

$$\mathcal{L}_2' = -\frac{1}{2}e^{2\sigma}\rho_{ab}F_{+-}{}^a F_{+-}{}^b + e^\sigma(D_+\sigma)(D_-\sigma) - \frac{1}{2}e^\sigma \rho^{ab}\rho^{cd}(D_+\rho_{ac})(D_-\rho_{bd}), \tag{4.3}$$

where $D_\mu\sigma$, $D_\mu\rho_{ab}$, and $F_{+-}{}^a$ are

$$D_\pm\sigma = \partial_\pm\sigma - A_\pm^a \partial_a\sigma, \tag{4.4a}$$

$$D_\pm\rho_{ab} = \partial_\pm\rho_{ab} - [A_\pm, \rho]_{ab}, \tag{4.4b}$$

$$F_{+-}{}^a = \partial_+ A_-^a - \partial_- A_+^a - [A_+, A_-]^a. \tag{4.4c}$$

Under the infinitesimal variations

$$\delta y^a = \xi^a(y, u, v), \qquad \delta x^\mu = 0, \qquad (\partial_a \xi^a = 0), \tag{4.5}$$

the fields transform as

$$\delta\sigma = -[\xi, \sigma] = -\xi^a \partial_a\sigma, \tag{4.6a}$$

$$\delta\rho_{ab} = -[\xi, \rho]_{ab} = -\xi^c \partial_c\rho_{ab} - (\partial_a\xi^c)\rho_{cb} - (\partial_b\xi^c)\rho_{ac}, \tag{4.6b}$$

$$\delta A_+^a = -D_+\xi^a = -\partial_+\xi^a + [A_+, \xi]^a, \tag{4.6c}$$

$$\delta A_-^a = -\partial_-\xi^a, \tag{4.6d}$$

which shows that it *is* a linear realization of the area-preserving diffeomorphisms. The geometric picture of the action principle (4.3) is now clear: it is equipped with the natural bundle structure, where the gauge fields are the connections valued in the Lie algebra associated with the area-preserving diffeomorphisms of $N_2$. Thus the action principle (4.3) provides a field theoretical realization of $w_\infty$-gravity [8, 9] in a linear and geometric way, with the built-in area-preserving diffeomorphisms as the local gauge symmetry.

With this picture of $w_\infty$-geometry at hands, we may construct as many different realizations of $w_\infty$-gravity as one wishes. The simplest example would be a single real scalar field representation, which we may write

$$\mathcal{L}_2'' = -\frac{1}{2} F_{+-}{}^a F_{+-}{}^a + (D_+\sigma)(D_-\sigma), \tag{4.7}$$

where we used $\delta_{ab}$ in the summation, and $D_\pm\sigma$ and $F_{+-}{}^a$ are as given in (4.4a) and (4.4c). By choosing the gauge $A_-{}^a = 0$ and eliminating the auxiliary field $A_+{}^a$ in terms of $\sigma$ using the equations of motion of $A_+{}^a$, we recognize (4.7) a single real scalar field realization of $w_\infty$-gravity. In presence of the auxiliary field $A_+{}^a$, (4.7) provides an example of the *linearized* realization of $w_\infty$-gravity for a single real scalar field. It would be interesting to see if the representation (4.7) is related to the ones constructed in the literatures [8, 9].

## 5.  Discussion

In this review, we examined space-times of 4-dimensions from a (1+1)-dimensional point of view. That general relativity admits such a description is rather surprising, even though the action principle in general appears rather formal. For the algebraically special class of space-times, however, the (1+1)-dimensional action principle, as we have shown here, is formulated as the Yang-Mills type gauge theories interacting with matter fields, where the infinite dimensional group of diffeomorphisms of the 2-surface becomes the 'internal' gauge symmetry. The (Hamiltonian) constraint conjugate to the lapse function appears partly as polynomial. The non-polynomial part is a typical non-linear sigma model type in (1+1)-dimensions, where such models often admit exact solutions. We also discussed the so-called $w_\infty$-gravity as special cases of the algebraically special class of space-times. The detailed study of the $w_\infty$-gravity and its geometry in terms of the fibre bundle will be presented somewhere else.

We wish to conclude with a few remarks. First, one might be interested in finding exact solutions of the Einstein's equations in this formulation. Various two (or more) Killing reductions of the Einstein's equations have been known for sometime which led to the discovery of many exact solutions to the Einstein's equations, by making the system essentially two (or lower) dimensional. In our formulation, the Einstein's equations are already put into a two dimensional form without such assumptions. This might be useful in finding new solutions of the Einstein's equations, which possess no Killing symmetries[5].

Second, we need to find the constraint algebras for the algebraically special class of space-times explicitly in terms of the variables we used here. As we have shown here, the splitting of the metric variables into the gauge fields and the 'matter fields' is indeed suitable for the description of general relativity as Yang-Mills type gauge theories in (1+1)-dimensions. It remains to study the constraint algebras in detail to see if the ordering problem in the constraints of general relativity becomes manageable in terms of these variables.

Lastly, that the Lie algebra of $SU(N)$ for large $N$ can be used as an approximation of the infinite dimensional Lie algebra of the area-preserving diffeomorphisms of the 2-surface has been suggested as a way of 'regulating' the area-preserving diffeomorphisms. In connection with the problem regarding the regularization of quantum gravity in this formulation, one might wonder as to whether it is also possible to approximate the diffeomorphism algebras of the 2-surface in terms of finite dimensional Lie algebras in a certain limit. There seem to be many interesting questions to be asked about general relativity in this formulation.

# References

[1] Q.H. Park, *Phys. Lett. 238B (1990), 287.*

[2] C.W. Misner, *Phys. Rev. D18 (1978), 4510;*
V.E. Belinskii and V.E. Zakharov, *Sov. Phys. JETP 48 (1978), 985;*
D. Maison, *Phys. Rev. Lett. 41 (1978), 521.*

[3] E. Witten, *Phys. Rev. D44 (1991), 314;*
S.B. Giddings and A. Strominger, *Phys. Rev. Lett. 67 (1991), 2930;*
J.A. Harvey and A. Strominger, *Quantum aspects of black holes,* lectures presented at the 1992 Trieste Spring School on String Theory and Quantum Gravity, September 1992, and references therein.

[4] Y.M. Cho, Q.H. Park, K.S. Soh, and J.H. Yoon, *Phys. Lett. B286 (1992), 251.*

[5] D. Kramer, H. Stephani, E. Herlt, and M. MacCallum, *Exact Solutions of Einstein's Field Equations,* Cambridge University Press, 1979.

[6] J.H. Yoon, *Algebraically special class of space-times and (1+1)-dimensional field theories,* SNUTP-92-98, submitted to Phys. Lett. B.

[7] W. Kundt, *Z. Phys. 163 (1961), 77.*

---

[5]Interestingly, there *are* exact solutions of the Einstein's equations which possess no *space-time* Killing symmetry, known as the Szekeres' dust solutions [5]. For the vacuum Einstein's equations, however, no such solutions are known, at least to the author.

[8]  E. Bergshoeff, C.N. Pope, L.J. Roman, E. Sezgin, X. Shen, and K.S. Stelle, *Phys. Lett. B243 (1990), 350;*
     K.S. Schoutens, A. Sevrin, and P. van Nieuwenhuizen, *Phys. Lett. B251 (1990), 355;*
     E. Sezgin, *Area-preserving diffeomorphisms, $w_\infty$ algebras and $w_\infty$ gravity,* lectures presented at the Trieste Summer School in High Energy Physics, July 1991;
     J.-L. Gervais and Y. Matsuo, *Phys. Lett. B274 (1992), 309.*

[9]  I. Bakas, *Phys. Lett. B228 (1989), 57;*
     C.M. Hull, *Phys. Lett. B269 (1991), 257.*

[10] Y.M. Cho, *J. Math. Phys. 16 (1975), 2029;*
     Y.M. Cho and P.G.O. Freund, *Phys. Rev. D12 (1975), 1711.*

[11] Y.M. Cho, *Phys. Rev. Lett. 67 (1991), 3469.*

[12] R. Wald, *Phys. Rev. D33 (1986), 3613 ;*
     S. Deser, J. McCarthy, and Z. Yang, *Phys. Lett. 222 (1989), 61;*
     C.N. Pope and P.K. Townsend. *Phys. Lett. B225 (1989), 245;*
     M.P. Blencowe, *Class. Quantum. Grav. 6 (1989), 443.*

[13] A.M. Polyakov, *Mod. Phys. Lett. A2 (1987), 893;*
     V.G. Knizhnik, A.M. Polyakov, and A.B. Zamolochikov, *Mod. Phys. Lett. A3 (1988), 819.*

[14] K. Kuchař, *Canonical quantization of gravity,* in Relativity, Astrophysics and Cosmology, ed. W. Israel, 1973.

[15] E. Witten, *Comm. Math. Phys. 92 (1984), 455;*
     V. Baluni, *Phys. Lett. B90 (1980), 407.*

[16] D.J. Gross and E. Witten, *Phys. Rev. D21 (1980), 446;*
     I. Bars, *Phys. Lett. 245 (1980), 35;*
     A. Jevicki and B. Sakita, *Phys. Rev. D22 (1980), 467;*
     A.A. Migdal, *Phys. Rep. 102 (1983), 199.*

[17] Y.M. Cho, *Phys. Rev. D14 (1976), 2521;*
     A. Trautman, *Rep. Math. Phys. 1 (1970), 29.*

# Coalescence of Primal Gravity Waves to Make Cosmological Mass Without Matter

*Daniel E. Holz* [*]    *Warner A. Miller* [†]    *Masami Wakano* [‡]
*John A. Wheeler* [§]

## Abstract

We propose primal-chaos black holes (PCBHs) as candidates for the missing mass. Beginning with a discussion of the mystery of the missing mass, in its various formulations, we motivate PCBHs as "dark matter." Envisioning black hole production from colliding gravity waves, we develop a model of time symmetric, axially symmetric gravity waves by making use of the Brill methodology. Through numerical spectral-element techniques, the geometry of space is determined. We discuss trapped surfaces as the signatures of impending collapse to a black hole, and are thereby able to identify, through numerical relaxation, which geometries will undergo gravitational collapse. We are thus able to determine the critical wave amplitude at which black hole production from imploding gravity waves begins. We conclude with a brief discussion of observational limits.

## 1. Introduction

"I just can't understand it. All the young men I know are retiring." So exclaimed Mrs. Niels Bohr in a post-war visit to Princeton on seeing Paul Dirac look from floor to ceiling and back again to floor in a desperate effort to answer her question, "Who is there now at Cambridge? Is Robert Frisch still there?"

"Frisch is retiring. I cannot remember who else is there, except me."

Any thought of Brill retiring is foreign to anyone who sees him in action, as vigorous now as he was in his Princeton undergraduate (A.B. 1954) and graduate (Ph. D. 1959) days. And love for the big view and passion for accurate calculation are as

---

[*]Department of Physics, University of Chicago.
[†]Theoretical Astrophysics Group, Theoretical Division, Los Alamos National Laboratory.
[‡]Department of Fundamental Sciences, Kyoto University
[§]Department of Physics, Princeton University

strong today as they were then. Nobody who looks at his list of publications with an imaginative eye will fail to spot domains of exploration to which he has opened inviting doors. Into one of them, the quantum state of a spin-endowed object, we can take a brief look-see without entering before we undertake here a fuller examination of another, the proper subject of this paper, the mass of a gravity wave.

## 2.   Neutrinos and Spin

Electrons and neutrinos and other elementary objects present us with that mysterious and ever-alluring phenomenon that Wolfgang Pauli called "non-classical two-valuedness" and the rest of us call "spin 1/2." Thanks to Dieter Brill, [1,2] we now possess a radial wave equation for the radial factor in the wave function of a neutrino moving under the influence of a black hole, source of the only field we can point to with the ability to find an electrically neutral particle in orbit. This analysis, done before the days when black hole evaporation was recognized, [3] brought to light the existence under some circumstances of an effective potential barrier separating the regions where the neutrino undergoes something like Kepler motion under the attraction of the black hole from the region of small $r$-values where the neutrino has been caught and is falling down in to the throat of the black hole. However interesting it may be to examine further the consequences of such capture processes, it would seem even more interesting to look insightfully into the question, "How come spin 1/2?" We turn to Hermann Weyl and Élie Cartan for rays of light. In his 1924 book, *Was ist Materie*, Weyl [4] pointed out that space may be multiply connected and that through what Misner and I came to call a "wormhole" [5] can't thread electric lines of force, or, as Weyl put it, "One cannot say, here *is* charge, but only that this closed surface cutting through the field includes charge.

Weyl thus made it clear that Euclidean topology is far from the only physically natural topology for space. A subsequent look at the consequences of quantum fluctuations for geometry made it appear compelling to consider the wormhole, not as an isolated anomaly, but a microscopic portion of a foam-like structure [6] pervading all of space, gravitational collapse going on and being undone everywhere and all the time.

Every wormhole, Élie Cartan [7] had pointed out as long ago as 1928, doubles the options for understanding the geometry according to the choice one makes for the triad structure one lays down on the 3-geometry. [8]

What happens when quantum considerations are added to this concept of space geometry with this, that or the other triad structure or "spinor structure?" Then the superspace breaks up from a singly-connected domain to an entity double sheeted, illustrated in [8, p. 289]. Then the remarkable knot-representation of a solution of the WDW "Schrödinger equation" discovered by Ashtekar [9], finds the knot structure pierced as it were by a knitting needle, thereby altering the knots that come into consideration in an easily picturable way as kindly pointed out to us by Princeton graduate student, Chetan Nayak. It is difficult to imagine any more natural point of origin for "Pauli's non-classical two-valuedness." At this point the moment has come to close the door on "spin without spin," and turn to the:

## 3. Mystery of the Missing Mass

The mystery of the missing mass is presently one of the most active fields of cosmological research [10]. From the study of galactic luminosity functions and rotation curves, it has been found that dark mass represents about 90% of the dynamical mass present in spiral galaxies [11,12]. The virial theorem argues that about 97% of the mass in galactic clusters must be in the form of dark mass [11]. These observations yield values of $\Omega$, the ratio of the density of the universe to the critical density required for closure, of $\Omega_{galaxies} \geq 0.03$–$0.10$, and $0.1 \leq \Omega_{clusters} \leq 0.3$ [12,13].

As we move to larger scales (such as the infall of the Virgo cluster [12] or distortions of the Hubble flow [14,15,16,17]), one finds that the density approaches the critical density, $\Omega \approx 1$.

There is strong support in the astrophysics community for the Einstein-Friedmann cosmology, and the inflationary scenario [12,18,19]. Although both models predict a value of $\Omega \approx 1$, current observations indicate that $\Omega_{visible} \leq 0.01$ [11,13,19]. Taken at face value, this survey suggests that the mass observed is less than one percent of the mass that might be there.

Primordial nucleosynthesis appears to be a trustworthy way to understand conditions as they were in the first few seconds of the universe, at temperatures over $10^{10}$ $K$. Observations of the abundance ratios of $^2H$ and $^3H$ to H argue for a constraint on the baryon density [12,20,10,21]

$$\Omega_{baryons} \approx 0.05 \pm 0.03. \qquad (1)$$

This condition excludes purely baryonic sources (dust, brownies, stars, baryonic black holes) from being prime contributors to the cosmic density ($\Omega \approx 1$). In other words, today's "standard model", appears according to present evidence to require that at least 88% of the universe is mass without being matter. A mystery of missing matter? No! What we have is a mystery of missing *mass*.

## 4. Dark Matter Candidates

We need to find an overwhelming non-baryonic source of mass. There are a number of popular candidates, including both hot and cold dark mass theories. The hot dark mass (HDM) theory involves mass which was moving near the speed of light around the epoch of galaxy formation. The primary components of this theory are low mass neutrinos, with $m_\nu c^2 \approx 25 eV$. HDM candidates have problems clumping on small scales, and for the most part are unable to bring about galaxy formation, as the structure that is produced is generally over-developed [12,10]. The cold dark mass (CDM) theory proposes mass which was moving slowly during galaxy formation, thereby lending itself to clumping on small scales. The primary components of the CDM theory are weakly-interacting massive particles (WIMPS), neutralinos, and axions. There are many detailed treatments of these dark mass candidates, including [11,12,13,22]. To avoid an absolutely homogeneous universe, and initiate the formation of structure, the mass needs to be seeded. The primary seeding models are random density fluctuations, and topological defects, such as cosmic "strings" and "textures" [12,10].

It is important to note that all of the popular CDM theories require new fundamental physics, in the form of WIMPS, shadow matter, and the like. The foremost "traditional physics" candidates for the missing mass are black holes and brownies. The main difficulties with these candidates is that they appear to presume baryons, and are therefore unable to constitute a significant portion of the missing mass.

We are animated by two concerns: that we find non-baryonic mass, and that this mass require no new physics. To begin such a search, we examine the universe during the first few seconds, before nucleosynthesis gets into full swing. Any mass originating at this time must be predominantly non-baryonic.

Analyses of the approach to crunch of the universe show that the effective energy density of long-wavelength gravitational radiation increases as $1/R^6$ ($R$ is the radius of the universe), while that of electromagnetic radiation increases as $1/R^4$, and matter density increases as $1/R^3$ [23,24]. Thus, long-wavelength gravitational radiation may be expected to dominate at early times.

We picture an early universe with a chaotic, energetic gravity wave spectrum; waves of all amplitudes and frequencies colliding "higgledy-piggledy" throughout space [23,25,26,27]. To apply this tremendous energy spectrum to the CDM scenario, we must localize it. Our goal is to transform the gravitational radiation into compact, well-defined, massive objects. General relativity offers us the perfect "box" for containing energy: a black hole. Gravitational waves represent energy, and having a high enough concentration of energy in a confined space will lead to gravitational collapse to a black hole. With a strong, chaotic gravitational wave spectrum in the early universe, the conditions needed for such collapse may become fairly likely. We call objects created in this way primal-chaos black holes, as they are formed in the first few seconds, and are therefore non-baryonic. In our search for "mass without matter," we cannot help but turn to primal-chaos black holes as promising candidates.

# 5.   Background

All information regarding the fate of a time-symmetric gravitational wave is contained within its particular distortion of space. A distinctive feature of this distortion is signaled by the presence of non-trivial extremal surfaces. These extremal surfaces are characterized by the convergence of all emitted perpendicular null-geodesics. In simpler terms, light is trapped, and cannot escape to infinity. Accordingly, these surfaces are called "trapped surfaces." If a wave's 3-geometry contains a trapped surface, then, from a theorem due to Roger Penrose, [28][29, §34.6] it is inevitable that the geometry collapse to a black hole.

We take a scenario of gravity waves coming in from infinity and converging upon the origin, combining to form a single, localized disturbance. Using Brill's model characterizing axially symmetric waves at the moment of time symmetry, we catch a glimpse of the localized wave just as it reaches its pinnacle. From this fleeting vision, we are able to determine the fate of the wave; the presence of a trapped surface signals impending collapse to a black hole. By tinkering with the amplitude parameter, we will be able to determine the critical amplitude at which trapped surfaces first appear, and black holes start to form. There will be three major scenarios. In the first, the gravity waves converge upon the origin, the localized wave rises to its summit, then recedes, the waves effectively pass through each other, and return to infinity. In the second scenario, the localized wave reaches its apex, and due to the tremendous amount of energy confined about the origin, collapses to a black hole. At a maximum value of the amplitude, the gravitational energy will be so great as to curve space up into closure, yielding a closed universe animated by a pure gravity wave. The third scenario mediates between the previous two, representing the critical wave strength at which collapse to a black hole becomes inevitable. At this value, the waves are teetering between explosion and collapse. These three scenarios are depicted in the embedding diagrams of Figure 1.

We now determine the metric for a specific gravity wave implosion scenario, and then look for trapped surfaces. In so doing, we will determine the value of $A_{crit}$.

# 6.   The Brill Methodology

We utilize a model, first suggested by Weber and Wheeler, [30] and further developed by Brill, [31,32] which characterizes axially symmetric, time-symmetric gravity waves. We begin with the following assumptions:

1. Take a gravity wave of the lowest multipolarity, with no fancy tailoring. This is a quadropole wave, of order $L = 2$. Assume the wave has no rotation.
2. Take the gravity wave to be at the moment of time symmetry, such that all time derivatives of space vanish: the extrinsic curvature of space is zero. This allows a slicing of the spacetime geometry with a spacelike hypersurface of zero extrinsic

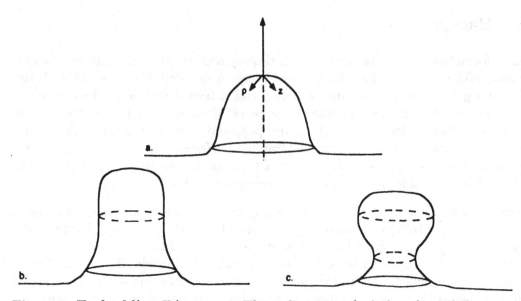

Figure 1. **Embedding Diagrams.** Three diagrams, depicting three different geometrical situations, for the case of a spherically symmetric metric. a.) For low amplitudes, the deformation to space will be slight. A small "bump" in an otherwise flat space. b.) As we increase the amplitude, we hit a critical value, $A_{crit}$, at which the first trapped surface appears (dashed curve). The geometry has just barely gone "vertical," and an extremal surface is produced. c.) As we increase $A$ beyond $A_{crit}$, we find that the geometry possesses *two* extremal surfaces. The inner surface represents the point at which an expansion of the surface begins to *decrease* the surface area. The outer surface marks the critical point at which surface area transitions to the more familiar state of increasing with expansion of the surface. As we go to even higher amplitudes, the throat pinches shut, and we have a closed universe animated by a pure gravity wave.

curvature, a "snapshot." The moment of time symmetry can be thought of as the moment of maximum impact.

**3.** Idealize to pure geometrodynamics, considering a gravity wave in otherwise empty (flat) spacetime.

We take the metric as a conformal transformation from an arbitrary base metric, $ds_1^2$,

$$ds^2 = \psi^4 ds_1^2, \qquad (2)$$

where $\psi$ is the conformal factor. To avoid "bag of gold" singularities, [33] we require that $\psi$ be non-zero everywhere, or by convention, positive.

Following the methodology developed by Brill, [31,32,33][29, §21.10] we utilize an axially symmetric base metric,

$$ds^2 = \psi^4 \left[ e^{2Aq(\rho,z)} \left( dz^2 + d\rho^2 \right) + \rho^2 d\varphi^2 \right], \qquad (3)$$

where $\rho$, $z$, and $\varphi$ are the familiar cylindrical coordinates, $A$ is the amplitude of the wave, and $q(\rho, z)$ represents the "distribution of gravitational wave amplitude," or more commonly, the "form factor." Due to the singularity of the cylindrical coordinate system as $\rho$ approaches 0 (along the $z$–axis), we are required to impose certain boundary conditions on this "form factor" to ensure non-singular metrics. Defining $r = \sqrt{\rho^2 + z^2}$, these boundary conditions are

$$q(\rho, z)|_{\rho=0} = 0, \tag{4}$$

$$\left.\frac{\partial q(\rho,z)}{\partial \rho}\right|_{\rho=0} = 0, \tag{5}$$

$$\left.\frac{\partial \psi(\rho,z)}{\partial \rho}\right|_{\rho=0} = 0, \tag{6}$$

$$q = O\left(\tfrac{1}{r^2}\right)\Big|_{r\to\infty}, \tag{7}$$

$$\psi = 1 + \tfrac{m^*}{2r} + O\left(\tfrac{1}{r^2}\right)\Big|_{r\to\infty}. \tag{8}$$

Of special note is boundary condition (8), which dictates that the gravity wave represents positive energy, causing the metric to go asymptotically to the standard Schwarzschild form. $m^*$ represents the effective mass of the gravitational wave, measured in cm. Plugging equation (8) into equation (3), we have, for $r \to \infty$,

$$ds^2 = \left(1 + \frac{m^*}{2r}\right)^4 \left[e^{2Aq(\rho,z)}\left(dz^2 + d\rho^2\right) + \rho^2 d\varphi^2\right]. \tag{9}$$

Although space is completely devoid of baryons, far away from the "active zone" the geometry behaves exactly as if it were under the influence of matter, with mass given by $m^*$. This is an example of mass without matter! The amplitude of the gravity wave, $A$, uniquely fixes the effective mass. For $A = 0$ we have a flat space metric, and $m^* = 0$. As we increase $A$, we eventually hit the maximum amplitude at which space curls itself up into closure. Before this value, however, we expect to pass $A_{crit}$, at which we are teetering between explosion and collapse.

Making use of the assumptions and boundary conditions, Brill shows that the conformal factor satisfies the "wave equation" [31,32]

$$\nabla^2\psi + A\phi\psi = 0, \tag{10}$$

with

$$4\phi = \frac{\partial^2 q}{\partial \rho^2} + \frac{\partial^2 q}{\partial z^2}. \tag{11}$$

$\nabla^2$ is the cylindrical Laplacian with respect to flat space. The wave equation allows us to solve for the geometry of space, given $q(\rho, z)$ and $A$, and is the essential tool in constructing our gravity waves.

## 7.  A Particular Wave

The heart of the wave is given by specifying the form factor, $q(\rho, z)$. We choose, in accordance with the boundary conditions (4)—(8), a particularly simple form, $q = \rho^2 e^{-(\rho^2+z^2)}$, and the equation for the full metric becomes

$$ds^2 = \psi^4 \left[ e^{2A\rho^2 e^{-(\rho^2+z^2)}} \left(dz^2 + d\rho^2\right) + \rho^2 d\varphi^2 \right]. \tag{12}$$

By plugging the form factor, $q(\rho, z)$, into equation (11), we solve for $\phi$:

$$\phi = \left(\frac{1}{2} - 3\rho^2 + \rho^4 + \rho^2 z^2\right) e^{-(\rho^2+z^2)}. \tag{13}$$

Plugging this into the Brill wave equation (10), we end up with the governing wave equation for the conformal factor

$$\frac{1}{\rho} \frac{\partial}{\partial \rho} \left( \rho \frac{\partial \psi}{\partial \rho} \right) + \frac{\partial^2 \psi}{\partial z^2} + A \left(\frac{1}{2} - 3\rho^2 + \rho^4 + \rho^2 z^2\right) e^{-(\rho^2+z^2)}\psi = 0, \tag{14}$$

where we have dropped the $\frac{1}{\rho^2}\frac{\partial^2 \psi}{\partial \varphi^2}$ term, due to axial symmetry. The solution of this equation yields the conformal factor throughout space, and thus the metric. This equation is fundamental in constructing the gravity wave, as it links the abstract description of the wave, $q(\rho, z)$, to its physical space manifestation. We note that this is a Schrödinger-type equation, with a zero-energy "effective potential".

## 8.  Solving for the Geometry

As there is no known analytical solution to equation (14), the best way to approach the problem is numerically. As a first try, we applied a relaxation method using standard "difference equation" techniques [34,35]. We found that the equations are not well suited to such techniques, as we were plagued by convergence problems about the axes [34]. A detailed account of convergence issues relevant to this type of equation can be found in [36]. As an alternative to "differencing," we turn to the spectral element method.

The spectral element method for solving equations combines the generality of a finite element method with the accuracy of spectral techniques. We divide up space into grid elements, and in each of the grid elements we use spectral techniques. Instead of approximating derivatives with difference equations, the spectral method approximates equation (14) with an iterative version,

$$\nabla^2 \psi^{(n+1)} = A \left(\frac{1}{2} - 3\rho^2 + \rho^4 + \rho^2 z^2\right) e^{-(\rho^2+z^2)}\psi^{(n)}. \tag{15}$$

The method represents a solution as a truncated series of smooth functions. In a sense, one does a fourier transform on the desired solution, approximating the results by functions as opposed to specific values at isolated points. The method is global with respect to each element, as opposed to a difference equation's local viewpoint. For well-behaved, smooth solutions, this can be a much more "intelligent" way of solving problem, as it hones in on the large scale solution quickly, without worrying needlessly about details at every single point of the mesh. The spectral element method is very sensitive to boundary conditions, and special care must be taken to avoid pathologies. For background, and a much more detailed explanation, refer to [37,38,39,40].

The spectral method has been extensively developed by the Applied and Computational Mathematics Program of Princeton University. Using computer code created by Ananias Tomboulides and George Karniadakis, we solve equation (14) on a quarter-circle grid, using Dirichlet boundary conditions along the circular edge, and Neumann boundary conditions along the axes. In our particular case, we demand higher precision about the origin, so as to be able to focus on the region of activity. To achieve this, we use a square grid out to $r = 3$ cm, and then a quarter-circle grid out to an edge of $r = 100$ cm.

For each value of the amplitude, $A$, we vary the Dirichlet boundary value along the perimeter circle until our solution satisfies the boundary condition (8) demanding asymptotic flatness, with an uncertainty at each point, $\delta\psi$, given by $\delta\psi \leq 1 \times 10^{-5}$. From our solutions we are able to infer the mass value $m^*$ corresponding to the given amplitude, and arrive at a value for the conformal factor of such a mass throughout the grid.

As a test of the numerical values, we make use of the following equation for $m^*$, valid for small $A$, [41,33]

$$m^* = \frac{A^2}{8\pi^2} \iint \frac{\phi(x)\phi(x')}{r_{12}} d^3x\, d^3x', \tag{16}$$

where $\phi$ is given by equation (13). This equation can be solved, giving us

$$m^* = 0.0372\, A^2. \tag{17}$$

Our numerically derived mass at $A = 0.5$ was $m^* = 0.0092$ cm. Equation (17) gives us a value of $m^* = 0.0093$ cm. *Phenomenal* agreement! Using *Mathematica* we were able to do a quadratic fit on the first few mass values

$$m^* = -0.00112733 + 0.006184A + 0.0289413A^2$$
$$\approx 0.03\, A^2 \tag{18}$$

Equation (18) is consistent with the theoretical predictions of equation (17), and hence verifies the numerics.

We have constructed the gravity waves! Our next task is to look for black holes.

## 9.   Trapped Surfaces as Geodesics

To search for trapped surfaces, we start with our metric (12). We are looking for an extremal surface within this metric. We require that this surface be smooth and closed [28]. Due to axial symmetry, we need only consider the $\rho$–$z$ plane, and treat arcs within this plane as surfaces of revolution about the $z$-axis. Our extremal surface will be such a surface of revolution, designated as a path from $P_1$ to $P_2$ in the $\rho$–$z$ plane. The total surface area, $S$, of this surface is given by the integral of the product of the two $ds$ components,

$$S = \int_{P_1}^{P_2} \int_0^{2\pi} ds(\varphi) ds(\rho, z),$$

$$S = \int_{P_1}^{P_2} 2\pi \psi^4 e^{A\rho^2 e^{-\left(\rho^2 + z^2\right)}} \rho \left(d\rho^2 + dz^2\right)^{\frac{1}{2}}. \tag{19}$$

We are interested in a minimal surface,

$$\delta S = 0. \tag{20}$$

We notice that the metric is even in both $\rho$ and $z$, and therefore our solution for $\psi$ will be symmetric about both axes. The problem of finding a trapped surface is thus reduced to the problem of finding an extremal arc in a quadrant of the $\rho$–$z$ plane.

This extremal problem is mathematically identical to finding geodesics of the metric

$$ds^2 = 4\pi^2 \psi^8 \rho^2 e^{2A\rho^2 e^{-\left(\rho^2 + z^2\right)}} \left(d\rho^2 + dz^2\right) \tag{21}$$

in the first quadrant of the $\rho$–$z$ plane. To ensure a closed, smooth surface, we add the boundary constraints that the geodesics hit the $\rho$ and $z$ axes perpendicularly. Using the geodesic equation, [29, §8.5] this problem reduces to the set of equations:

$$\ddot{\rho} + \frac{1}{2g} \left(g_{,\rho} \dot{\rho}^2 + 2g_{,z} \dot{\rho}\dot{z} - g_{,\rho}\dot{z}^2\right) = 0, \tag{22}$$

$$\ddot{z} + \frac{1}{2g} \left(-g_{,z}\dot{\rho}^2 + 2g_{,\rho}\dot{\rho}\dot{z} + g_{,z}\dot{z}^2\right) = 0, \tag{23}$$

where $_{,x}$ denotes partial differentiation with respect to the variable $x$, $\dot{x}$ denotes partial differentiation with respect to the parametrization variable $\lambda$ (distance along the curve), and the "pseudo-metric", $g(\rho, z)$, is given by

$$g = 4\pi^2 \psi^8 \rho^2 e^{2A\rho^2 e^{-\left(\rho^2 + z^2\right)}}. \tag{24}$$

Equations (22) and (23) parametrize the path of a geodesic through the pseudo-metric. To ensure that our geodesic hit the axes perpendicularly, we further require

$$\rho(\lambda_0) = \dot{z}(\lambda_0) = 0, \tag{25}$$

$$z(\lambda_1) = \dot{\rho}(\lambda_1) = 0, \tag{26}$$

where $\lambda_0$ and $\lambda_1$ are the endpoints of the geodesic.

## 10. Looking for Trapped Surfaces

We know the value of $\psi(\rho, z)$, and hence $g(\rho, z)$, everywhere. We are ready to use equations (22) and (23), with boundary conditions (25) and (26), to search for trapped surfaces. We define two new variables, $a = \dot{\rho}$ and $b = \dot{z}$, and rewrite the two second order equations as four coupled first order linear differential equations

$$a = \dot{\rho}, \quad \dot{a} = -\tfrac{1}{2g}\left(g_{,\rho}a^2 + 2g_{,z}ab - g_{,\rho}b^2\right), \tag{27}$$

$$b = \dot{z}, \quad \dot{b} = -\tfrac{1}{2g}\left(-g_{,z}a^2 + 2g_{,\rho}ab + g_{,z}b^2\right). \tag{28}$$

The boundary conditions are rewritten

$$\rho(\lambda_0) = b(\lambda_0) = 0, \tag{29}$$

$$z(\lambda_1) = a(\lambda_1) = 0. \tag{30}$$

These equations can be solved using the relaxation method outlined in *Numerical Recipes in C* [42, Sections 16.3 and 16.4]. This involves the iterative multidimensional Newton's method, coupling matrix equations at adjoining pairs of points. We parametrize a curve as an initial-guess at an arbitrary number of points, and then "relax" to a solution which solves the equations (27)–(28) at pairs of adjoining points. We fix the first and last point of the curve in accordance with the boundary conditions (29) and (30).

For our purposes, we use a curve defined by 801 points, and specify a convergence of $1 \times 10^{-2}$. The number of iterations required will vary from roughly 500 to 2,000. Starting at $A = 0.50$, we work our way upwards looking for trapped surfaces. The first one appears at $A = 7.5$, corresponding to an effective mass $m^* = 1.546$ cm. This is the solid arc in Figure 2, shown projected onto the pseudo-metric factor, $g(\rho, z)$.

For larger values of $A$, two trapped surfaces appear. For $A = 10.00$ ($m^* = 2.914$ cm), the numerically determined trapped surfaces are shown as the arcs of Figures 3 and 4, the former being projected upon the conformal factor $\psi(\rho, z)$, and the latter projected upon the pseudo-metric factor, $g(\rho, z)$. In Figure 5 we have taken advantage of symmetry about the axes to plot the trapped surfaces projected onto $g(\rho, z)$ over the whole $\rho$–$z$ plane, instead of just a quadrant.

Figures 4 and 5 are particularly physically revealing. We can motivate the location of the trapped surfaces with the aid of a "mental rubber-band." The maximal surface is given by the path along which a rubber-band is most stretched. We expect this path to start at the top of the peak along the $\rho$-axis, come down the ridge, and then cross rapidly to the $z$-axis. We indeed find an extremal surface along this path. The minimal surface is the surface at which the rubber-band is least stretched. This is the surface we are most familiar with in our common interactions with rubber bands! Placing a rubber band in the maximal position,

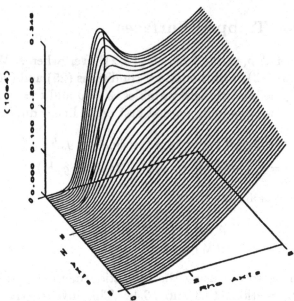

**Figure 2. A critically trapped surface,** projected onto $g(\rho, z)$, corresponding to $A = 7.5$, and $m^* = 1.546$. Recall that the curve is symmetric about both axes, and that the resulting ellipse is actually an ellipsoid of revolution about the $z$ axis. The curve is critical in the sense that it is at the peak of the pseudo-metric. For a slightly lower amplitude, this peak will become diminished, and the curve will no longer be "held" back: it will collapse upon the origin.

we see that it does one of two things: (1) it may collapse to the origin, or (2) it may collapse to the other side of the peak, looping about the peak and hugging it. Although the second case is only a local minimum, it will satisfy the equations as being a geodesic through the surface, as geodesics are only concerned with the local curvature of space. Thus, the two geodesics of Figure 4 and Figure 5 are in precisely the locations we expect them to be.

We have found trapped surfaces, and hence a black hole, within our pure gravity wave geometry!

## 11.   Eppley's calculations

In 1977 Eppley numerically solved a very similar set of equations, both arriving at a conformal factor, and looking for trapped surfaces [36]. Although this paper does much the same thing, there are some crucial differences:

(1) Eppley's motivation was primarily numeric; this was an interesting problem through which to develop and test numerical methods. We propose that primal-chaos black holes formed in this way may have cosmological significance.

(2) Eppley used $q = \frac{\rho^2}{\rho^2 + z^2}$, as opposed to our $q = \rho^2 e^{-(\rho^2 + z^2)}$. Our choice gives a

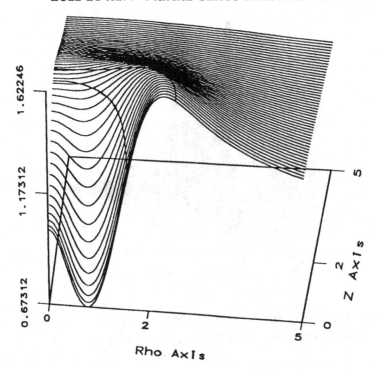

**Figure 3. Two trapped surfaces, projected onto** $\psi(\rho, z)$**, corresponding to** $A = 10.0$**, and** $m^* = 2.914$ cm.

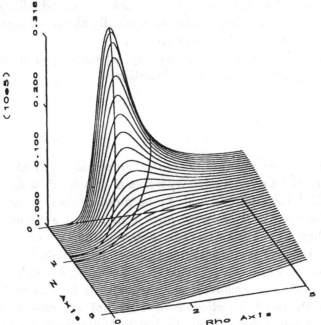

**Figure 4. Two trapped surfaces, projected onto** $g(\rho, z)$**, corresponding to** $A = 10.0$**, and** $m^* = 2.914$ cm.

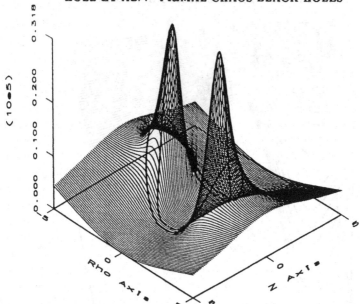

Figure 5. **Two trapped surfaces,** projected onto $g(\rho, z)$, corresponding to $A = 10.0$, and $m^* = 2.914$ cm. Shown for the whole $\rho$–$z$ plane.

more well-defined quadrupole moment, and therefore a clearer picture of the wave. (3) Eppley solved for $\psi(\rho, z)$ using difference equations, and his paper outlines convergence problems he had with such an approach. As already discussed, we found similar problems with difference equations, especially about the axes, and have had great success with the spectral element method.
(4) Eppley found only one (the minimal) trapped surface, as opposed to the two predicted by theory, and found in our calculations.

## 12.   Generality

From Section 10., we conclude that the critical amplitude at which black holes start to form in our model for colliding gravity waves is $A_{crit} = 7.5$, with a corresponding mass of $m^* = 1.546$ cm. However, this by no means indicates that, for imploding axially symmetric waves of the form we have proposed, we are *limited* to black hole masses greater than $m^* = 1.546$. The governing equations which we have been using can all be written inclusive of a scale factor [33,32], redefining $\rho' = \rho/a$, $z' = z/a$. The same physical scenario can be made to occur on scales both much larger and much smaller than the one we have been working with, by varying this factor. The essential point of the last few sections is to show that a mechanism to create black holes from gravity waves not only exists, but in fact must occur for gravitational waves of sufficient strength.

Questions could be raised as to the "generality" of our method. Brill's methodology applies to the particular case of axially-symmetric gravity waves. The collapse to a black hole is an essentially stable phenomenon. For amplitudes much greater than $A = A_{crit}$, we expect a black hole to form even if additional factors, such as rotation or other asymmetries, are thrown in. At the heart, what we are dealing with is sufficient energy density to induce gravitational collapse. Different initial configurations will lead to quantitatively different collapses (for example, Shapiro and Teukolsky's footballs [43]), but the collapse itself is inevitable.

We have shown that primal-chaos black holes, products of colliding gravity waves, are viable physical objects. It is clear that these objects are not composed of baryons, and therefore can close the universe without affecting the success of nucleosynthesis. We are forced to take PCBH's seriously, and explore their consequences!

## 13.   Observational Limits

We can make some rough estimates of the sizes of possible PCBHs created by the mechanisms described in this paper. In the first place, we have assumed that the black hole is produced in the first ten seconds of the universe, before nucleosynthesis gets into full swing. We want these black holes to be causally connected, and therefore we expect their horizons to be less than ten seconds of light-travel time across. This gives an upper limit to the ensuing mass of

$$M_{max} = \frac{c^2}{2G} r_{Schwarzschild} \tag{31}$$

$$= 3.372 \times 10^{28} grams \tag{32}$$

$$= 2.504 \text{ cm.} \tag{33}$$

We can also find a lower limit to the ensuing mass. Due to Hawking radiation [3,44], black holes with masses significantly less than $M_{min} \approx 10^{15}$ grams $= 7 \times 10^{-14}$ cm will have evaporated by now. Furthermore, black holes produced with roughly this mass will be evaporating during the present day, becoming increasingly observable towards the end of their lifetimes. Density limits can be put on the number of black holes with masses around this minimum mass by searching the cosmic and $\gamma$-ray background for their Hawking emission. This has been done, [45] with a resulting density bound on black holes of

$$\Omega \leq 7.6(\pm 2.6) \times 10^{-9} h_0^{1.95 \pm 0.15}, \tag{34}$$

(where $H_0 = h_0/100 \ kms^{-1} Mpc^{-1}$), for the mass range

$$4.3 \times 10^{14} grams < M < 7.0 \times 10^{14} grams. \tag{35}$$

A density limit on the number of burst from exploding black holes was also done, [45] with the results

$$\# \text{ of bursts} < 0.04 \ pc^{-3} year^{-1}. \tag{36}$$

For masses greater than $\approx 10^{14}$ *grams*, observational limits are scarce. In fact, for the mass range

$$1 \times 10^{14} grams < M < 1 \times 10^{25} grams, \tag{37}$$

there has not yet been proposed any convincing method of detection [46]. For higher mass black holes, in the range

$$1 \times 10^{25} grams < M < 1 \times 10^{32} grams, \tag{38}$$

a number of important experiments involving gravitational microlensing are currently being undertaken [47,48].

Studies of the gravitational radiation spectrum of the early universe have to date mostly focused on present-day backgrounds. Preliminary calculations involving the spectrum of relic gravitational waves have been made [24][23, Chapter 17]. Calculations have also been made for the particular case of the inflation model [12, Section 8.4]. These calculations predict a background spectrum of relic gravitons of $T_{gravitons} \approx 1 \ K$. Extrapolating these models back to early times poses some serious theoretical difficulties [12].

There is much discussion of cosmological processes which produce gravitational radiation in the early universe. These include gravitational radiation from collapsing vacuum bubbles [46], from cosmological phase transitions [49], from colliding vacuum bubbles [50], and from cosmic strings and textures [51].

Primal chaos black holes are a versatile candidate for the missing mass. They can be of arbitrary mass (the actual masses will depend on the early epoch gravity wave spectrum), are intrinsically non-baryonic, and are the epitome of "dark" [52,53]. In fact, studies have been done in the particular case of planetary mass black holes, given by masses in the range $10^{15} grams < M < M_{\odot}$, where $M_{\odot}$ is the mass of the sun. These planetary mass black holes have been shown to make very good dark matter candidates [54,55].

## 14.   Conclusion

Present day cosmological theories hold that roughly 99% of the mass of the universe has not yet been detected, and that 90% of this mass is of a non-baryonic form. This is an intriguing state of affairs, and much research is being done to shed light on the situation. However, all current candidates for the missing mass require new physics, in the form of axions, WIMPs, shadow matter, and the like. In this paper we have presented a versatile candidate for the dark mass based soundly upon general relativity. Although the feasibility of this proposal is crucially dependent upon the spectrum of gravity waves in the early universe, we note that primal chaos black holes have not been ruled out observationally.

Primal-chaos black holes, formed by colliding gravity waves, are compelling candidates to help solve the mystery of the missing mass, and deserve further consideration.

## 15. Acknowledgments

We would like to recognize the extensive help received from the Applied and Computational Mathematics group at Princeton University, and in particular, from Ananias Tomboulides. Two of us, W.A.M. and D.E.H., acknowledge partial support for this research from the Air Force Office of Scientific Research.

# References

[1] D.R. Brill and J.A. Wheeler. Interaction of neutrinos and gravitational fields. *Rev. Mod. Phys* **29**,465 (1957).

[2] D.R. Brill and J.A. Wheeler. Interaction of neutrinos and gravitational fields. *Rev. Mod. Phys* **33**,623 (1961).

[3] S.W. Hawking. Particle creation by black holes. *Comm. Math. Phys.* **43**,199 (1975).

[4] H. Weyl. *Was ist Materie.* Springer, Berlin (1924). p. 57.

[5] C.W. Misner and J.A. Wheeler. Classical physics as geometry. Gravitation, electromagnetism, unquantized charge, and mass as properties of curved empty space. *Ann. Phys.* **2**,525 (1957).

[6] J.A. Wheeler. On the nature of quantum geometrodynamics. *Ann. Phys.* **2**,604 (1957). see also *Gravitation.*

[7] É. Cartan. *Leçons sur la Géométrie des Espaces de Riemann.* Gauthier-Villars, Paris, France (1928 and 1946). see also his *Theory of Spinors*, M.I.T. Press, Cambridge (1966).

[8] J.A. Wheeler. Superspace and quantum geometrodynamics. In C. DeWitt and J.A. Wheeler, editors, *Battelle Rencontres: 1967 Lectures in Mathematics and Physics.* W.A. Benjamin, Inc., New York, (1968).

[9] A. Ashtekar. New Hamiltonian formulation of general relativity. *Phys. Rev. D* **36**,1587 (1987).

[10] D.N. Schramm. Dark matter and cosmology. *Nucl. Phys. B (Proc. Suppl.)* **28A**,243 (1992).

[11] P. Galeotti and D.N. Schramm, editors. *Dark Matter in the Universe*. NATO ASI Series. Kluwer Academic Publishers, The Netherlands (1990).

[12] E.W. Kolb and M.S. Turner. *The Early Universe*. Addison-Wesley, Redwood City, CA (1990).

[13] M.S. Turner. Dark matter in the Universe. *preprint* **FERMILAB-Conf-90/230-A** (November 1990).

[14] M. Davis and P.J.E. Peebles. *Ap. J.* **267**,465 (1983).

[15] E. Bertshinger. Large-scale motions in the universe: A review. In J.M. Alimi, A. Blanchard, A. Bouquet, F. Martin de Volnay, and J. Tran Thanh Van, editors, *Particle Astrophysics: The Early Universe and Cosmic Structures*, page 411, Gif-sur-Yvette, (1990). Editions Frontières.

[16] A. Dekel. Streaming velocities and formation of large-scale structure. In J.D. Barrow, editor, *Proc. 15th Texas Symposium on Relativistic Astrophysics*, Singapore, (1991). World Scientific. in press.

[17] N. Kaiser. Streaming velocities and formation of large-scale structure. In J.D. Barrow, editor, *Proc. 15th Texas Symposium on Relativistic Astrophysics*, Singapore, (1991). World Scientific. in press.

[18] P.J.E. Peebles, D.N. Schramm, E.L. Turner, and R.G. Kron. The case for the relativistic hot big bang cosmology. *Nature* **352**,769 (1991).

[19] M.S. Turner. The best-fit universe. In K. Sato, editor, *Proceedings of the IUPAP Conference on Primordial Nucleosynthesis and the Early Evolution of the Universe*, Kluwer, Dordrecht, (1991). preprint: FERMILAB-Conf-90/226-A.

[20] D.N. Schramm. Experimentally testing the standard cosmological model. In *25th International Conference on High Energy Physics*, Singapore, (1990). preprint: FERMILAB-Conf-90/241-A.

[21] D.N. Schramm. Tests of the particle physics-physical Cosmology interface. *preprint* **FERMILAB-Conf-93/022-A** (1993).

[22] D.N. Schramm. Cosmological structure formation. In *Proceedings of PASCOS-91: The Second International Symposium on Particles, Strings, and Cosmology*, (September 1991). preprint: FERMILAB-Conf-91/266-A.

[23] Y. Zel'dovich and I. Novikov. The structure and evolution of the universe. volume 2 of *Relativistic Atrophysics*, Chicago, (1983). The University of Chicago Press.

[24] L.P.Grishchuk. Primordial gravitational waves (October 1990). A talk given at L. Gratton's meeting, Rome.

[25] V.A. Belinsky, I.M. Khalatnikov, and E.M. Lifshitz. Oscillatory approach to a singular point in the relativistic cosmology. *Adv. Phys.* **19**,252 (1970).

[26] I.M. Khalatnikov and E.M. Lifshitz. General cosmological solutions of the gravitational equations with a singularity in time. *Phys. Rev. Lett.* **24**,76 (1970).

[27] L.P. Grishchuk and Yu.V. Sidorov. *Class. and Quant. Grav.* **6**,L155 (1989).

[28] S.W. Hawking and G.F.R. Ellis. *The Large Scale Structure of Spacetime*. Cambridge University Press, London (1973).

[29] C.W. Misner, K.S. Thorne, and J.A. Wheeler. *Gravitation*. W. H. Freeman and Company, New York (1973).

[30] J. Weber and J.A. Wheeler. *Rev. Mod. Phys.* **29**,509 (1957).

[31] D.R. Brill. *Time-Symmetric Solutions of the Einstein Equations: Initial Value Problem and Positive Definite Mass*. PhD thesis, Princeton University, (1959).

[32] D.R. Brill. *Ann. Phys.* **7**,466 (1959).

[33] J.A. Wheeler. Geometrodynamics and the issue of the final state. In C. DeWitt and B. DeWitt, editors, *Relativity, Groups and Topology*. Gordon and Breach, Science Publishers, New York, (1964).

[34] D.E. Holz. *Junior Paper, Princeton University Physics Deparment* (Spring 1991). Advisor: J.A. Wheeler.

[35] J. Stoer and R. Bulirsch. *Introduction to Numerical Analysis*. Springer Verlag, New York (1980).

[36] K. Eppley. Evolution of time-symmetric gravitational waves: Initial data and apparent horizons. *Phys. Rev. D* **16**(6),1609 (1977).

[37] S.A. Orszag. Spectral methods for problems in complex geometries. *J. Comp. Phys.* **37**,70 (1980).

[38] A.T. Patera. A spectral element method for fluid dynamics: Laminar flow in a channel expansion. *J. Comp. Phys.* **54**,468 (1984).

[39] K.Z. Korczak and A.T. Patera. An isoparametric spectral element method for solution of the navier-stokes equations in complex geometry. *J. Comp. Phys.* **62**,361 (1986).

[40] Y. Maday and A.T. Patera. Spectral element methods for the Navier-Stokes equations. *ASME State of the art surveys in Computational Mechanics* (1987).

[41] H. Araki. *Ann. Phys.* **7**,456 (1959).

[42] W.H. Press, B.P. Flannery, S.A. Teukolsky, and W.T. Vetterling. *Numerical Recipes in C.* Cambridge University Press (1988).

[43] S. Shapiro and S. Teukolsky. *Phys. Rev. Lett.* **66**,994 (1991).

[44] K.S. Thorne, R.H. Price, and D.A. Macdonald. *Black Holes: The membrane paradigm.* Yale University Press, New Haven (1986).

[45] F. Halzen, E. Zas, J.H. MacGibbon, and T.C. Weekes. Gamma rays and energetic particles from primordial black holes. *Nature* **353** (October 1991).

[46] L.J. Hall and S.D.H. Hsu. Cosmological production of black holes. *Phys. Rev. Lett.* **64**(24),2848 (1990).

[47] B. Paczynski. *Ap. J.* **304**,1 (1986).

[48] B. Paczynski. Experiments by B. Sadoulet at CfPA. *private communication* (1992).

[49] A. Kosowsky, M.S. Turner, and R. Watkins. Gravitational waves from cosmological phase transitions. *preprint* **FERMILAB-Pub-91/333-A** (1991). submitted to *Phys. Rev. Lett.*

[50] A. Kosowsky, M.S. Turner, and R. Watkins. Gravitational waves from colliding vacuum bubbles. *preprint* **FERMILAB-Pub-91/323-A** (1991). submitted to *Phys. Rev. D.*

[51] B. Allen and E.P.S. Shellard. Gravitational radiation from cosmic strings. *preprint* **WISC-MILW-91-TH-12** (1991).

[52] D.N. Schramm, Brian Fields, and Dave Thomas. Quark matter and cosmology. In *Quark Matter '91*, Gatlinburg, Tennessee, (1992). preprint: FERMILAB-Conf-92/20-A.

[53] J. Silk. *The Big Bang.* W.H. Freeman and Co., New York (1989). p. 133.

[54] M. Crawford and D.N. Schramm. Spontaneous generation of density perturbations in the early universe. *Nature* **298**,538 (1982).

[55] K. Freese, R. Price, and D.N. Schramm. *Ap. J.* **275**,405 (1983).

# Curriculum Vitae

## DIETER R. BRILL

University of Maryland

## EDUCATION

B.A., 1954, Princeton University
M.A., 1956, Princeton University
Ph.D., 1959, Princeton University

## UNIVERSITY POSITIONS

1958–60 Instructor, Physics Department, Princeton University
1961–62 Instructor, Physics Department, Yale University
1962–67 Assistant Professor, Physics Department, Yale University
1967–70 Associate Professor, Physics Department, Yale University
1970–　 Professor, Physics, University of Maryland, College Park

## VISITOR

1993 (spring term) Institute for Theoretical Physics, University of California, Santa Barbara
1992 (fall term) Max Planck Institute, Garching and Jena (Germany)
1984-85 Max Planck Institute, Garching (Germany)
1977 (fall term) University of Texas at Austin
1977 (January-March) Department of Mathematics, Oregon State University
1977 (April-May) Collège de France, Paris (France)
1975 (spring term) Max Planck Institute, Munich (Germany)
1972 (fall term) Max Planck Institute, Munich (Germany)
1961 (summer) International School "Enrico Fermi" (Italy)
1960-61 University of Hamburg (Germany)

## HONORS AND AWARDS

Visiting Scientist, Institute for Theoretical Physics, University of California, Santa
    Barbara (1993)
Sponsored by Humboldt-Foundation under the Senior US Scientist Award (1972,
    1975, 1984 and 1992)
A. Schild Memorial Lecturer, University of Texas (1977)
Recipient, G. Budé Medal, Collège de France (1977)
Flick Exchange Fellow, University of Hamburg (1960-61)
Proctor Fellow, Princeton University (1957)

## PROFESSIONAL ACTIVITIES

American Physical Society fellow
International Society on General Relativity and Gravitation member
Associate Editor, American Journal of Physics, 1976 - 1978
Editorial Board, Einstein Centenary Volume of the International Society on
    General Relativity and Gravitation
International Advisory Committe, Second Marcel Grossmann meeting
Local Organizing Committee, 10th Texas Symposium on Relativistic Astrophysics
    (1980)
Co-organizer, Aspen Summer Workshop (1981)
Co-organizer, Conference on Asymptotic Structure of Mass and Space-Time
    Geometry, Oregon State University (1984)
Nominating Committe of the GRG Society
Member, US delegation for planning Conference at Indian institute of Science,
    Bangalore (1984)
Organizer, Symposium on Kaluza-Klein Theories, 11th International Conference of
    the GRG Society
Scientific Organizing Committee, International Conference on Gravitation and
    Cosmology at Goa, India (1987)
Chairman, Scientific Council, Maryland Academy of Sciences (1991-92)
Program Committee, International Mach Conference (1991-93)

## INVITED LECTURER

Invited participant, *Elftes Bremer Universitätsgespräch*, University of Bremen,
    Germany (1992)
Invited Lecture, Springer Verlag, New York (1991)

Invited Speaker, 3rd Regional Conference on Mathematical Physics, Islamabad
(Pakistan), 1989
Featured speaker at annual Sigma Pi Sigma banquet, University of Cincinnati
(1986)
Lecturer, Fünfzehnte Jenaer Relativitätstagung, Georgenthal, Germany, 1984
Lecturer, Workshop on Gauge Theories, Gravitation, and the Early Universe,
Ahmedabad, India, 1984
Invited plenary lecturer, 9th International GRG Conference, 1982
Les Houches Summer School of Theoretical Physics, 1982
Einstein Centennial Symposium, Physical Research Institute, Ahmedabad (India)
1979.
Alfred Schild Memorial Lectures, Austin, Texas, 1978
Invited Lecturer, Conference on Quantum Theory and the Structure of Space and
Time, Feldafing, Germany (1974)
Lecturer, Banff Summer School (1972)
Lecturer, International School "Enrico Fermi", Varenna (1961)
Guest Lecturer, State University of Iowa (1961)

## BIOGRAPHICAL LISTINGS

Who's Who in the East
Dictionary of International Biography
American Men and Women of Science
Who's Who in Science and Engineering

# Physics Ph.D. Theses
# Supervised by Dieter R. Brill

*Yale University:*

1   Jeffrey M. Cohen 1965 *Rotating Masses and Their Effects on Inertial Frames*

2   Robert H. Gowdy 1968 *Quantization of Gravitational Field Fluctuations*

3   Lawrence E. Thomas 1970 *The State Functional and the Phenomenon of Time in Quantized Gravitation*

*University of Maryland:*

4   Paul L. Chrzanowski 1973 (Prof. Charles Misner, co-supervisor) *Gravitational Synchrotron Radiation*

5   Frank J. Tipler 1976 *Causality Violation in General Relativity*

6   Lee A. Lindblom 1978 *Fundamental Properties of Equilibrium Stellar Models*

7   James A. Isenberg 1979 (Prof. Charles Misner, principal supervisor) *The Construction of Spacetimes from Initial Data*

8   John C. Dell 1981 *Metric and Connection in Einstein and Yang-Mills Theory*

9   Steve M. Lewis 1982 *Regge Calculus: Applications to Classical and Quantum Gravity*

10   Mark D. Matlin 1991 *On Some Features of the Five-Dimensional Witten Bubble Spacetime*

11   Jong Hyuk Yoon 1991 *Dimensional Reductions in Gravity Theories*

# Batchelor's Theses

*Princeton University:*

12    James B. Hartle 1960 *Gravitational Geons*

13    John C. Graves 1960 *Singularities in Solutions of the Field Equations of General Relativity*

*Yale University:*

14    Bjarne L. Everson 1967 *Bound States in the Schwarzschild-Kruskal Geometry*

# List of Publications

## DIETER R. BRILL

### BOOKS AND THESIS

A  *Time-symmetric Solutions of the Einstein Equations: Initial Value Problem and Positive Definite Mass* (Ph.D. Thesis, Princeton University, May 1959)

B  *Seeing the Light: Optics in Nature, Photography, Color, Vision and Holography*, (textbook, 446 pp., with D. Falk, and D. Stork), Harper & Row (1986)

C  *Ein Blick ins Licht* (textbook, with D. Falk and D. Stork), Springer Birkhäuser (1989)

### JOURNAL ARTICLES

1  "Interaction of Neutrinos and Gravitational Fields" (with J. A. Wheeler) *Rev. Mod. Phys.* **29**, 465 (1957).

2  "On the Positive Definite Mass of the Bondi-Weber-Wheeler time-symmetric gravitational waves" *Ann. Phys.* **7** 466 (1959).

3  "Oscillatory Character of the Reissner-Nordström Metric for an Ideal Charged Worm-hole" (with J. C. Graves) *Phys. Rev.* **120**, 1507 (1960).

4  "Significance of Electromagnetic Potentials in the Quantum Theory in the Interpretation of Electron Interferometer Fringe Observations" (with F. G. Werner) *Phys. Rev. Letters* **4**, 344 (1960).

5  "Derivation of the Speed of Electromagnetic Waves in Terms of Dielectric Constant, Magnetic Permeability, and Ratio of Charge Unit" (with F. G. Werner) *Am. J. Phys.* **28**, 126 (1960).

6 "Krümmmung der leeren Raum-Zeit als Baumaterial der physikalischen Welt—eine Einschätzung" (with J. A. Wheeler) *Physikalische Blätter* 8, 354 (1963).

7 "Interaction Energy in Geometrostatics" (with R. Lindquist) *Phys. Rev.* **131**, 471–476 (1963).

8 "Electromagnetic Fields in a Homogeneous Nonisotropic Universe" *Phys. Rev.* **133**, B845–48 (1964).

9 "Method of the Self-Consistent Field in General Relativity and Its Application to the Gravitational Geon"(with J. B. Hartle) *Phys. Rev.* **135** B271–B278 (1964).

10 "General Relativity: Selected Topics of Current Interest" *Il Nuovo Cimento Supplement* **2**, No. 1, 1–56 (1964).

11 "Cartan Frames and the General Relativistic Dirac Equation"(with J. M. Cohen) *J. Math. Phys.* **7**, 238–243 (1966).

12 "Rotating Masses and Their Effects on Inertial Frames" (with J. J. Cohen), *Phys. Rev.* **143**, 1011 (1966).

13 "Erweiterte Gravitationstheorie, Mach'sches Prinzip und Rotierende Massen" *Zeitschrift für Naturforschung* **22a**, 1336 (1967).

14 "Positive Definiteness of Gravitational Field Energy" (with S. Deser) *Phys. Rev. Letters* **20**, 75–78 (1968).

15 "Sign of Gravitational Energy" (with S. Deser and L. Faddeev) *Phys. Letters*, **26A**, 538–9 (1968).

16 "Further Examples of Machian Effects of Rotating Bodies in General Relativity" (with J. M. Cohen) *Nuovo Cimento* **56B**, 209 (1968).

17 "Variational Methods and Positive Energy in General Relativity" (with S. Deser) *Ann. of Phys.* **50**, 548 (1968).

18 "Quantization of General Relativity" (with R. Gowdy) *Reports on Progress in Physics,* **33**, 413 (1970).

19  "Gravitational Synchrotron Radiation in the Schwarzschild Geometry" (with Breuer, Chrzanowski, Hughes, Misner and Pereira), *Phys. Rev. Letters* **28**, 1998 (1972).

20  "Solutions of the Scalar Wave Equation in a Kerr Background by Separation of Variables" (with Chrzanowski, Pereira, Fackerell and Ipser), *Phys. Rev.* **D5**, 1913 (1972).

21  "Instability of Closed Spaces in General Relativity" (with S. Deser) *Commun. Math. Phys.* **32**, 291 (1973).

22  "Gravitations-Synchrotronstrahlung" *Nova Acta Leopoldina* **39**, 291 (1974).

23  "Inertial Effects in the Gravitational Collapse of a Rotating Shell" (with L. Lindblom) *Phys. Rev.* **D10**, 3151 (1974).

24  "Isolated Maximal Surfaces in spacetime" (with F. Flaherty) *Comm. Math. Phys.* **50**, 157 (1976).

25  "Maximizing Properties of Extremal Surfaces in General Relativity" (with F. Flaherty) *Ann. Inst. H. Poincaré* **28**, 335 (1978).

26  "K-Surfaces in the Schwarzschild Space-Time and the Construction of Lattice Cosmologies" (with J. Cavallo and J. Isenberg) *J. Math. Phys.* **21**, 2789–2796 (1980).

27  "The Laplacian on Asymptotically Flat Manifolds and the Specification of Scalar Curvature" (with M. Cantor) *Compositio Mathematica* **43**, 317 (1981).

28  "Barrier Penetration and Initial Values in Kaluza-Klein Theories" *Foundations of Physics* **16**, 637 (1986).

29  "Joint Linearization Instabilities in General Relativity" with C. V. Vishveshwara), *J. Math. Phys.* **27**, 1813 (1986).

30  "Local Linearization Stability" (with O. Reula and B. Schmidt), *J. Math. Phys.* **28**, 1844 (1987).

31  "Geodesic Motion in a Kaluza-Klein Bubble Spacetime" (with M. Matlin), *Phys. Rev.* **D39**, 3151 (1989).

32 "States of Negative Energy in Kaluza-Klein Theories" (with H. Pfister), *Physics Letters* **B228** (1989) pp. 359–362.

33 "Inflation from Extra Dimensions" (with J. H. Yoon), *Class. and Quantum Grav.* **7**, 1253 (1990).

34 "Negative Energy in String Theory" (with G. Horowitz), *Physics Letters* **B262**, 437 (1991).

35 "Splitting of an Extremal Reissner-Nordström Throat via Quantum Tunneling" *Phys Rev.* **D46**, 1560 (1992).

36 "Spell it Nordström" (with T. Dray), *GRG Journal* (to appear).

## ARTICLES FROM BOOKS/CONFERENCES

1[B] "Time-Symmetric Gravitational Waves" *Les Théories Relativistes de la Gravitation*, Edition CNRS, Paris (1962).

2[B] "Experiments on Gravitation" (with B. Bertotti and R. Krotkov) Chapter 1 in *Gravitation, An Introduction to Current Research*, John Wiley, N.Y. (1962).

3[B] "Review of Jordan's Extended Theory of Gravitation" *Proceedings of the International School of Physics "Enrico Fermi"* XX Course, Academic Press, New York (1962).

4[B] "Geons and Gravitational Collapse" Chapter 5 in Perspectives in *Geometry and Relativity*, B. Hoffmann, Editor, Indiana University Press, August 1966.

5[B] "Positive Definiteness of Gravitational Field Energy" (with S. Deser and L. Faddeev) *Proceedings of 5th International Congress Gravitation and Relativity*, Tbilisi, USSR.

6[B] "Isolated solutions in General Relativity" in *Gravitation: Problems and Prospects*, Naukova Dumka, Kiev, p. 17 (1972).

7[B] "A simple Derivation of the General Redshift Formula" in *Methods of Local and Global Differential Geometry in General Relativity*, lecture notes in Physics **14**, Farnsworth, Fink, Porter and Thompson, editors, Springer-Verlag, Heidelberg 1972.

8<sup>B</sup> "Thoughts on Topology Change" in *Magic Without Magic*, J. Klauder editor, W. H. Freeman, p. 309 (1972).

9<sup>B</sup> "Observational Contacts of General Relativity" in *Relativity Astrophysics and Cosmology*, W. Israel, editor, D. Reidel, Dodrecht-Holland 1973.

10<sup>B</sup> "Geometrodynamics" article for *Encyclopaedic Dictionary of Physics*, Pergammon Press LTD, 1975.

11<sup>B</sup> "Quantum Cosmology" in *Quantum Theory and the Structure of Space and Time*, L. Castell, M. Brieschner and C. F. von Weizsäcker, editors, Carl Hauser Verlag, Munich 1975.

12<sup>B</sup> "Extremal Surfaces in Open and Closed Spaces," *Proceedings of the First 'Marcel Grossmann' meeting on Recent Progress of the Fundamentals of General Relativity*, R. Ruffini, ed., North-Holland Elsevier (1977).

13<sup>B</sup> "Comments on the Topology of Nonsingular Stellar Models" (with L. Lindblom) in *Essays on General Relativity: A Festschrift for Abraham Taub*, F. Tipler, editor, Academic press (1980).

14<sup>B</sup> "Positivity of Total Energy in General Relativity" in *Gravitation, Quanta, and the Universe*, R. Prasanna, editor, Wiley Eastern Ltd. (1980).

15<sup>B</sup> "The Positive Mass Conjecture" (with Pong Soo Jang) in *General Relativity and Gravitation; One Hundred Years After the Birth of Albert Einstein*, A. Held, editor, Vol. 1, Plenum Press, pp. 173–193 (1980).

16<sup>B</sup> "Hypersurfaces of Constant Mean Curvature" in *Proceedings: Einstein Centenary Symposium*, K. Kondo, ed., Duhita publishers, Nagpur, India, pp. 213–225 (1980).

17<sup>B</sup> "Linearization Stability" Third A. Schild Memorial Lecture, in *Spacetime and Geometry*, Matzner and Shepley, editors, University Press, pp. 59–81 (1982).

18<sup>B</sup> "Surfaces of Constant Mean Curvature in Schwarzschild, Reissner-Nordström, and Lattice Spacetimes" (with J. Cavallo and J. Isenberg), *Proceedings of the Second Marcel Grossmann meeting*, R. Ruffini, editor, North-Holland (1982).

19[B] "The Positive Energy Program" *Proceedings of the Ninth International Conference on General Relativity and Gravitation*, E. Schmutzer, ed., Cambridge University Press, pp. 229–237 (1983).

20[B] "On Spacetimes Without Maximal Surfaces" *Proceedings of the Third Marcel Grossmann meeting*, Hu Ning, editor, Science Press and North-Holland Publ. Co., pp. 79–87 (1983).

21[B] "Kaluza-Klein Theories" pp. 320–325 in *General Relativity and Gravitation-11* (M.A.H. MacCallum, editor), Cambridge University Press (1987).

28=22[B] "Barrier Penetration and Initial Values in Kaluza-Klein Theories" in *Between Quantum and Chaos*, W. H. Zurek, A. van der Merwe, and W. A. Miller, editors, Princeton University Press (1988).

23[B] "Workshop on Quantum Gravity and New Directions" (with L. Smolin), in *Highlights in Gravitation and Cosmology* Iyer, Khembavi, Vishveshwara and Narliker (editors), Cambridge University Press (1989).

24[B] "Kaluza Klein Theories" in *IIIrd Regional Conference on Mathematical Physics*, F. Hussain and A. Qadir, editors, World Scientific (1990).

25[B] "Euclidean Einstein-Maxwell Theory" in *Festschrift for L. Witten*, World Scientific

26[B] "A Pictoral History of some Gravitational Instantons" in *Festschrift for Charles W. Misner*, Cambridge University Press